The Biology of *Frankia* and Actinorhizal Plants

The Biology of *Frankia* and Actinorhizal Plants

Christa R. Schwintzer and John D. Tjepkema
Department of Botany and Plant Pathology
University of Maine
Orono, Maine

Academic Press, Inc.
Harcourt Brace Jovanovich, Publishers

San Diego New York Boston London Sydney Tokyo Toronto

Front cover photographs:
Upper left: Figure 1(a), page 5. *Upper right:* Figure 1, page 63.
Lower left: Figure 6, page 69. *Lower right:* Figure 11, page 74.

This book is printed on acid-free paper. ∞

Academic Press, Inc.
San Diego, California 92101

United Kingdom Edition published by
Academic Press Limited
24–28 Oval Road, London NW1 7DX

Library of Congress Cataloging-in-Publication Data

The Biology of Frankia and actinorhizal plants / [edited by] Christa R. Schwintzer and John D. Tjepkema.
p. cm.
ISBN 0-12-633210-X (alk. paper)
1. Frankia. 2. Actinorhizal plants. 3. Actinorhiza.
I. Schwintzer, Christa R. II. Tjepkema, John D.
QR82.F7B56 1990
589.9'2--dc20 89-18425
CIP

Printed in the United States of America
90 91 92 93 9 8 7 6 5 4 3 2 1

This book is dedicated to the memory of Emeritus Professor George Bond, F.R.S. 1906–1988

George Bond
1906–1988

George Bond was appointed Assistant Lecturer in the Botany Department of the University of Glasgow in 1927 and held the Hooker Chair of Botany when he retired in 1976. His initial research concerned the distribution and function of the root endodermis, and he started researching nitrogen fixation in legumes in 1932. His work ranged from the transfer and excretion of fixed nitrogen from the nodules to the effects of pO_2 on the fixation process and the utilization of host carbohydrates by nodule bacteria.

His work on nitrogen fixation was interrupted by the war years, when he served as a technical officer with the Ministry of Food. His research into nitrogen fixation in actinorhizal plants started shortly after his return to Glasgow, when he realized how little was then known of woody plant symbioses with *Frankia*. His first study was of *Myrica gale*, a prevalent shrub in western Scotland. He extended his studies to at least 30 other actinorhizal species, concentrating on the physiology and biology of nitrogen fixation. He was primarily responsible for promoting research in symbiotic nonlegume N fixation for a quarter of a century and, consequently, is viewed by actinorhizal plant researchers today as the "father" of present-day research in this field. His scientific excellence was given well-deserved recognition by his election as a Fellow of the Royal Society of London in 1973.

Professor Bond was of a modest and retiring nature. His was a selfless personality, and he would always do all that he possibly could to advise and to help those less experienced than himself. He possessed a quiet sense of humor and could be drawn to recount many amusing anecdotes from his past experiences.

One of his particular wishes was to see greater exploitation of actinorhizal plants for human benefit. It is, therefore, fitting that five chapters on this topic are included in this book. His experience, wisdom, and integrity of character are missed by all who knew him.

C. T. Wheeler

Contents

8. Oxygen Regulation and Hemoglobin

Warwick B. Silvester, Sharon L. Harris, and John D. Tjepkema

9. Spore-Positive and Spore-Negative Nodules

Christa R. Schwintzer

10. The Molecular Genetics of *Frankia*

Beth C. Mullin and Chung Sun An

11. Micropropagation, Tissue Culture, and Genetic Transformation of Actinorhizal Plants and *Betula*

Armand Séguin and Maurice Lalonde

Contributors

Numbers in parentheses indicate the pages on which the authors' contributions begin.

Chung Sun An (196), Department of Botany, Seoul National University, Seoul 151, Korea

Dwight D. Baker (3), School of Forestry & Environmental Studies, Yale University, New Haven, Connecticut 06511

Larry F. Benoit (281), Department of Environmental Horticulture, University of California, Davis, California 95616

David R. Benson (107), Department of Molecular and Cell Biology, University of Connecticut, Storrs, Connecticut 06268

Alison M. Berry (61, 281), Department of Environmental Horticulture, University of California, Davis, California 95616

Jean Bousquet (239), Centre de recherche en biologie forestière, Faculté de foresterie et de géomatique, Université Laval, Ste-Foy, Québec, Canada G1K 7P4

Kermit Cromack, Jr. (343), Department of Forest Science, Oregon State University, Corvallis, Oregon 97331

Jeffrey O. Dawson (299), Department of Forestry, University of Illinois, Urbana, Illinois 61801

H. G. Diem (317), Laboratoire de Biotechnologie des Systèmes Symbiotiques Forestiers Tropicaux, (ORSTOM/CTFT/CNRS), 94736 Nogent sur Marne Cedex, France

Yvon R. Dommergues (317), Laboratoire de Biotechnologie des Systèmes Symbiotiques Forestiers Tropicaux, (ORSTOM/CTFT/CNRS), 94736 Nogent sur Marne Cedex, France

Sharon L. Harris (157), Department of Biological Sciences, University of Waikato, Private Bag, Hamilton, New Zealand

David E. Hibbs (343), Department of Forest Science, Oregon State University, Corvallis, Oregon 97331

Kerstin Huss-Danell (129), Department of Plant Physiology, University of Umeå, S-901 87 Umeå, Sweden

Maurice Lalonde (216, 239), Centre de recherche en biologie forestière, Faculté de foresterie et de géomatique, Université Laval, Ste-Foy, Québec, Canada G1K 7P4

Hubert A. Lechevalier (35), Waksman Institute of Microbiology, Rutgers University, Piscataway, New Jersey 08855

Mary P. Lechevalier (35), Waksman Institute of Microbiology, Rutgers University, Piscataway, New Jersey 08855

I. M. Miller (365), Department of Biological Sciences, Wright State University, Dayton, Ohio 45435

Beth C. Mullin (196), Department of Botany, The University of Tennessee, Knoxville, Tennessee 37996

Anton Quispel (17), Department of Plant Molecular Biology, University of Leiden, Nonnensteeg 3, 2311 VJ Leiden, The Netherlands

Armand Séguin (216), Centre de recherche en biologie forestière, Faculté de foresterie et de géodésie, Université Laval, Ste-Foy, Québec, Canada G1K 7P4

Nancy A. Schultz (107), Department of Molecular and Cell Biology, University of Connecticut, Storrs, Connecticut 06269

Christa R. Schwintzer (3, 178), Department of Botany & Plant Pathology, University of Maine, Orono, Maine 04469

Warwick B. Silvester (157), Department of Biological Sciences, University of Waikato, Private Bag, Hamilton, New Zealand

Leslie A. Sunell (61), Department of Environmental Horticulture, University of California, Davis, California 95616

John D. Tjepkema (157, 264), Department of Botany & Plant Pathology, University of Maine, Orono, Maine 04469

John G. Torrey (83), Harvard Forest, Harvard University, Petersham, Massachusetts 01366

C. T. Wheeler (365), Department of Botany, The University of Glasgow, Glasgow G12 8QQ, Scotland

Lawrence J. Winship (264), School of Natural Science, Hampshire College, Amherst, Massachusetts 01002

Preface

Actinorhizal plants rival legumes in the amount of nitrogen that they fix on a global basis, yet knowledge of their biology and uses is for the most part very recent. Much of this knowledge has accumulated since the first confirmed isolation of *Frankia*, the nitrogen fixing microsymbiont, approximately ten years ago. This book is the first comprehensive review of *Frankia* and the actinorhizal plants and for the first time brings together in one volume the widely scattered information on these topics. It reviews present knowledge on all aspects from molecular genetics through ecology to practical applications, describes methods used in research and practical applications, and is a guide to the literature.

The book begins with an introductory chapter, which gives an overview of *Frankia* and the actinorhizal plants and provides enough background to permit readers with little knowledge of these topics to read the remaining chapters in any order desired. The second chapter provides an overview of developments in this field prior to the first confirmed isolation of *Frankia*. Next is a series of authoritative chapters on the biology of *Frankia*, the symbiosis, and actinorhizal plants. Although methods used in research and in practical applications are included throughout the book, they are given special emphasis in the middle section. The final section of the book concerns the ecology and current and potential uses of actinorhizal plants in both the temperate regions and the tropics.

This work is intended as a reference text and handbook of methods for a wide audience, including established workers and students of *Frankia* and actinorhizal plants, specialists and students in other areas of nitrogen fixation including the *Rhizobium*–legume symbiosis, soil microbiologists, plant physiologists, ecologists, general biologists, foresters, specialists in land reclamation, and managers requiring an authoritative overview of this rapidly developing field.

In planning this treatise we extended invitations to prepare chapters to distinguished scientists throughout the world. We wish to thank these eminent contributors for their scholarly contributions, their enthusiasm for this joint venture, and their patient cooperation in revising chapters.

Christa R. Schwintzer
John D. Tjepkema

1 Introduction

Dwight D. Baker
School of Forestry and Environmental Studies
Yale University
New Haven, Connecticut, U.S.A.

Christa R. Schwintzer
Department of Botany and Plant Pathology
University of Maine
Orono, Maine, U.S.A.

This chapter provides an overview of *Frankia* and the actinorhizal plants and is intended to provide enough background to permit readers with little previous knowledge of these topics to read the remaining chapters in any order desired.

Actinorhizal root nodules are nitrogen-fixing symbioses involving the actinomycete *Frankia* and roots of dicotyledonous plants belonging to 8 plant families and 25 genera (Table I). Most are capable of high rates of nitrogen fixation comparable to those found in legumes. Although they are taxonomically diverse, the actinorhizal plants have some common features. All are perennial dicots and all except *Datisca*, which has herbaceous shoots, are woody shrubs or trees (Tjepkema *et al.*, 1986). Well-known genera include *Alnus* (alder), *Myrica* (e.g., bayberry), *Hippophae* (sea buckthorn), *Elaeagnus* (e.g., autumn olive), and *Casuarina* (beef wood).

These symbioses were known as nonleguminous nitrogen-fixing plants until 1978 when the term actinorhizal plants was coined at a meeting at Harvard Forest in Petersham, Massachusetts (Tjepkema and Torrey, 1979). The term is formed from the roots *actino* for the actinomycete *Frankia* and *rhiza* for the plant root bearing the symbiosis. It is

Table I Currently Identified Actinorhizal Families and Genera[a]

Family	Genus	Number of species
Betulaceae	*Alnus*	47
Casuarinaceae	*Allocasuarina*	54
	Casuarina	16
	Ceuthostoma	2
	Gymnostoma	18
Coriariaceae	*Coriaria*	16
Datiscaceae	*Datisca*	2
Elaeagnaceae	*Elaeagnus*	38
	Hippophae	2
	Shepherdia	2
Myricaceae	*Comptonia*	1
	Myrica	28
Rhamnaceae	*Ceanothus*	31
	Colletia	4
	Discaria	5
	Kentrothamnus	1
	Retanilla	2
	Talguenea	1
	Trevoa	2
Rosaceae	*Cercocarpus*	4
	Chamaebatia	1
	Cowania	1
	Dryas[b]	3
	Purshia	2
	Rubus[c]	2

[a]Compiled from Bond (1983) and Torrey and Berg (1988).
[b]Although three species were originally reported to be nodulated, recent reports have confirmed nodulation only in *Dryas drummondii*.
[c]Originally reported as an actinorhizal plant, *Rubus* more recently has been discounted as a truly nodulated species (Stowers, 1985).

analogous to mycorrhizal, a much older term used to describe symbioses between fungi and plant roots. But it should be noted that "actinorhizal" is spelled with only one "r." In describing actinorhizal root nodules the frankiae (strains of *Frankia* collectively) inhabiting the nodule are called the "microsymbiont" or the "endophyte." At the 7th International Meeting on *Frankia* and Actinorhizal Plants in Storrs, Connecticut, in August 1988, a majority of *Frankia* workers decided to discontinue use of the term "endophyte" because it implies that frankiae are plants and not bacteria.

It is ironical that plant growth is often limited by the availability of nitrogen when approximately 80% of the atmosphere is composed of dinitrogen gas. This irony is explained by the fact that dinitrogen is chemically very stable and cannot be used directly by plants. However, atmospheric nitrogen is the ultimate source of all nitrogen used by organisms. Before dinitrogen can be used it must be fixed, that is, split and combined with other chemical species. This process requires large amounts of energy and can occur via either biological nitrogen fixation or chemical fixation. Because of the large amounts of energy required, the cost of chemically produced nitrogenous fertilizers is closely linked to the price of petroleum.

Among living organisms, only certain prokaryotic microorganisms (bacteria) have the enzyme nitrogenase, which is responsible for the fixation of dinitrogen. This enzyme is very sensitive to oxygen and thus must be protected from oxygen. On the other hand, large amounts of ATP are consumed by nitrogenase and are provided by respiration, which in turn requires oxygen. Chapters 6 and 7 give details about how nitrogenase functions in *Frankia* and Chapter 8 describes how frankiae have solved the problem of protecting nitrogenase from oxygen while allowing respiration to occur.

I. Frankiae

The microsymbionts within the root nodules of actinorhizal plants were first described in the second half of the nineteenth century. Initially they were believed to be eukaryotic fungi but later were found to be actinomycetes (Chapter 2). Our understanding of the actinomycetes has also evolved through time and it is now known that they are filamentous bacteria and not fungi.

A. Pure Culture

Unlike the rhizobia, which were cultured *in vitro* successfully in the late 1800s, frankiae have been cultured only since 1978 (Chapter 2). Pommer (1959) isolated a *Frankia* strain earlier from *Alnus glutinosa* but his work was disputed and the cultures were lost (Chapter 3). Even today strains of *Frankia* have not been pure-cultured from some genera of actinorhizal plants. In particular strains that are infective on the genera of the Rosaceae and Rhamnaceae are lacking. Difficulties associated with isolating frankiae seem to be their slow growth rates and unidentified nutritional requirements (Chapter 3; Baker, 1989).

From studies of strains of frankiae that have been isolated in pure culture, we have some understanding of the physiology and morphology of these bacteria. Frankiae, once isolated, usually grow well on simple media containing major salts and a single carbon source, often intermediates of the citric acid cycle. Some strains seem to require complex additives, but the majority of strains do not (Chapter 3).

B. Morphology and Physiology

Frankia differentiates into three different cell types. All three can be observed in both pure culture (*in vitro*) and symbiotically (*in planta*). Vegetative cells are filamentous and form a mycelium on which the other two cell types are borne (Fig. 1). Sporangia (Fig. 1a) containing numerous ovoid spores are produced routinely *in vitro* if certain nutritional requirements are met, but production of sporangia *in planta* is observed in only some nodules on some genera (Chapter 9).

Vesicles are thick-walled spherical structures borne on short lateral branches of the vegetative mycelium (Fig. 1b) and are produced if the *Frankia* is fixing nitrogen. The shape of vesicles produced *in planta* is determined by the host plant (Lalonde, 1979; Baker and Seling, 1984). For example, vesicles produced by the same strain of *Frankia* are spherical in *Alnus* but club-shaped in *Myrica*. The enzyme nitrogenase has been localized in these specialized structures *in vitro* (Meesters *et al.*, 1985, 1987) and *in planta* (Sasakawa *et al.*, 1988). Vesicles are analogous to heterocysts in cyanobacteria because they restrict the diffusion of oxygen to the site of nitrogenase activity.

Frankia can be cultured in the presence of very high concentrations of oxygen and still retain nitrogenase activity. The vesicles of cultures grown in high concentrations of oxygen bear more layers of wall material (Chapter 8), indicating that the microorganism is morphologically as well as physiologically responsive to its environment.

Frankiae are heterotrophic, aerobic, and sometimes microaerophilic. They are able to utilize a variety of carbon sources including short-chain fatty acids like propionate and citric acid cycle intermediates like succinate. Some strains also use sugars. Several frankiae degrade cellulose and some have pectinase activity. Trehalose and glycogen are major storage compounds (Chapter 6). Assimilation of dinitrogen follows the nitrogenase—glutamine synthetase (GS)–glutamate synthase (GOGAT) pathway (Chapter 6).

C. Molecular Genetics

Understanding of the molecular genetics of *Frankia* is still very limited in part because frankiae have been difficult subjects because of their multicellular nature, slow growth rates, and lack of a vector to introduce and

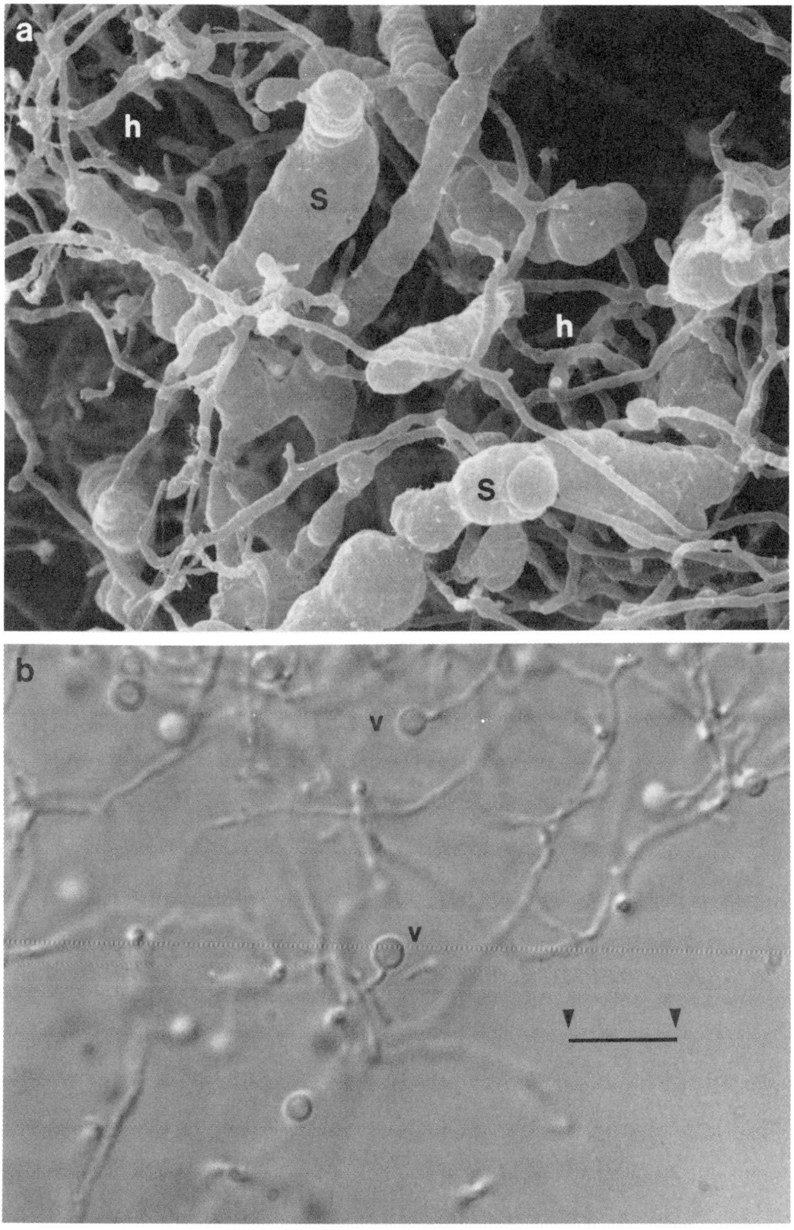

Figure 1 Microscopic views of *Frankia* pure cultures. (a) A scanning electron micrograph showing hyphae (h) and sporangia (s). (b) A Nomarski light micrograph showing numerous vesicles (v), the specialized structure for nitrogen fixation. Bar = 10 μm.

replicate DNA within *Frankia* cells. However, great genetic diversity has been shown to exist among frankiae by DNA–DNA hybridization of total cellular DNAs (Chapter 10). Moreover, plasmids have been found in some frankiae. But relatively few *Frankia* strains have been shown to have plasmids and to date no functions have been assigned to these plasmids (Chapter 10). Several of the *nif* genes have been identified in *Frankia* but progress in identifying *nod* genes has been limited (Chapter 10).

D. Taxonomy

Frankia belongs to the family Frankiaceae together with the closely related genus *Geodermatophilus* (Chapter 3). Although there has been a recent proposal to assign some frankiae to species (Lalonde *et al.*, 1988), most workers in the field prefer not to do this until more reliable criteria for species definition have been developed (Chapter 3). By agreement among the workers in the field, frankiae are designated as *Frankia* spp. and given an acronym of three letters followed by up to 10 numbers, the first four of which represent the plant host genus and species from which the strain was isolated (Lechevalier, 1983). Although it has not been possible to identify species, two broad groups of frankiae designated groups A and B have been identified (Chapter 3). Group A is very heterogeneous and contains strains that may not be infective for the host from which they were isolated. Further, most of these strains grow relatively rapidly and can be maintained in slant culture. Group B is far more homogeneous. These strains grow relatively slowly, cannot be maintained on slants, and are infective for their original host plant.

E. Ecology

Very little is known about the ecology of frankiae independent of their host plants (Baker, 1988). It is almost certain that frankiae are able to grow saprophytically in soil, because at least some frankiae can be cultured on relatively simple media *in vitro*, where they make substantial growth, fix dinitrogen, and sporulate. In addition there is considerable indirect evidence that frankiae can grow in soil, including the presence of infective *Frankia* particles in soils that either were never known to support host plants or had been devoid of hosts for many years (Rodriguez-Barrueco, 1968; Wollum *et al.*, 1968; Benecke, 1969; Bermudez de Castro *et al.*, 1976; van Dijk, 1984; Huss-Danell and Frej, 1986; Weber, 1986; Smolander and Sundman, 1987). Further it is intriguing that more infective *Frankia* particles occur in soils under birch, a close relative of alder, than under alder itself (van Dijk, 1984; Smolander and Sundman, 1987).

II. Actinorhizal Root Nodule Symbioses

Actinorhizal root nodules are clusters of modified lateral roots in which the vascular tissue is central and the bacterially infected cells are found in the cortex (Fig. 2a). Depending on the host species, the lobes may be loosely clustered as in *Comptonia peregrina* or tightly packed as in *Alnus* spp. (Chapter 4). In most species the apex of the nodule lobes is terminal but in the Myricaceae and *Casuarina* spp. *sensu stricto* uninfected roots, termed nodule roots, arise from the apices of the nodule lobes (Chapter 8). In contrast legume root nodules are morphologically tumors with vascular tissue peripheral to the infected cells (Fig. 2b).

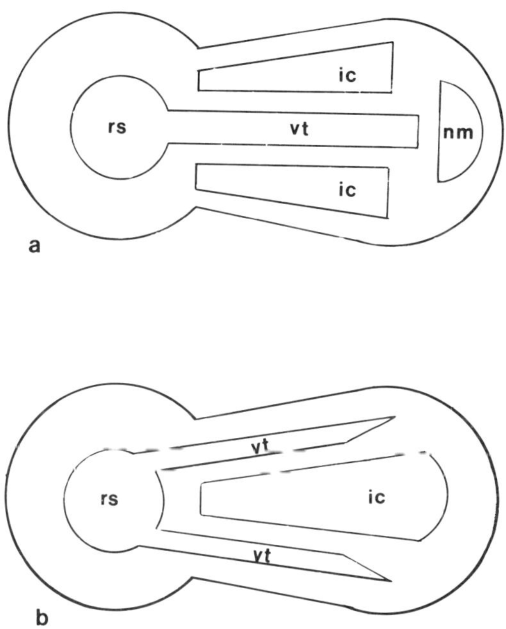

Figure 2 Conceptual diagrams of the anatomy of root nodules of actinorhizae and legumes. (a) Anatomy of an actinorhizal nodule: vascular tissue (vt) is centralized, attached to the root stele (rs), and surrounded by microsymbiont-infected cortical cells (ic). In most actinorhizal nodules, a nodule meristem (nm) exists at the tip of the nodule. (b) Anatomy of a legume nodule: vascular tissue (vt) surrounds the microsymbiont-infected cells (ic) and attaches to the root stele (rs).

When frankiae occur within actinorhizal nodules, they are filamentous and surrounded by a host cell membrane plus a cell-wall-like material called the encapsulation. Frankiae are usually differentiated into three cell types similar to those seen in *in vitro* cultures. Vegetative cells are filamentous and vesicles, the site of nitrogen fixation, are enlarged structures formed at the ends of branch hyphae (Fig. 3a). Some nodules also contain sporangia filled with spores. In contrast the rhizobia within legume root nodules are modified single cells, termed bacteroids (Fig. 3b), and are surrounded only by a host cell membrane.

Root infection and nodule development of both actinorhizal and legume symbioses have some similarities. In many genera of actinorhizal and leguminous plants the infection process occurs via the penetration of deformed root hairs by the bacteria. A nodule meristem develops in response to the invasion of the root system and the bacteria grow to fill some of the cells of this new structure (Chapter 4).

In some genera of actinorhizal plants and in some rhizobium-induced symbioses the infection of the root occurs via an intercellular mechanism rather than a root-hair-mediated one. Members of the Elaeagnaceae are infected in this manner by *Frankia*. In essence the symbiotic bacteria gain entry to the root by growing through the intercellular spaces between adjacent epidermal cells. Once inside the root cortex, they continue to colonize the intercellular spaces until at some point in development they penetrate individual cortical cells of the newly formed nodule (Chapter 4).

III. Distribution of Actinorhizal Plants

Actinorhizal plants are found on every continent except Antarctica. Silvester (1977) provided maps showing the current distribution of many of the actinorhizal genera. This distribution reflects both the native distributions of these genera and the regions where they have been introduced. Introductions have been common for horticultural purposes (Chapter 18) as well as for land reclamation and forestry. The ranges of *Casuarina* and *Elaeagnus* in particular have been greatly expanded through these activities.

Table II lists the major actinorhizal genera by continent. Africa has only one known indigenous actinorhizal genus. This is probably due to a shortage of habitat favoring actinorhizal plants and may also reflect incomplete exploration of the flora. North and South America have the greatest diversity of native actinorhizal genera because of the presence of endemics in isolated desert regions.

The actinorhizal plants are found primarily in the temperate zone.

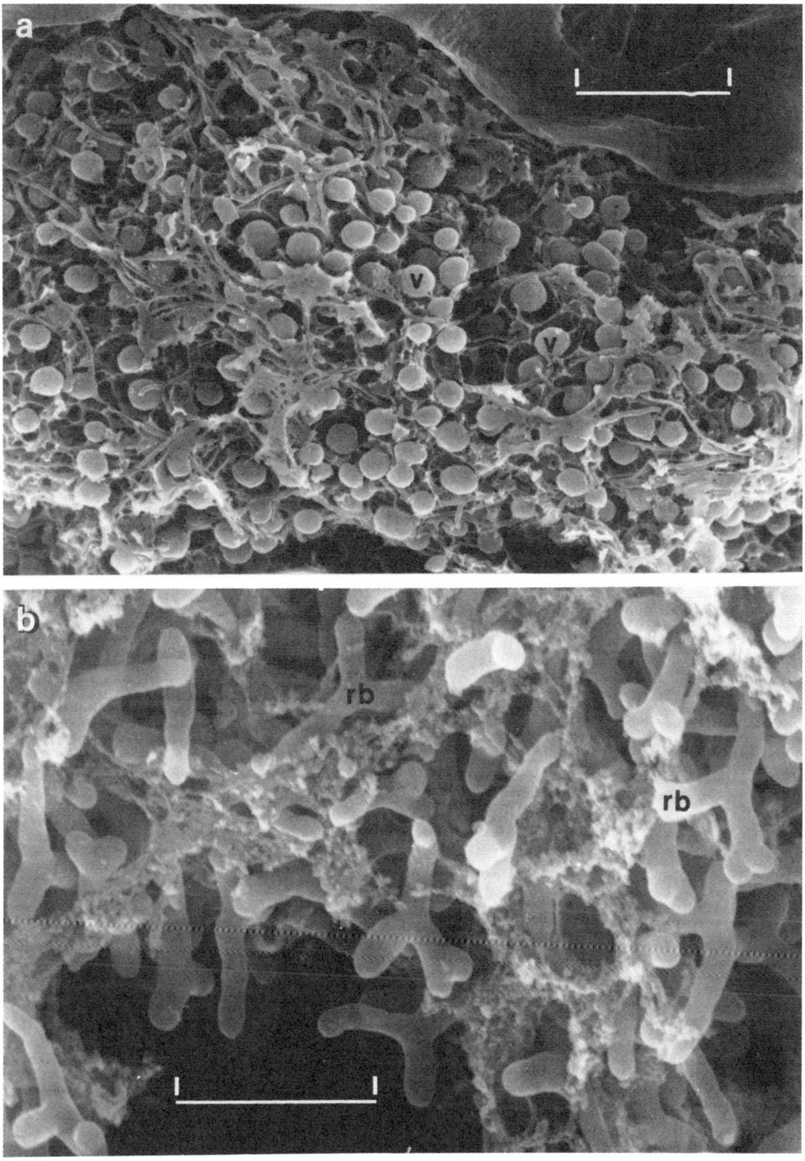

Figure 3 Scanning electron micrographs of root nodule microsymbionts. (a) *Frankia* within a root nodule cell of *Elaeagnus*. The filamentous nature of the endophyte can be seen. Vesicles (v) are visible protruding out of the mass of plant cell cytoplasm and the vegetative mycelium. Bar = 10 μm. (b) Rhizobial bacteroids (rb) within a root nodule cell of *Vicia*. Note that the bacteroids are Y-shaped and single-celled. Bar = 5 μm.

Table II Distribution of Actinorhizal Plants by Continent[a]

Continent	Native genera	Introduced genera
North America	*Alnus, Ceanothus, Cercocarpus, Chamaebatia, Comptonia, Coriaria, Cowania, Datisca, Dryas, Elaeagnus, Myrica, Purshia, Shepherdia*	*Casuarina, Elaeagnus* spp.
South America	*Alnus, Colletia, Coriaria, Discaria, Kentrothamnus, Myrica, Retanilla, Talguenea, Trevoa*	*Casuarina, Elaeagnus*
Africa	*Myrica*	*Casuarina, Elaeagnus, Gymnostoma*
Eurasia	*Alnus, Coriaria, Datisca, Dryas, Elaeagnus, Hippophae, Myrica*	*Casuarina, Gymnostoma*
Oceania (including Australia)	*Allocasuarina, Casuarina, Ceuthostoma, Coriaria, Discaria, Gymnostoma, Myrica*	*Alnus, Elaeagnus, Purshia*

[a]Complied by D.D.B. based on personal observations during his travels and consultation of numerous national floras.

Only some species of the Casuarinaceae and the Myricaceae can be considered truly tropical. Some species of *Alnus* and *Elaeagnus* are also found in the tropical zone but these species are restricted to the higher elevations of montane regions, where the climate is essentially temperate. Actinorhizal plants are especially important in high-latitude countries such as Scandinavia, Canada, and New Zealand, where conditions are not favorable for legumes but actinorhizal plants are abundant and capable of vigorous growth (Silvester, 1977).

Ecologically, the majority of actinorhizal plants are pioneers on nitrogen-poor, open sites. In late glacial and postglacial times (Silvester, 1977), they were abundant and widely distributed in Europe and North America, where they colonized nitrogen-poor glacial deposits and accelerated soil development by adding nitrogen and organic matter. They were displaced as increasing soil nitrogen permitted non-nitrogen-fixing plants to flourish (Silvester, 1976). In modern times actinorhizal plants are much less abundant, being found in a variety of nitrogen-poor sites, including sandy and gravelly sites, shores of streams and lakes, wetlands, and exposed raw mineral soils. Many of the environments occupied by actinorhizal plants are relatively harsh. For example, the chaparral and desert environments occupied by actinorhizal members of the Rosaceae (e.g., *Cercocarpus, Purshia,* and *Chamaebatia*) and Rhamnaceae (e.g., *Ceanothus, Kentrothamnus,* and *Colletia*) in the New World are subject to severe water stress and large temperature fluctuations. Moreover, the riparian environments favored by many species of *Alnus* and *Myrica* throughout North America and Europe are subject to stresses associated with periodic high water and anaerobiosis. Consequently actinorhizal plants adapt to a variety of stresses in addition to their ability to fix nitrogen (Chapters 16 and 18).

IV. Practical Uses of Actinorhizal Plants

Actinorhizal plants have economic potential as timber and fuel wood and in forestry, biomass production, land reclamation, and amenity planting. In Western Europe actinorhizal plants are used primarily for reclamation of industrial wastelands and for land stabilization. *Alnus glutinosa* has been used extensively in the reclamation of mine spoils in Britain and *Alnus, Elaeagnus,* and *Hippophae* have been widely used for land stabilization (Chapter 18). Other uses include use of *Alnus* for timber or biomass production, use of *Alnus incana* to improve the fertility of degenerated forest soils and peat bog bottoms, and use of a wide variety of actinorhizal plants in horticulture for amenity planting. In Eastern Europe and China, *Hippophae rhamnoides* is cultivated for its fruits for human consumption (Chapter 18).

In the Pacific Northwest in North America the primary interest in utilization of actinorhizal plants focuses on *Alnus rubra,* a valuable pulp and timber species. In addition to being managed for pulp wood and timber, this tree is an important alternative crop in areas infected with conifer root diseases and can be used to improve the growth of associated species by increasing the amounts of nitrogen and organic matter in soils (Chapter 17). In highland areas of Central and South America, *Alnus acuminata* is used in pastures to provide shade and shelter for livestock while producing fuelwood and timber (Chapter 15).

Although there are few actinorhizal plants in the tropics and subtropics, several species of Casuarinaceae play an important role as a source of timber and fuelwood (Chapter 16). They are also planted extensively as windbreaks and to stabilize dunes against wind erosion. Like other actinorhizal plants they are valuable for reclamation of mine spoils and rehabilitation of wastelands. Finally, they are highly valued for landscaping, where they provide shade and contribute to the beautification of parks and cities (Chapter 16).

References

Baker, D. D. (1988). Opportunities for autecological studies of *Frankia,* a symbiotic actinomycete. *In* "Biology of Actinomycetes" (Y. Okami, T. Beppu, and H. Ogawara, eds.), pp. 271–276. Jpn. Sci. Soc. Press, Tokyo.

Baker, D. D. (1989). Methods for the isolation, culture and characterization of the

Frankiaceae: Soil actinomycetes and symbionts of actinorhizal plants. *In* "Isolation of Biotechnological Organisms from Nature" (D. P. Labeda, ed.), p. 213–236. McGraw-Hill, New York.

Baker, D., and Seling, E. (1984). *Frankia:* New light on an actinomycete symbiont. *In* "Biological, Biochemical, and Biomedical Aspects of Actinomycetes" (J. Ortiz-Ortiz, L. F. Bojalil, and V. Yakoleff, eds.), pp. 563–574. Academic Press, Orlando, Florida.

Benecke, U. (1969). Symbionts of alder nodules in New Zealand. *Plant Soil* **30,** 145–149.

Bermudez de Castro, F., Miguel, C., and Rodriguez-Barrueco, C. (1976). A study of the capacity of soil to induce nodules of *Alnus glutinosa* (L.) Gaertn. and *Myrica gale* L. with special reference to the specificity of the endophytes. *Ann. Microbiol. (Paris)* **127A,** 307–315.

Bond, G. (1983). Taxonomy and distribution of non-legume nitrogen-fixing systems. *In* "Biological Nitrogen Fixation in Forest Ecosystems: Foundations and Applications" (J. C. Gordon and C. T. Wheeler, eds.), pp. 55–87. Nijhoff/Dr. W. Junk, The Hague.

Huss-Danell, K., and Frej, A.-K. (1986). Distribution of *Frankia* in soils from forest and afforestation sites in northern Sweden. *Plant Soil* **90,** 407–418.

Lalonde, M. (1979). Immunological and ultrastructural demonstration of nodulation of the European *Alnus glutinosa* (L.) Gaertn. host plant by an actinomycetal isolate from North American *Comptonia peregrina* (L.) Coult. root nodules. *Bot. Gaz. (Chicago), Suppl.* **140,** S35–S43.

Lalonde, M., Simon, L., Bousquet, J., and Seguin, A. (1988). Advances in the taxonomy of *Frankia*: Recognition of species *alni* and *elaeagni* and novel subspecies *pommerii* and *vandijkii. In* "Nitrogen Fixation: Hundred Years After" (H. Bothe, F. J. de Bruijn, and W. E. Newton, eds.), pp. 671–680. Fischer, Stuttgart.

Lechevalier, M. P. (1983). Cataloging *Frankia* strains. *Can. J. Bot.* **61,** 2964–2967.

Meesters, T. M., van Genesen, S. T., and Akkermans, A. D. L. (1985). Growth, acetylene reduction activity and localization of nitrogenase in relation to vesicle formation in *Frankia* strains Cc1.17 and Cp 1.2. *Arch. Microbiol.* **143,** 137–142.

Meesters, T. M., Van Vliet, M. W., and Akkermans, A. D. L. (1987). Nitrogenase is restricted to the vesicles in *Frankia* strain $EAN1_{pec}$. *Physiol. Plant* **70,** 267–271.

Pommer, E. H. (1959). Über die Isolierung des Endophyten aus den Wurzelknöllchen *Alnus glutinosa* Gaertn. und über erfolgreiche Re-Infektionsversuche. *Ber. Dtsch. Bot. Ges.* **72,** 138–150.

Rodriguez-Barrueco, C. (1968). The occurrence of the root nodule endophyte of *Alnus glutinosa* and *Myrica gale* in soils. *J. Gen. Microbiol.* **52,** 189–194.

Sasakawa, H., Hiyoshi, T., and Sugiyama, T. (1988). Immuno-gold localization of nitrogenase in root nodules of *Elaeagnus pungens* Thunb. *Plant Cell Physiol.* **29,** 1147–1152.

Silvester, W. B. (1976). Ecological and economic significance of non-legume symbioses. *In* "Proceedings of the First International Symposium on Nitrogen Fixation, Pullman 1974" (W. E. Newton and C. J. Nyman, eds.), pp. 489–506. Washington State Univ. Press, Pullman, Washington.

Silvester, W. B. (1977). Dinitrogen fixation by plant associations excluding legumes. *In* "A Treatise on Dinitrogen Fixation" (R. Hardy and W. Silver, eds.), Vol. 4, pp. 141–190. Academic Press, New York.

Smolander, A., and Sundman, V. (1987). *Frankia* in acid soils of forests devoid of actinorhizal plants. *Physiol. Plant.* **70,** 297–303.

Stowers, M. D. (1985). Further studies on the nodulating potential of *Rubus elipticus* by the actinomycete *Frankia. In* "Nitrogen Fixation Research Progress" (H. J. Evans, P. J. Bottomley, and W. E. Newton, eds.), p. 702. Nijhoff, The Hague.

Tjepkema, J. D., and Torrey, J. G. (1979). Symbiotic nitrogen fixation in actinomycete-nodulated plants. Preface. *Bot. Gaz. (Chicago), Suppl.* **140,** i–ii.

Tjepkema, J. D., Schwintzer, C. R., and Benson, D. R. (1986). Physiology of actinorhizal nodules. *Annu. Rev. Plant Physiol.* **37,** 209–232.

Torrey, J. G., and Berg, R. H. (1988). Some morphological features for generic characterization among the Casuarinaceae. *Am. J. Bot.* **75,** 864–874.

van Dijk, C. (1984). "Ecological Aspects of Spore Formation in the Frankia–Alnus Symbiosis," Ph.D. thesis. State Univ., Leiden, The Netherlands.

Weber, A. (1986). Distribution of spore-positive and spore-negative nodules in stands of *Alnus glutinosa* and *Alnus incana* in Finland. *Plant Soil* **96,** 205–213.

Wollum, A. G., Youngberg, C. T., and Chichester, F. W. (1968). Relation of previous timber and stand age to nodulation of *Ceanothus velutinus*. *For. Sci.* **14,** 114–118.

2

Discoveries, Discussions, and Trends in Research on Actinorhizal Root Nodule Symbioses before 1978

Anton Quispel
Department of Plant Molecular Biology
University of Leiden
2311 VJ Leiden, The Netherlands

The alder, whose fat shadow nourisheth—
Each plant set neere him long flourisheth.
William Browne, 1613

These lines, published in *Britannia's Pastoral Book 1* (song 2, lines 357–358), were first cited in the context of nitrogen fixation by Goldman (1961). They indicate that the beneficial effects of alder trees on the surrounding vegetation, presently explained by their ability to fix nitrogen, have been known and appreciated for centuries. This chapter will survey the observations and research that led from these vague indications to our present knowledge as discussed in the other chapters of this book. Such a review cannot avoid being colored by our present insight.

Three periods will be recognized: (1) the classical period, comprising the last half of the nineteenth century in which the base was laid for our present insight after many mistakes and controversies; (2) an intermediary period, comprising the first part of the twentieth century in which this basal knowledge was deepened and extended without important new discoveries; and (3) the modern period, beginning about 1950 and continuing until 1978, when the first pure culture of *Frankia* became available. The years after 1978 are too recent to permit an objective historical overview.

Each of these periods demands a different approach. The observations

The Biology of *Frankia* and Actinorhizal Plants

of lasting importance made during the first period can only be identified after a comprehensive survey of all relevant observations. Moreover, the early discussions of actinorhizal nodules, known as nonleguminous root nodules until 1978, are best understood in the context of the exciting developments occurring in the research on other nitrogen-fixing organisms (for a review, see Quispel, 1988). The second period contains some excellent contributions, as well as a number of publications that added little to our knowledge, so here it is possible to be selective. For the third period a comprehensive discussion of all publications, even those of high standards, would lead to considerable duplication of other chapters. Here we will restrict ourselves to the main lines of research and the trends that contributed the most to rapid progress.

I. The Classical Period until the End of the Nineteenth Century

The first description of root nodules in the alder (*Alnus glutinosa*) was given by Meyen (1829), who, during a study of parasitic plants, discovered that roots of alder, when growing in the proximity of water or in deep shadows, showed a remarkable tendency to produce large nodules. A microscopic examination of these nodules led him to the remarkable conclusion that they were parasitic plants growing on the roots.

A better description was given by Schacht (1853) but in a paper in 1860 (cited by Woronin, 1866) he considered these nodules to be examples of abnormal growth, and von Jager (1860) suggested that they were caused by insect bites. Meanwhile Gasparrini (1851) described root nodules in *Ceanothus intermedius*, while Oersted (1865) was the first to describe the root nodules of *Hippohaë rhamnoides*.

The classic work of Woronin (1866) produced a great step forward. He described the anatomical structure of the root nodules of *Alnus glutinosa* as consisting of a central cylinder surrounded by a thick parenchymatous cortex. In this cortex he observed intercellular hyphae passing through the cell walls from cell to cell, while inside the cells the tips of these hyphae formed round vesicular swellings, which he considered as fungal spores (Fig. 1). On the basis of a presumed resemblance to the parasitic fungus *Schinzia cellulicola*, he proposed the name *Schinzia alni* for the root nodule "parasite" of alder.

This identification was highly disputed. Gravis (1879) concluded from certain intracellular structures in alder root nodules that a myxomycete, *Plasmodium*, was present, which was confirmed by Möller (1885) for *Alnus* and by Warming (1876) for Elaeagnaceae. Correspondence between Gravis and Woronin (1885) led to the suggestion that these root

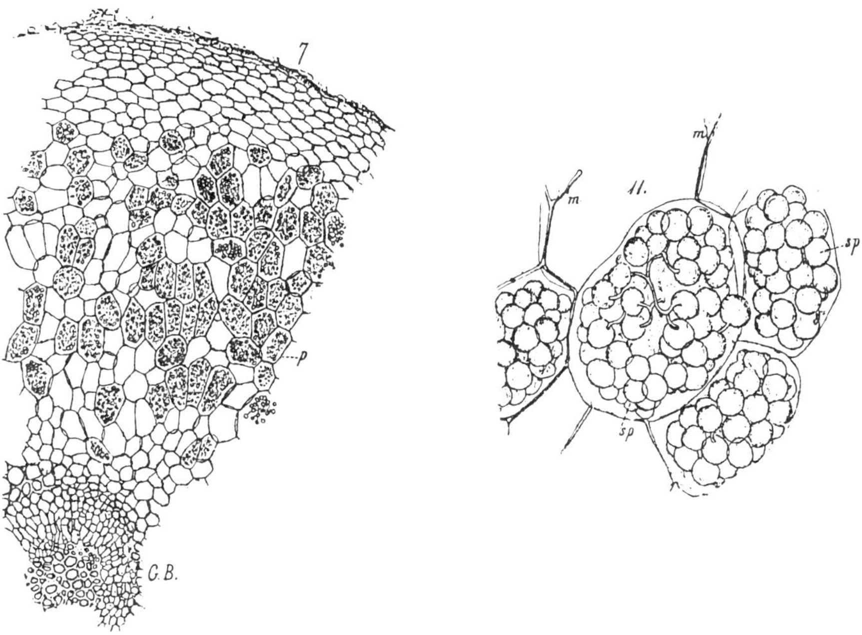

Figure 1 Reproduction of two figures of portions of root nodules of *Alnus glutinosa* from Woronin (1866). 7. Part of a nodule cross section with cells in the region "p" containing "the parasitic fungus *Schinzia alni.*" The microsymbiont is now known to be the actinomycete *Frankia.* 11. Enlargement of four infected cells showing hyphae and "spores" of the "fungus." The "spores" are now known to be vesicles.

nodules contained both a *Schinzia* and a *Plasmodium.* The question remained whether both were essential for root nodule formation or whether the *Plasmodium* was merely a companion.

These discussions of the root nodules of *Alnus* and the Elaeagnaceae cannot be considered in isolation from the studies of leguminous root nodules in the same period. Woronin (1866) in his classical paper made extensive comparisons between alder nodules and those of lupines. The quite different cytological contents of the two types of nodules convinced him that they contained different types of microorganisms. Subsequent studies, however, showed the presence of hyphae in leguminous root nodules as well (see Quispel, 1988). From our present vantage point such observations are easy to understand. The formation of infection threads by deposits of cell-wall material in leguminous as well as in actinorhizal root nodule infections can give the impression of fungal hyphae. The masses of spores in certain actinorhizal root nodules, or merely bad fixation techniques, may have given the impression of plasmodia.

The situation became still more complicated when Frank (1887), primarily on the basis of cytological studies by his student Brunchorst (1885), denied the microbial character of the contents of leguminous root nodules, considered the bacteroids as mere protein deposits, and extended this denial to the nodules of *Alnus* and Elaeagnaceae. In a subsequent study Brunchorst (1886–1888, 1887) himself was so struck by the difference between the cytology of leguminous and actinorhizal root nodules that he became convinced that in the actinorhizal nodules a microorganism was indeed involved. This microorganism formed hyphae with swollen sporangia from which spores were formed. Since this microorganism differed substantially from known myxomycetes or hyphal fungi he proposed a new name: *Frankia subtilis.*

A few years later Frank (1891) returned to his original opinion that a fungus was present in alder root nodules so that the nodules had to be regarded as "Pilzkammer" or *Mycodomatia.* Möller (1890), after dismissing his original observation of plasmodia as the result of bad fixation techniques, accepted the conclusions of Brunchorst though he failed to see septae in the hyphae and considered the microsymbiont to be a one-celled hyphal fungus. His description of the formation of spores from the vesicles through subsequent cell-wall formations may be explained by confusion of vesicles and sporangia, a confusion that persisted far into the twentieth century. Some cytological differences between the structures of the microsymbiont in *Alnus* and *Myrica* made him propose a new species, *Frankia Brunchorstii.*

The magnificent experiments of Hellriegel and Wilfarth (1888) demonstrating symbiotic nitrogen fixation in leguminous root nodules gave an explanation for the well-known agricultural difference between nitrogen-using and nitrogen-accumulating plants. Many observations and experiments (e.g., Mayer, 1895; Dinger, 1895) suggested that plants like alder belonged to the nitrogen-accumulators but proof was obtained in the experiments of Hiltner (1895), which were based on the experimental principles of Hellriegel and Wilfarth (1888). In these experiments the growth of *Alnus glutinosa* was compared in N-free soils with or without the addition of nitrates, and with or without infection with root nodule suspensions leading to the formation of root nodules. After one year it was clear that the plants could not develop on N-free soils or nutrient solutions unless nitrates had been added or the plants had formed root nodules. Although these results could still be criticized because no determinations of N-content were available, Hiltner preferred to continue to grow his plants and not destroy them for N-determinations. The continuing good growth of nodulated plants for eight years without any additions of nitrogenous compounds was considered to be sufficient evidence to conclude that nodulated plants could use atmospheric nitrogen (Nobbe and Hiltner, 1904). These later experiments also included Elaeagnaceae.

Since these authors had recently concluded that in leguminous plants the fixation of N_2 was correlated with the formation of the bacteroids, a comparison with the development of the microsymbiont of *Alnus* root nodules could be predicted. This study led to the conclusion that the microsymbiont was not a fungus but was related to the bacteria (Hiltner, 1898).

While these European studies were based on species of *Alnus, Myrica, Elaeagnus,* and *Hippophaë,* in other parts of the world other genera were being examined. Beal (1890) and Atkinson (1892) worked on *Ceanothus* and proposed a new species, *Frankia ceanothi,* based on cytological comparisons with *Alnus.* Schimper (1891) and Janse (1897) discovered root nodules in *Casuarina* species. The latter author described the cell contents as filled with hyphae and bacteroidlike forms.

II. The Intermediary Period from the Beginning of the Twentieth Century until 1950

At the beginning of the twentieth century the foundation was laid for our understanding of the form and function of the actinorhizal root nodules. There was a general consensus about the structure of the nodules and the work of Hiltner had made it very probable that these nodules functioned in a similar way as those of leguminous plants. Yet there remained several discrepancies concerning the nature of the microsymbionts and the interpretation of the vesicles as sporangia, spores, or degenerative structures. Although the bacteria of leguminous root nodules had been isolated (Beyerinck, 1888) these bacteria appeared to be restricted to Leguminosae and did not appear to infect other types of plants (Kordes, 1925). The very different cytological structures in the cells of leguminous and actinorhizal root nodules definitely indicated other types of microorganisms.

Shibata (1902) made an important contribution to the cytology of the root nodules of *Alnus* and *Myrica.* He observed hyphae primarily near the meristems and reported that they seemed to divide into small rods in somewhat older cells. He described the vesicles as degenerative forms and noted the resorption of all hyphae and vesicles in the oldest parts of the nodules. He was the first to consider the microsymbiont of *Alnus* as a *Mycobacterium,* and the microsymbiont of *Myrica* as an actinomycete. A later paper by Shibata and Tahara (1917) included *Elaeagnus, Coriaria,* and *Casuarina* and gave evidence for an actinomycetelike organism in all nodules. In this work a clear distinction was made between the round vesicles of *Alnus* and *Elaeagnus* and the club-shaped forms in *Myrica* and *Coriaria,* while no real vesicles were observed in *Casuarina.*

We do not need to discuss the results of most authors who described root nodules of *Alnus, Elaeagnus,* or *Casuarina* species since they did not add important new information to the already existing knowledge. Some confusion still remained about the presence of bacterialike cells. Peklo (1910) used the word "bacteroids" and distinguished between their presence within cells and in the intercellular spaces. He also distinguished between nodules with or without such intercellular bacteroids. Arcularius (1928) too used the word "bacteroids" and described them as being present together with the actinomycetous hyphae. In *Casuarina cunninghamiana,* MacLuckie (1923) found such bacteroids primarily in the degenerative zone. Miehe (1918) introduced the word "Rhizothamnia" for the root nodules of *Casuarina,* which he defined as perennial, continuously growing, metamorphosed roots. The descriptions of alder nodules by Krebber (1932) left little doubt that the microsymbiont belonged to the actinomycetes. Finally, the work of Schaede (1933, 1938, 1962) can be considered as the culmination of the descriptive cytological studies of actinorhizal root nodules. His very careful microscopic studies, fine photographs, and thorough interpretation put an end to many speculations and misinterpretations. The only question that remained unsolved was related to the nature of the bacterialike cells for which he again used the word "bacteroids" without suggesting any relationship to the bacteroids of the leguminous root nodules. There remained uncertainty whether they belonged to the definitely actinomycetelike hyphae, either as a phase of development or as the result of a "dissymbiotic disturbance," or represented a second microorganism.

Although the actinomycetous character of the hyphae was now accepted by virtually all further workers in this field, definite proof that the microsymbionts belonged to the actinomycetes could not yet be obtained. All descriptions were made on the forms found within nodules, which could be markedly affected by the surrounding host cells. The same applied to conclusions about specific differences between the microsymbionts based on different forms of the vesicles (round or club-shaped). Better indications for specific differences were obtained in cross-inoculation experiments that showed that all *Alnus* species could be nodulated by infection with crushed nodules from other *Alnus* species, but not by crushed nodules from the Elaeagnaceae, while on the other hand *Elaeagnus* and *Hippophaë* could be mutually infected but inoculation with crushed nodules from *Alnus* species on roots of these plants remained without success (Roberg, 1934b, 1938). Similar experiments led to conclusions about specific differences between the microsymbionts from *Casuarina* species and *Alnus* (Miehe, 1918). All *Casuarina* species could be mutually infected (Mowry, 1933).

Definite proof that the microsymbionts really belonged to the actinomycetes could only be obtained after successful isolation and cultivation. Of the many efforts to isolate and cultivate the microsymbionts only those with negative results can be taken seriously (e.g., Krebber, 1932). All those presumed isolations where subsequent inoculation experiments failed to lead to any nodule formation or those experiments where such inoculations were not even tried can be forgotten. This also includes isolations where nodules were formed many months after inoculation (e.g., von Plotho, 1941) since such late nodules certainly are merely the result of spontaneous infections (Bouwens, 1943).

Although the fixation of atmospheric nitrogen had been made highly probable by the experiments of Hiltner, some doubts still remained and other theories about symbiotic functions, for example, based on prevailing ideas about mycorrhizae, still could play a role (Miehe, 1918). A new impetus for the acceptance of the N_2-fixation hypothesis was given by the repetition and extension of Hiltner's experiments in water cultures of *Alnus, Myrica,* and the Elaeagnaceae by Krebber (1932) and by Roberg (1934a). Virtanen and Saastomoinen (1933, 1936) demonstrated increases of nitrogenous substances in the surroundings of alders and the excretion of amino acids by their roots. Good growth of nodulated plants on N-free substrates was shown for *Casuarina* by Aldrich-Blake (1932) and by Mowry (1933) and for *Coriaria* by Kataoka (1930).

III. The Modern Period from 1950 to 1978

Notwithstanding the publication of some excellent papers on actinorhizal root nodules, the first half of our century did not show real progress and nitrogen fixation by actinorhizal plants was still a rather neglected field of biological research. This gradually changed after World War II and the number of studies on actinorhizal root nodules is still increasing exponentially. There are three different reasons for this increased interest. Progress in science depends on influential scientists whose publications draw attention to an important topic. In this respect we owe much to the large number of influential papers by G. Bond and collaborators from the University of Glasgow from 1954 to 1976. A second reason for modern progress was the stimulating effect of the International Biological Program in which Bond took an important part. Finally, the development of many important new methods and techniques, including the electron microscope, use of isotopes, and the acetylene reduction technique, made possible new approaches to studies of symbiotic nitrogen fixation.

A. The Work of G. Bond

Space limitations do not allow us to cite all the publications of Bond and his students. We must restrict ourselves to those publications in which a new approach was begun and refer to some of Bond's many review articles in which all individual papers are cited (1959, 1963, 1967a, 1974).

After some research with leguminous plants, Bond began his work with actinorhizal plants with a description of the root nodules of *Myrica gale* in which he compared growth and N-content of inoculated and uninoculated plants on N-free nutrient solutions (Bond, 1949). This was soon followed by an estimation of their contribution to the N-economy of natural vegetation (Bond, 1951). A feature that was studied in detail was the negative geotropism of the rootlets that grow from the tips of the root nodules. The work on *Myrica* was soon followed by similar work on N_2 fixation in *Alnus glutinosa* and *Hippophaë rhamnoides* in which the effect of combined nitrogen on nodulation and N_2 fixation was compared with similar effects in the Leguminosae. While he still relied on the classical methods of growth and determinations of total N in these first experiments, from 1954 onward he applied the $^{15}N_2$ method to actinorhizal root nodules and thus unequivocally proved their ability to fix N_2 (Bond *et al.*, 1954). In experiments with excised root nodules it was shown that these structures really were the sites of N_2 fixation (Bond and MacConnell, 1955), while subsequent studies showed that the fixed N was immediately transported upward through the xylem. These studies were extended by biochemical analyses and showed that in *Alnus* the main transport amino acid was citrulline (first found in the root nodules of alder by Miettinen and Virtanen in 1952). Experiments with plants in the laboratory were extended to the field and estimates were made of the importance of N_2 fixation in natural vegetation. In many cases the contributions of actinorhizal plants exceeded the estimated values for leguminous plants and thereby focused attention on the important role that actinorhizal plants play in the N-economy of ecosystems.

Initially Bond's work was restricted to plants native in Britain, but subsequent work extended his observations to plants from other regions of the world, including *Casuarina, Ceanothus, Shepherdia,* and *Coriaria.* Cross-inoculation experiments led to a greater refinement of the different infection groups. In some cross-inoculations noneffective nodules were formed (Bond, 1967b). Successful inoculation with soil suspensions showed that the infective organisms must be present in soils, an observation that later led to the discovery of infective organisms in soils where actinorhizal plants had been absent for many years (Rodriguez-Barrueco and Bond, 1976). Finally, we want to mention his many contributions on the effects of ammonium and nitrates, oxygen concentration, pH, and diurnal light–dark rhythms on nodule formation and activity.

B. The International Biological Program (IBP)

The International Biological Program (IBP) was established by the International Council of Scientific Unions (ICSU) in 1964 as a counterpart to the International Geophysical Year. The aim of IBP was defined as "the biological basis of productivity and human welfare" and one of the main items concerned biological productivity in agriculture and natural communities, including the contribution of biological nitrogen fixation. The program on nitrogen fixation consisted of six parts, one of which was directed to the importance of actinorhizal plants. In this respect it was again Bond (1976) who played a central role in collecting data from all over the world. This survey has greatly stimulated subsequent research on plants with N_2-fixing root nodules.

Moreover the IBP stimulated research on N_2 fixation and organized conferences on N_2 fixation in which research on the nodules of actinorhizal plants was included (Lie and Mulder, 1971; Nutman, 1976). At the same time that the synthesis meetings of IBP emphasized the importance of biological nitrogen fixation for the N-economy of the world, the oil crisis and increasing prices of commercial N-fertilizers showed the necessity for more and better applications of biological systems. This practical impetus together with the fascinating possibilities of modern biology greatly stimulated research on all aspects of biological nitrogen fixation.

C. New Research Methods and Their Results

1. Studies on Cytology and Ultrastructure

Great progress was made possible by the application of electron microscopy while the traditional methods of light microscopy led to new conclusions through the application of new cytochemical methods. The first electron microscopic studies of actinorhizal root nodules were made by Becking *et al.* (1964) with *Alnus glutinosa* and by Silver (1964) with *Myrica cerifera*. In both plants the hyphae were thin and had an actinomycete-like character. Most spectacular were the remarkable structures of the vesicles of *Alnus* nodules with their incomplete cell-wall septae. Such cell-wall septae were seen in the club-shaped structures of *Myrica* as well.

In *Alnus* nodules the so-called "bacteroids" were seen but Becking *et al.* (1964) preferred to describe them as bacterialike cells or later, based on their typical form, as "polyhedral-shaped cells." It was suggested that they were formed by fragmentation of tightly interlaced hyphae. Further studies by Gardner (1965, 1976) showed that the septae in the vesicles increased during their maturation, which led to the erroneous conclusion that the bacterialike cells were formed from the vesicles. The absence of vesicles in noneffective root nodules (Becking, 1965; Mian *et al.*, 1976) led to the conclusion that they might have a special function in N_2 fixation. More direct evidence for a special metabolic activity of the vesicles, which made them especially suitable as sites for nitrogen reduction, was obtained by Akkermans (1971), who observed after application of tetrazolium salts that the vesicles were the main sites of tetrazolium reduction. Application of neo-tetrazolium led to the formation of crystals of formazan inside the vesicles.

Progress leading to the correct interpretation of the bacterialike cells began with the discovery by light microscopy of the so-called "Spindels" or "granulated bodies" by Käppel and Wartenberg (1958) and their interpretation as intermediary stages in the formation of the bacterialike cells. The electron microscopic studies of van Dijk and Merkus (1976) showed conclusively that although the "Spindels" superficially resembled the vesicles, they could be distinguished by their normal cell walls as compared with the incomplete and irregular septae in the vesicles. The "Spindels" or granulated bodies thus were real sporangia that formed the bacterialike cells as spores. The vesicles had nothing to do with sporangia.

Many ultrastructural studies described the still insufficiently identified striated bodies and the multilayered cell walls, especially of the vesicles. All intracellular stages appeared to be surrounded by encapsulations that contained pectin, as shown by cytochemical techniques, and were continuous with the host cell wall (Lalonde and Devoe, 1976; Lalonde and Knowles, 1975).

2. Morphogenesis of Root Nodules

Pommer (1956) and Taubert (1956) simultaneously published their results of studies on infection and nodulation in *Alnus glutinosa*. Taubert showed the first pictures of infections through root hairs. Both authors described the two-phase nodulation process. First the cells of the cortex appeared to be infected, then lateral root primordia developed that were infected from the cortical cells and then formed the ultimate rhizothamnia (nodule lobes). This process was confirmed and further studied by Angulo Carmona (1974). A comparison of the pattern of this lateral root induction with the normal pattern led to the conclusion that the primordia for the nodule lobes were induced by the infections in the cortex. This two-step nodulation process was also observed in *Casuarina* (Torrey, 1976) and *Comptonia* (Callaham and Torrey, 1977). Time-lapse photography permitted a still more detailed analysis of the nodulation process in the latter plant (Bowes *et al.*, 1977).

Only a few observations are available on the role of plant hormones in this nodulation process. Angulo Carmona (1974) suggested that auxins are involved since application of indole acetic acid induces root primordia. This effect is in accordance with the observed effect of this substance on nodule formation (Pizelle, 1970). Fully developed nodules of alder have a high level of indole acetic acid (Dullaart, 1970), gibberellic acid (Henson and Wheeler, 1977a), and cytokinin synthesis (Henson and Wheeler, 1977b). Cytokinin may induce pseudonodules (Rodriguez Barrueco and Bermudez de Castro, 1973). On the other hand, Silver *et al.* (1966) explained the negative geotropism of nodule roots in *Myrica* by high levels of indole acetic acid oxidase and consequently very low content of this auxin in the nodules.

3. N_2 Fixation Studies

The availability of the stable isotope ^{15}N in N_2 was a great advance in the study of nitrogen fixation because it made possible unequivocal proof of N_2 fixation. This method was first applied to actinorhizal root nodules by Virtanen *et al.* (1954) in *Alnus* and by Bond *et al.* (1954) in *Alnus, Myrica,* and *Hippophaë*. The acetylene reduction method was first applied in *Myrica asplenifolia* (*Comptonia peregrina*) and *Alnus incana* spp. *rugosa* by Stewart *et al.* (1967).

The results with $^{15}N_2$ gave unquestionable proof of N_2 fixation. The albeit highly improbable suggestion that contaminating rhizosphere organisms and not the microsymbionts within the nodules were responsible for N_2 fixation could be excluded after good growth of alder plants under axenic conditions (Quispel, 1954a). Subsequent experiments led to the localization of the N_2 fixation: N_2 fixation in excised nodules showed that these were indeed the structures where N_2 fixation took place (Bond, 1957; Ziegler and Hüser, 1963). A further step in this localization was taken when van Straten *et al.* (1977) succeeded in demonstrating reduction of acetylene after addition of dithionite and low levels of ATP to microsymbiont clusters prepared according to Akkermans (1971). The inhibition of the synthesis of nitrogenase by chloramphenicol showed that this synthesis is in a prokaryote and thus in the microsymbiont (Skeffington and Stewart, 1976).

Egle and Munding (1951) demonstrated that nodules of *Alnus, Hippophaë,* and *Myrica* contain more hemes than do the roots, but they did not succeed in demonstrating the presence of hemoglobin. This was achieved for *Casuarina* nodules with a microspectroscopic method by Davenport (1960). Application of the acetylene reduction method together with studies on transport of ^{14}C, applied as $^{14}CO_2$ to the leaves of alder, showed that N_2 fixation is dependent on a supply of carbohydrates from the leaves (Wheeler and Bowes, 1974). This explains the diurnal and seasonal periodicity of nitrogenase activity as demonstrated by Pizelle (1975).

4. Field Studies

Total N-determinations in plants and soils had already given some indication of the contribution of actinorhizal N_2 fixation to the N-economy in the field (Virtanen, 1957; Crocker and Major, 1955; Goldman, 1961). The first applications of $^{15}N_2$ in field experiments were reported by Stevenson (1958) and of the acetylene reduction method by Stewart and Pearson (1967). A critical evaluation of both methods, directly in the field or after excision of the nodules, was made by Akkermans (1971) for *Alnus glutinosa* and *Hippophaë rhamnoides.* Various studies led to extrapolations of annual rates of nitrogen fixation, including alder in groves (Akkermans, 1971) or in woods (Tarrant and Trappe, 1971), *Hippophaë* on sand dunes (Akkermans, 1971), *Myrica* vegetation (Sprent *et al.*, 1978), *Dryas drummondii* following recession of glaciers (Lawrence *et al.*, 1967), *Ceanothus* vegetation (Delwiche *et al.*, 1965), *Purshia* soils (Wagle and Vlamis, 1961), and *Casuarina* on tropical dunes (Dommergues and Mangenot, 1970). Extensive data on N_2 fixation by actinorhizal plants in temperate forests and natural vegetation were collected in a symposium in 1979 (Gordon *et al.*, 1979). These studies, followed by many more recent publications, focused attention on the importance of these plants for the N-economy in natural vegetation, forestry, and reclamation of barren soils.

5. Studies of the Microsymbionts

Although Hawker and Fraymouth defended the myxomycetelike character of the microsymbionts as late as 1951, all other investigators followed the conclusions of Schaede and considered the microsymbionts to be actinomycetes. However, some uncertainty remained because all taxonomic conclusions had to be based on observations of microsymbionts that had been subjected to the influences of surrounding host cells. For practical reasons Becking (1970) proposed to reestablish the old name *Frankia* Brunchorst as the genus name for the microsymbionts although the original author of this name was convinced that he had applied this name to a fungus (Brunchorst, 1886–1888). Becking's further assignment of frankiae to species based on the form of the vesicles was premature and distinctions based on cross-inoculation experiments have also proved unreliable (see Chapter 3). A very important distinction was discovered by van Dijk (1978) when his cross-inoculation experiments revealed that in *Alnus glutinosa* the differences between nodules with spores (designated sp^+) and without spores (sp^-) are based on differences between the strains of the microsymbiont.

All efforts to isolate and cultivate the microsymbionts were unsuccessful for many years. Quispel (1954a) demonstrated that disinfected and peeled root nodules of alder, which after six weeks of incubation on a rich nutrient agar did not show any growth of microorganisms, were still infective. This suggested that the microorganisms that had grown out of the other nodules had nothing in common with the real infective microsymbiont. Moreover, this method permitted preparation of inoculant that was axenic according to the usual microbiological criteria. After a study of the relationship between the amount of inoculant and the number of root nodules obtained, this relationship could be used for a more sensitive, although very time-consuming, method for measuring the eventual increase of the number of infective particles during incubation of inoculated nutrient solutions (Quispel, 1954b, 1955). A definite increase could be observed, especially during the second week of incubation, but only in those nutrient media to which an alcoholic extract of alder roots had been added. After further purification of these extracts, the reproducibility of these experiments was increased. It was shown that the active substance in these extracts was a neutral lipid. In some highly infective nutrient media the presence of actinomycete hyphae was shown (Quispel, 1960).

Meanwhile Pommer (1959) published the results of isolation experiments from root nodules of *Alnus glutinosa* in which an actinomycete was isolated and cultivated on a simple nutrient medium containing glucose and asparagine. Unfortunately his cultures were lost (see Chapter 3). The still remaining irreproducibility in the experiments of Quispel could only be resolved after distinction between sp^+ and sp^- nodules was made. Increases in the number of infective particles were obtained with sp^- nodules and were far more difficult with the sp^+ type. This observation and further improvements in the technique and the nutrient solutions led to more reproducible experiments in which a correlation was found between infectivity and the amount of hyphae in the incubated media, indicating that these hyphae indeed belonged to the cultivated microsymbiont (Quispel and Tak, 1968). At the same time Callaham *et al.* (1978), after using Quispel's method for the preparation of axenic inoculant, succeeded in isolating and cultivating the microsymbiont from root nodules of *Comptonia* on rather simple media without the addition of lipids. The typical sporangia, soon also observed in Quispel's cultures of the *Alnus* microsymbiont, were identical to the sporangia inside the nodule (van Dijk and Merkus, 1976) and were in all respects similar to the sporangia described by Pommer (1959).

We thus can conclude that the first cultures of *Frankia* were obtained in 1959 although we had to wait till 1978 before cultures became available for further research. This research has included the isolation of many more strains from a variety of host plants and the elucidation of their growth requirements, the evaluation of the still existing differences in the success of isolations, experiments on infection and nodulation, and finally the beginning of molecular genetics of *Frankia*. These topics fall outside the scope of this historical introduction and will be discussed in subsequent chapters.

References

Akkermans, A. D. L. (1971). "Nitrogen Fixation and Nodulation of *Alnus* and *Hippophaë* under Natural Conditions," thesis. Univ. Leiden, The Netherlands.

Aldrich-Blake, R. N. (1932). On the fixation of atmospheric nitrogen by bacteria living symbiotically in root nodules of *Casuarina equisetifolia*. *Oxford For. Mem.* **14,** 1–20.

Angulo Carmona, A. F. (1974). La formation des nodules fixateurs d'azote chez *Alnus glutinosa* (L.) Vill. *Acta Bot. Neerl.* **23,** 257–303.

Arcularius, J. (1928). Cytologische Untersuchungen an einigen endotrophen Mykorrhizen. *Zentralbl. Bakteriol., Parasitenkd., Infektionskr. Hyg., Abt. 2, Naturwiss.: Allg. Landwirtsch. Tech. Mikrobiol.* **74,** 191–207.

Atkinson, G. F. (1892). The genus *Frankia* in the United States. Bull. Torrey Bot. Club **19,** 171–177.

Beal, W. J. (1890). Root-galls in *Ceanothus americanus*. *Bot. Gaz. (Chicago)* **15,** 232.

Becking, J. H. (1965). *In vitro* cultivation of alder root-nodule tissue containing the endophyte. *Nature (London)* **207,** 885–887.

Becking, J. H. (1970). *Frankiaceae* fam. nov. (Actinomycetales) with one new combination and six new species of the genus *Frankia* Brunchorst 1886. *Int. J. Syst. Bacteriol.* **20,** 201–220.

Becking, J. H., de Boer, W. E., and Houwink, A. L. (1964). Electron microscopy of the endophyte of *Alnus glutinosa*. *Antonie van Leeuwenhoek* **30,** 343–376.

Beyerinck, M. W. (1888). Die Bacterien der Papilionacean Knöllchen. *Bot. Ztg.* **46,** 725–735, 741–750, 757–772, 781–790, and 797–804.

Bond, G. (1949). Root nodules of bog myrtle or sweet gale (*Myrica gale* L.) *Nature (London)* **163,** 730.

Bond, G. (1951). The fixation of nitrogen associated with the root nodules of *Myrica gale* L., with special reference to its pH relation and ecological significance. *Ann. Bot. (London)* **15,** 447–459.
Bond, G. (1957). Isotopic studies of nitrogen fixation in non-legume root nodules. *Ann. Bot. (London)* **21,** 513–521.
Bond, G. (1959). Fixation of nitrogen in non-legume root-nodule plants. *In* "Utilization of Nitrogen and Its Compounds by Plants" (H. K. Porter, ed.), pp. 59–72. Cambridge Univ. Press, Cambridge, England.
Bond, G. (1963). The root nodules of non-leguminous angiosperms. *Symp. Soc. Gen. Microbiol.* **13,** 72–91.
Bond, G. (1967a). Fixation of nitrogen by higher plants other than legumes. *Annu. Rev. Plant Physiol.* **18,** 107–126.
Bond, G. (1967b). Nitrogen fixation in some non-legume root nodules. *Phyton (Buenos Aires)* **24,** 57–66.
Bond, G. (1974). Root-nodule symbioses with actinomycete-like organisms. *In* "The Biology of Nitrogen Fixation" (A. Quispel, ed.), pp. 342–378. Elsevier/North-Holland Publ., Amsterdam.
Bond, G. (1976). The results of the IBP survey of root-nodule formation in non-leguminous angiosperms. *In* "Symbiotic Nitrogen Fixation in Plants" (P. S. Nutman, ed.), pp. 443–474. Cambridge Univ. Press, Cambridge, England.
Bond, G., and MacConnell, J. T. (1955). Nitrogen fixation in detached non-legume root nodules. *Nature (London)* **176,** 606.
Bond, G., Fletcher, W. W., and Ferguson, T. P. (1954). The development and function of the root nodules of *Alnus, Myrica* and *Hippophaë*. *Plant Soil* **5,** 309–323.
Bouwens, H. (1943). On the symbiont of *Alnus glutinosa, Alnus incana* and *Hippophaë rhamnoides*. *Antonie van Leeuwenhoek* **9,** 107–114.
Bowes, B., Callaham, D., and Torrey, J. G. (1977). Time lapse photographic observations of morphogenesis in root nodules of *Comptonia peregrina*. *Am. J. Bot.* **64,** 516–525.
Brunchorst, J. (1885). Über die Knöllchen an den Leguminosenwurzeln. *Ber. Dtsch. Bot. Ges.* **3,** 241–257.
Brunchorst, J. (1886–1888). Über einige Wurzelanschwellungen, besonders diejenigen von *Alnus* und den *Elaeagnaceen*. *Unters. Bot. Inst. Tubingen* **2,** 150–177.
Brunchorst, J. (1887). Die Struktur der Inhaltskörper in den Zellen einiger Wurzelanschwellungen. *Bergens Mus. Aarsb.* 235.
Callaham, D., and Torrey, J. G. (1977). Prenodule formation and primary nodule development in roots of *Comptonia* (Myricaceae). *Can. J. Bot.* **55,** 2306–2318.
Callaham, D., Del Tredici, P., and Torrey, J. G. (1978). Isolation and cultivation *in vitro* of the actinomycete causing root nodulation in *Comptonia*. *Science* **199,** 899–902.
Crocker, R. L., and Major, J. (1955). Soil development in relation to vegetation and surface age at Glacier Bay, Alaska. *J. Ecol.* **43,** 427–448.
Davenport, H. E. (1960). Haemoglobin in the root nodules of *Casuarina cunninghamia*. *Nature (London)* **186,** 653–654.
Delwiche, C. C., Zinke, P. J., and Johnson, C. M. (1965). Nitrogen fixation by *Ceanothus*. *Plant Physiol.* **40,** 1045–1047.
Dinger, R. (1895). De els een stikstofverzamelaar. *Landbouwkd. Tijdschr.* pp. 167–192.
Dommergues, Y., and Mangenot, F. (1970). "Ecologie Microbienne des Sols," pp. 158–187 and 595–656. Masson, Paris.
Dullaart, J. (1970). The auxin content of root nodules and roots of *Alnus glutinosa* (L.) Vill. *J. Exp. Bot.* **21,** 975–984.
Egle, K., and Munding, H. (1951). Über den Gehalt an Häminkörpern in den Wurzelknöllchen von Nicht-Leguminosen. *Naturwissenschaften* **38,** 548–549.

Frank, B. (1887). Sind die Wurzelanschwellungen der Erlen und Elaeagnaceen Pilzgallen? *Ber. Dtsch. Bot. Ges.* **5,** 50–58.

Frank, B. (1891). Über die Aufverdauung von Pilzen abzielende Symbiosen mit endotrophen Mykorhizen begabten Pflanzen, sowieder der Leguminosen und Erlen. *Ber. Dtsch. Bot. Ges.* **9,** 244–253.

Gardner, I. C. (1965). Observations on the fine structure of the endophyte of the root nodules of *Alnus glutinosa* (L.) Gaertn. *Arch. Mikrobiol.* **51,** 365–383.

Gardner, I. C. (1976). Ultrastructural studies of non-leguminous root nodules. *In* "Symbiotic Nitrogen Fixation in Plants" (P. S. Nutman, ed.), pp. 485–495. Cambridge Univ. Press, Cambridge, England.

Gasparrini, G. (1851). Osservazioni sulla struttura die tubercoli spongiolary di alcune plante legumineuse. *Atti Accad. Sci. fys. mat. Napoli* **6,** 221–139.

Goldman, C. R. (1961). The contribution of alder trees (*Alnus tenuifolia*) to the primary productivity of Castle Lake, California. *Ecology* **42,** 282–288.

Gordon, J. C., Wheeler, C. T., and Perry, D. A. (eds.) (1979). "Symbiotic Nitrogen Fixation in the Management of Temperate Forests." For. Res. Lab., Oregon State Univ., Corvallis, Oregon.

Gravis, A. (1879). Observations anatomiques sur les excroissances des racines de l'aulne. *Bull. Soc. R. Bot. Belg.* **18,** 50–60.

Hawker, L. E., and Fraymouth, J. (1951). A re-investigation of the root-nodules of *Elaeagnus, Hippophaë, Alnus* and *Myrica,* with special reference to the morphology and life history of the causative organisms. *J. Gen. Microbiol.* **5,** 369–386.

Hellriegel, H., and Wilfarth, H. (1888). Untersuchungen über die Stickstoffnahrung der Gramineen und Leguminosen. *Beil. Z. Ver. Rübenzuckerind. D. Reiches,* pp. 1–234.

Henson, L. E., and Wheeler, C. T. (1977a). Hormones in plants bearing nitrogen-fixing root-nodules: Gibberellin-like substances in *Alnus glutinosa* (L.) Gaertn. *New Phytol.* **78,** 373–382.

Henson, L. E., and Wheeler, C. T. (1977b). Hormones in plants bearing nitrogen-fixing root nodules: Partial characterization of cytokinins from root nodules of *Alnus glutinosa* (L.) Gaertn. *J. Exp. Bot.* **28,** 1076–1086.

Hiltner, L. (1895). Über die Bedeutung der Wurzelknöllchen von *Alnus glutinosa* für die Stickstoffernährung dieser Pflanze. *Landwirtsch. Vers.-Stn.* **46,** 153–161.

Hiltner, L. (1898). Über Entstehung und physiologische Bedeutung der Wurzelknöllchen. *Forstl. Naturwiss. Z.* **7,** 415–423.

Janse, J. M. (1897). Les endophytes radicaux de quelques plantes Javanaises. *Ann. Jardin Bot. Buitenzorg* **14,** 53–201.

Käppel, M., and Wartenberg, H. (1958). Die Formenwechsel des *Actinomyces alni* Peklo der Wurzeln von *Alnus glutinosa* Gaertn. *Arch. Mikrobiol.* **30,** 46–63.

Kataoka, T. (1930). On the significance of the root-nodules of *Coriaria japonica* A.Gr. to the nitrogen nutrition of the plant. *Jpn. J. Bot.* **5,** 209–218.

Kordes, H. (1925). Kritische Besprechung der Frage "Impfung der Nichtleguminosen." *Z. Pflanzenernaehr. Dueng. B* **4B,** 382–394.

Krebber, O. (1932). Untersuchungen über die Wurzelknöllchen der Erle. *Arch. Mikrobiol.* **3,** 588–608.

Lalonde, M., and Devoe I. W. (1976). Origin of the membrane envelope enclosing the *Alnus crispa* var. *mollis* Fern. root nodule endophyte as revealed by freeze-etching microscopy. *Physiol. Plant Pathol.* **8,** 123–125.

Lalonde, M., and Knowles, R. (1975). Ultrastructure, composition and biogenesis of the encapsulation material surrounding the endophyte in *Alnus crispa* var. *mollis* root nodules. *Can. J. Bot.* **53,** 1951–1971.

Lawrence, D. B., Schoenicke, R. E., Quispel, A., and Bond, G. (1967). The role of *Dryas drummondii* in vegetation development following ice recession at Glacier Bay, Alaska, with special reference to its nitrogen fixation by root nodules. *J. Ecol.* **55,** 793–813.

Lie, T. A., and Mulder, E. G. (eds.) (1971). Biological nitrogen fixation in natural and agricultural habitats. *Plant Soil, Spec Vol.*, pp. 1–590.

MacLuckie, J. (1923). Studies in symbiosis. IV. The root nodules of *Casuarina cunninghamiana* and their physiological significance. *Proc. Linn. Soc. N.S.W.* **48,** 194–205.

Mayer, A. (1895). Wortelknolletjes geanalyseerd. *Landbouwkd. Tijdschr.* pp. 319–320.

Meyen, J. (1829). Über das Hervorwachsen parasitischer Gebilde aus den Wurzeln anderer Pflanzen. *Flora (Jena)* **12,** 49–64.

Mian, S., Bond, G., and Rodriguez-Barrueco, C. (1976). Effective and ineffective root nodules in *Myrica faya*. *Proc. R. Soc. London, Ser. B* **194,** 285–293.

Miehe, H. (1918). Anatomische Untersuchungen der Pilzsymbiose bei *Casuarina equesetifolia* nebst einigen Bemerkungen über das Mykorrhizenproblem. *Flora (Jena)* **111/112,** 431–449.

Miettinen, J. K., and Virtanen, A. L. (1953). The free amino acids in the leaves, roots and root nodules of the alder (*Alnus*). *Physiol. Plant.* **5,** 540–557.

Möller, H. (1885). *Plasmodiophora alni*. *Ber. Dtsch. Bot. Ges.* **3,** 102–105.

Möller, H. (1890). Beiträge zur Kenntnis der *Frankia subtilis* Brunchorst. *Ber. Dtsch. Bot. Ges.* **8,** 215–224.

Mowry, H. (1933). Symbiotic nitrogen fixation in the genus *Casuarina*. *Soil Sci.* **36,** 409–425.

Nobbe, F., and Hiltner, L. (1904). Über das Stickstoffsammlungsvermoĝen der *Erlen* und *Elaeagnacean*. *Naturwiss. Z. Forst- Landwirtsch.* **2,** 366–369.

Nutman, P. S. (ed.) (1976). "Symbiotic nitrogen fixation in plants." *Int. Biol. Programme* **7,** pp. 1–584. Cambridge Univ. Press, Cambridge, England.

Oersted, A. S. (1865). Smaa biologiske og morphologiske bidrag. 6. *Hippophaë rhamnoides*. *Bot. Tidsskr., Ser. 3* **1,** 108–110.

Peklo, J. (1910). Die pflanzlichen Aktinomykosen. *Zentralbl. Bakteriol., Parasitenkd., Infektionskr. Hyg., Abt. 2, Naturwiss.: Allg. Landwirtsch. Tech. Mikrobiol.* **27,** 451–579.

Pizelle, G. (1970). Effets comparés de l'acide β-indolyl-acétique et de l'azote nitrique sur la nodulation de l'aulne glutineux (*Alnus glutinosa*). *Bull. Acad. Soc. Lorraines Sci.* **9,** 174–178.

Pizelle, G. (1975). Variations saisionnières de l'activité nitrogénasiques des nodules d'*Alnus incana* (L.) Moench et d'*Alnus cordata* (Lois) Desj. *C. R. Hebd. Seances Acad. Sci. D.* **281,** 1829–1832.

Pommer, E. H. (1956). Beiträge zur Anatomie und Biologie der Wurzelknöllchen von *Alnus glutinosa* Gaertn. *Flora (Jena)* **143,** 603–634.

Pommer, E. H. (1959). Über die Isolierung des Endophyten aus den Wurzelknöllchen von *Alnus glutinosa* Gaertn. und über Erfolgreiche Reinfektionsversuche. *Ber. Dtsch. Bot. Ges.* **72,** 138–150.

Quispel, A. (1954a). Symbiotic nitrogen fixation in non-leguminous plants. 1. Preliminary experiments on the root-nodule symbiosis of *Alnus glutinosa* (L.) Gaertn. *Acta Bot. Neerl.* **3,** 495–511.

Quispel, A. (1954b). Symbiotic nitrogen fixation in non-leguminous plants. 2. The influence of inoculation density and external factors on the nodulation of *Alnus glutinosa* and its importance to our understanding of the mechanism of infection. *Acta Bot. Neerl.* **3,** 512–532.

Quispel, A. (1955). Symbiotic nitrogen fixation in non-leguminous plants. 3. Experiments on the growth *in vitro* of the endophyte of *Alnus glutinosa*. *Acta Bot. Neerl.* **4,** 380–396.

Quispel, A. (1960). Symbiotic nitrogen fixation in non-leguminous plants. 5. The growth requirements of the endophyte of *Alnus glutinosa*. *Acta Bot. Neerl.* **9,** 380–396.

Quispel, A. (1988). Hellriegel and Wilfarth's discovery of (symbiotic) nitrogen fixation hundred years ago. *In* "Nitrogen Fixation: Hundred Years After" (H. Bothe, F. J. de Bruijn, and W. E. Newton, eds.), pp. 3–12. Fischer, Stuttgart.

Quispel, A., and Tak, T. (1978). Studies on the growth of the endophyte of *Alnus glutinosa* (L.) Vill. in nutrient solutions. *New Phytol.* **81,** 587–600.

Roberg, M. (1934a). Weitere Untersuchungen über die Stickstoffernährung der Erle. *Ber. Dtsch. Bot. Ges.* **52,** 54–64.

Roberg, M. (1934b). Über den Erreger der Wurzelknöllchen von *Alnus* und den Elaeagnaceen *Elaeagnus* und *Hippophaë. Jahrb. Wiss. Bot.* **79,** 472–492.

Roberg, M. (1938). Über den Erreger der Wurzelknöllchen Europäischer Erlen. *Jahrb. Wiss. Bot.* **86,** 344–349.

Rodriguez-Barrueco, C., and Bermudez de Castro, F. (1973). Cytokinin-induced pseudonodules on *Alnus glutinosa. Physiol. Plant* **29,** 277–280.

Rodriguez-Barrueco, C., and Bond, G. (1976). A discussion of the results of cross-inoculation trials between *Alnus glutinosa* and *Myrica gale. In* "Symbiotic Nitrogen Fixation in Plants" (P. S. Nutman, ed.), pp. 561–565. Cambridge Univ. Press, Cambridge, England.

Schacht, H. (1853). Beitrag zur Entwicklungsgeschichte der Wurzel. *Flora (Jena)* **36,** 257–266.

Schaede, R. (1933). Über die Symbionten in den Knöllchen der Erle und des Sanddornes und die cytologischen Verhältnisse in ihnen. *Planta* **19,** 389–416.

Schaede, R. (1938). Die Actinomyceten-Symbiose von *Myrica gale. Planta* 29,32–46.

Schaede, R. (1962). "Die pflanzlichen Symbiosen," 3rd ed., pp. 42–54. Fischer, Stuttgart (originally published in 1943).

Schimper, A. F. W. (1891). Die indo-malayische Strandflora. *Bot. Mitt. Trop.* **3**, pp. 1–204, A. Fischer, Jena.

Shibata, K. (1902). Cytologische Studien über die endotrophen Mykorrhizen. *Jahrb. Wiss. Bot.* **37,** 643–684.

Shibata, K., and Tahara, M. (1917). Studien über die Wurzelknöllchen. *Jahrb. Wiss. Bot.* **31,** 157–182.

Silver, W. S. (1964). Root nodule symbiosis. I. Endophyte of *Myrica cerifera. J. Bacteriol.* **87,** 416–421.

Silver, W. S., Bendana, F. E., and Powell, R. D. (1966). Root nodule symbiosis. II. The relation of auxin to geotropism in roots and root nodules of non-legumes. *Physiol. Plant.* **19,** 207–210.

Skeffington, R. A., and Stewart, W. D. P. (1976). Evidence from inhibitor studies that the endophyte synthesizes nitrogenase in the root nodules of *Alnus glutinosa* (L.) Gaertn. *Planta* **129,** 1–6.

Sprent, J. I., Scott, R., and Perry, K. M. (1978). The nitrogen economy of *Myrica gale* in the field. *J. Ecol.* **66,** 657–668.

Stevenson, G. (1958). Nitrogen fixation by non-nodulated plants and by nodulated *Coriaria arborea. Nature (London)* **182,** 1523–1524.

Stewart, W. D. P., and Pearson, M. C. (1967). Nodulation and nitrogen fixation by *Hippophaë rhamnoides* in the field. *Plant Soil* **26,** 348–360.

Stewart, W. D. P., Fitzgerald, G. P., and Burris, R. H. (1967). *In situ* studies on N_2 fixation using the acetylene reduction technique. *Proc. Natl. Acad. Sci. U.S.A.* **58,** 2071–2078.

Tarrant, R. F., and Trappe, J. M. (1971). The role of *Alnus* in improving the forest environment. *Plant Soil, Spec. Vol.* pp. 335–348.

Taubert, H. (1956). Über den Infektionsvorgang und die Entwicklung der Knöllchen bei *Alnus glutinosa* Gaertn. *Planta* **48,** 135–156.

Torrey, J. G. (1976). Initiation and development of root nodules of *Casuarina* (Casuarinaceae). *Am. J. Bot.* **63,** 335–344.

van Dijk, C. (1978). Spore formation and endophyte diversity in root nodules of *Alnus glutinosa* (L.) Vill. *New Phytol.* **81,** 601–615.

van Dijk, C., and Merkus, E. (1976). A microscopical study of the development of a spore-like stage in the life cycle of the root-nodule endophyte of *Alnus glutinosa* (L.) Gaertn. *New Phytol.* **77,** 73–91.

van Straten, J., Akkermans, A. D. L., and Roelofsen, W. (1977). Nitrogenase activity of endophyte suspensions from root nodules of *Alnus, Hippophaë, Shepherdia* and *Myrica* spp. *Nature (London)* **266,** 257–258.

Virtanen, A. I. (1957). Investigations on nitrogen fixation by the alder. II. Associated culture of spruce and inoculated alder without combined nitrogen. *Physiol. Plant.* **10,** 164–169.

Virtaanen, A. I., and Saastomoinen, S. (1933). Über die Stickstoffbindung der Erle (*Alnus*). *Acta Chem. Fenn.* **6,** 57–58.

Virtanen, A. I., and Saastomoinen, S. (1936). Untersuchungen über die Stickstoffbindung der Erle. *Biochem. Z.* **284,** 72–85.

Virtanen, A. I., Moisio, Y., Allison, R. M., and Burris, R. H. (1954). Fixation of molecular nitrogen by excised nodules of the alder. *Acta Chem. Scand.* **8,** 1730–1731.

von Jäger, G. (1860). Über eine Krankhafte Veränderung der Blüthenorgane der Weintraube. *Flora (Jena)* **18,** 49–51.

von Plotho, O. (1941). Die Synthese der Knöllchen an den Wurzeln der Erle. *Arch. Mikrobiol.* **12,** 1–18.

Wagle, R. F., and Vlamis, J. (1961). Nutrient deficiencies in two bitterbrush soils. *Ecology* **42,** 745–750.

Warming, E. (1876). Små biologiske og morfologiske Bidrag. *Bot. Tidsskr.* 3^{die} *Räkke* **1,** 108–110.

Wheeler, C. T., and Bowes, B. G. (1974). Effects of light and darkness on nitrogen fixation by root nodules of *Alnus glutinosa* in relation to their cytology. *Z. Pflanzenphysiol.* **71,** 71–75.

Woronin, M. (1866). Über die bei der Schwarzerle (*Alnus glutinosa*) und bei der gewöhnlichen Gartenlupine (*Lupinus mutabilis*) auftretenden Wurzelanschwellungen. *Mem. Acad. Imp. Sci. St. Petersburg, Ser. 7* **T.10**(6), 1–13.

Woronin, M. (1885). Bemerkungen zu dem Aufsatze von Herrn H. Müller über Plasmodiophora *Alni. Ber. Dtsch. Bot. Ges.* **5,** 50–58.

Ziegler, H., and Hüser, R. (1963). Fixation of atmospheric nitrogen by root nodules of *Comptonia peregrina. Nature (London)* **199,** 508–509.

3

Systematics, Isolation, and Culture of *Frankia*

Mary P. Lechevalier and Hubert A. Lechevalier
Waksman Institute
Rutgers University
Piscataway, New Jersey, U.S.A.

I. Systematics

A. Systematics of Actinomycetes

Actinomycetes are filamentous, branching, gram-positive bacteria. Most are free-living saprophytes deriving sustenance from dead organic matter. They inhabit such substrates as soil, compost, and plant litter; some are also found in smaller numbers in air and water. A few are associated with animals and humans, both as a component of their natural flora or as pathogens. Plant pathogens (mostly *Streptomyces*) largely infect "root" crops such as beet, sweet potato, and potato (Lechevalier, 1981, 1988).

1. Criteria for Genus Assignment

Morphological structures formed by actinomycetes are the most complex among bacteria and serve to aid in assigning strains to broad groups. One may distinguish those forming (1) no spores (the vegetative hyphae may break down into fragments, however); (2) single spores; (3) pairs of spores; (4) short chains (less than 20 spores); (5) long chains of spores; (6) sporangia; (7) synnemata; (8) sclerotia; (9) vesicles; and (10) budding spores. Because dependence on morphological characters alone sometimes led to confusion and mistakes in classification, cell chemistry has been widely employed as an aid in distinguishing among groups. Compounds having taxonomic significance include the substituents of cell polymers such as the peptidoglycan; cell polysaccharides, both those covalently linked to the peptidoglycan and others more loosely associated with the cell; and deoxyribonucleic acid (G + C%). Other cell constituents used in this way include the cellular and membrane fatty acids, including specialized fatty acids such as mycolic acids, phospholipids (present in membranes), and menaquinones (respiratory quinones) (Lechevalier and Lechevalier, 1980, 1981; Minnikin and O'Donnell, 1983; H. A. Lechevalier, 1989). The correctness of the classification to genus on the basis of cell chemistry has been confirmed by the use of 16*S* RNA cataloging (Stackebrandt, 1986). This latter technique is thought to be reflective of phylogenetic relationships. Other criteria for generic placement include relation to oxygen (anaerobic vs. aerobic), fixation of nitrogen, phage sensitivity, and spore location, shape, arrangement, motility, and resistance to heat.

2. Criteria for Species Assignment

Once an actinomycete has been assigned to a given genus, classification to species is primarily dependent on DNA homology and physiological characteristics. The latter include acid production from, and utilization of, carbohydrates, utilization of organic acids, and presence or absence of enzymes degrading various substrates such as proteins, purines, pyrimidines, amino acids, nitrates, urea, and glycosides. Sensitivity to temperature, salt, pH, lysozyme, dyes, and antibiotics is also commonly used. Phenetic data from tests such as these have been used to construct dendrograms of relationships among different groups using numerical taxonomic techniques and large numbers of strains (Goodfellow *et al.*, 1985). Results from such tests also serve as data bases for probabilistic identifications of unknown stains through the use of computer programs written for the purpose. Because many tests (usually > 100) are involved and because slow-growing or fastidious strains often give anomalous results, this technique can only be applied to certain groups. Also, it is not a practical option for an investigator who wishes to identify a small number of strains in his own laboratory. Additional criteria for species classification may include morphological characteristics such as spore mass color, spore chain configuration, spore structure, spore surface, sporangial shape and disposition of the sporangiospores, lack of vegetative hyphae, and chemical characteristics such as menaquinones and whole cell sugar content.

Individual strains may be identified and grouped on the basis of serology and protein patterns or restriction patterns using gel electrophoresis (see Section I,B,1,b).

B. Systematics of *Frankia*

As discussed in Chapter 2, the nature of the endophyte that we now place in the genus *Frankia* remained controversial until the isolation of the endophyte of sweet fern (*Comptonia peregrina*) and the demonstration that the isolated actinomycete could reinfect the plant (Callaham *et al.*, 1978; Lalonde, 1978). Although Becking (1974), in the eighth edition of *Bergey's Manual of Determinative Bacteriology*, stated that "no species [of *Frankia*] has yet been grown on artificial media," Pommer (1959) had succeeded in isolating the endophyte of *Alnus glutinosa* and in reinfecting the host with the production of nodules. Pommer's publication was largely overlooked for the next twenty years consequent to his changing his field of research at the end of 1959. His strains were sent to Professors Ettlinger and Zähner in Zürich, who were interested in testing them for antibiotic production, and the strains were subsequently lost or discarded (E. Pommer, personal communication).

Becking (1970), after years of trying to isolate the endophytes, decided they were obligate symbionts and proposed a classification using host plant ranges similar to the system used for rhizobia. Of course, at that time, his cross-inoculation groups were determined using crushed nodule inocula; however, after pure cultures were finally at hand, it was found that the host specificities he described were not always correct. For example, *Frankia alni* strains, which are supposed to be specific for *Alnus*, are also capable of infecting species of *Comptonia* and *Myrica* (Dillon and Baker, 1982; Baker, 1987). Another problem became apparent after the isolation of the first strains: some of them, isolated from a given host, could not be demonstrated to form nodules on that host, but rather could be shown to infect plants of an entirely different family (Baker, 1987); thus the host range concept as a basis for the delimitation of species has been set aside.

1. Taxonomic Criteria for Frankiae

a. Morphology. Endophyte strains isolated from nodules of actinorhizal plants may usually be assigned to the genus *Frankia* on the basis of their characteristic morphology. Unique among actinomycetes, members of the genus are characterized by the formation, *in submerged liquid culture,* of round, cylindrical, or highly irregular sporangia filled with spores. These spores, when mature, are thick-walled with outer membranelike layers, smooth-surfaced, refractile, somewhat polygonal in shape, and nonmotile. Often, mature and young spores in different stages of development inhabit the same sporangium. In nitrogen-free media, or in some strains growing in media with a fixed source of nitrogen, "vesicles" are formed. These round to cylindrical, stipitate structures are the principal locus of nitrogenase, the enzyme responsible for nitrogen fixation (see Chapter 7). The hyphae of frankiae are poorly branched, and range in width from 0.5 to 2.0 μm. No aerial mycelium is formed. Cells may be colorless, yellow, orange, brown, pink, red, coral, greenish, or black, and soluble pigments are equally diverse. One of the most common types of pigments is a complex of red compounds of the benzo (α) naphthacene-8,13-dione type (Gerber and Lechevalier, 1984; Rickards, 1989). As in most actinomycetes, the capacity to produce a pigment may vary depending on the composition of the medium, or may be lost; thus, it is a character with only limited usefulness in modern taxonomy.

Both sporangia and vesicles are also formed within plant nodules (van Dijk, 1978; Normand and Lalonde, 1982; Lechevalier *et al.*, 1983; Holman and Schwintzer, 1987). In 9 of the 21 genera of actinorhizal plants, nodules can be found that contain many sporangia. This nodule type is referred to as Sp+ (or "P" type, positive for sporangia). Other nodules (the most common type; see Chapter 9) with few to no sporangia are referred to as Sp− (or "N"). The Sp+ character is reported to be transmissible since crushed Sp+ nodule inocula produce largely Sp+ nodules on fresh plants. The Sp+ endophyte is very difficult to isolate and grow *in vitro* (Normand and Lalonde, 1982; Torrey, 1987; M. P. Lechevalier, unpublished observations) and those that are isolated do not give consistently Sp+ nodules. Thus it is presently uncertain whether Sp+ endophytes have an inherent genetic makeup that gives them special characteristics, or if environmental factors play a role in production of sporangia *in planta*. Whether this characteristic will have future taxonomic significance remains to be determined.

b. Cell Chemistry. Frankiae have a cell-wall composition of type III, containing *meso*-diaminopimelic acid, alanine, glutamic acid, muramic acid, and glucosamine. This is the most common type among actinomycetes and the simplest regarding the number of substituents. *Frankia* polysaccharides may contain a variety of sugars. The sugars having the most significance include fucose, xylose, madurose (3-*O*-methyl-D-galactose), galactose, glucose, rhamnose, and 2-*O*-methyl-D-mannose, with xylose and fucose being the most common differentiating compounds.

Fucose has been found only in frankiae that are noninfective for their original host. Xylose has been found in both infective and noninfective strains. Some strains lack either sugar. Because fucose has been found in very few other actinomycetes (M. P. Lechevalier, unpublished observations), when it occurs, its presence could be considered diagnostic for the genus *Frankia*. 2-*O*-Methyl-D-mannose is also very common, and Lalonde and co-workers consider it to be genus marker (Mort *et al.*, 1983; St. Laurent *et al.*, 1987). In our experience, some strains lack this sugar in amounts detectable on paper or thin-layer chromatography, although it is possible it could be discerned using more sensitive techniques such as gas–liquid chromatography. Trehalose has also been reported in unhydrolyzed chloroform–methanol extracts of whole cells of a number of *Frankia* strains (Lopez *et al.*, 1983). This sugar also occurs in a wide variety of organisms including other actinomycetes (McBride and Ensign, 1987) and *Rhizobium* (Streeter, 1985) and thus had little taxonomic significance.

Lipid analysis shows that frankiae contain normal, unsaturated, and branched-chain fatty acids. This pattern is not uncommon among actinomycetes. In frankiae, major differences were found between the fatty acids acylated to the triglyceride fraction and the total cellular fatty acids, the unsaturated fatty acids being largely concentrated in the triglycerides (Lechevalier *et al.*, 1983). Not enough is known about other actinomycetes' fatty acid distribution to know whether this is unique to frankiae. Only minor quantitative differences were found in cells grown under nitrogen-fixing and fixed-nitrogen conditions (Lechevalier *et al.*, 1987). Anders *et al.* (1989), in examining the lipids of vesicles, found they were richer in neutral lipids than the vegetative cells, and that they contained large amounts of a long-chain fatty acid having 26 carbons. The long-chained, α-branched, β-hydroxylated fatty acids (mycolates) found in the actinomycete genera *Mycobacterium*, *Nocardia*, and *Corynebacterium*, and useful in their taxonomy, are lacking in frankiae (M. P. Lechevalier, unpublished observations). Their phospholipid pattern (PI) is of the simplest actinomycete type, containing diphosphatidyl glycerol, phosphatidyl inositol, and phosphatidyl inositol mannosides. The principal menaquinones are either MK9 (H_4) or MK9 (H_6) with lesser amounts of other MK9 derivatives (Lechevalier *et al.*, 1987; J.-S. Ruan, personal communication). Aside from the long-chain fatty acid reported by Anders *et al.* (1989), none of the data on the lipids of frankiae do more than enable us to declare that they have a lipid composition that is not unusual in actinomycetes. It will be of great interest to learn whether the unusual 26-carbon fatty acid of Anders will prove to be a taxonomic marker.

Gel electrophoresis of whole cell proteins demonstrated that the *Alnus–Myrica–Comptonia* host compatibility group can be divided into three groups. Isolates from *A. incana* ssp. *rugosa* were found in all three groups (Benson *et al.*, 1984). The mol% G + C of the DNA of frankiae ranges from 66 to 71 and no modified bases have been found (An *et al.*, 1983). This criterion by itself is thus of little taxonomic value.

c. DNA Homology. Based on the homology of *Frankia* DNA with DNA of strains of other actinomycetes having the same cell wall type, the closest relative to *Frankia* may be *Geodermatophilus* (An *et al.*, 1987). The latter actinomycete genus is a member, along with *Frankia* and *Dermatophilus*, of the so-called "actinomycetes dividing in more than one plane." Members of this group share the capacity to form their spores in more than one plane, a characteristic that is uncommon among the actinomycetes (M. P. Lechevalier, 1989). Interestingly, the geodermatophili were also pinpointed as being closely related to frankiae on the basis of 16*S* RNA cataloging (Stackebrandt, 1986; Hahn *et al.*, 1989a). Hahn *et al.* (1989b), in another study, found that, using a synthetic probe specific for the 16*S* rRNA of effective and infective *Frankia* strains from *Alnus*, they could distinguish between effective and ineffective *Alnus* strains but not isolates from *Comptonia, Colletia,* and *Elaeagnus*. The probe also hybridized with the 16*S* rRNA from some *Casuarina* isolates.

As in other actinomycetes (Crameri *et al.*, 1983), restriction endonuclease patterns of DNA may be strain-specific rather than species-specific (An *et al.*, 1985b; Dobritsa, 1985); however, recently some gel groupings have been correlated with phenotypic characters (Bloom *et al.*, 1989a,b).

d. Physiology and Serology Groups A and B. On the basis of DNA homology (An *et al.*, 1985a), serology (Baker *et al.*, 1981), and physiology (Lechevalier *et al.*, 1983), frankiae may be classed into two broad groups called A and B (Lechevalier and Lechevalier, 1989). In general, group A contains strains that are very diverse and approach, in phenotypic characteristics, saprophytic actinomycetes. They are not infective, under present experimental conditions, for the host from which they were originally isolated. Physiologically (Lechevalier *et al.*, 1983; Lechevalier and Ruan, 1984; Lechevalier and Lechevalier, 1989), group A strains are relatively aerobic and usually utilize a variety of monosaccharides and disaccharides with or without production of acid; many have proteolytic and/or amylolytic activities as well. Most grow relatively rapidly and can be maintained in slant culture. Both serology and DNA homology point up the diversity of this group.

Group B strains are far more homogeneous. They are relatively inactive physiologically, not utilizing (with few exceptions) carbohydrates, proteins, or starch. They grow well on carbon sources such as Tweens and certain organic acids (Blom *et al.*, 1980). They show a synergistic growth response when Tween 80 is added to media containing certain carbohydrates (i.e., their cell dry weight exceeds that of the sum of cell weights grown on the carbohydrate or the Tween in separate media) (Lechevalier *et al.*, 1983; Lechevalier and Ruan, 1984; Ruan *et al.*, 1986). They are usually microaerophilic, cannot be maintained on slants, and are infective for their original plant host. Attempts at using conventional bacteriological tests for grouping frankiae have met with little success (Lechevalier *et al.*, 1982, 1983; Horrière, 1984; Lechevalier and Lechevalier, 1989).

e. Host Compatibility Relative to Taxonomic Characters. As previously discussed, the allure of host plant specificity as a taxonomic criterion has been considerably diminished by the lack of infectivity of some strains for their original host plant. It is clear that the techniques presently used to demonstrate *Frankia* infectivity for plants growing in axenic culture do not emulate conditions in the field. Axenic techniques are used to fulfill Koch's postulates by demonstrating the inherent infective capacity of a given strain, but in the real world there are fungi, bacteria, actinomycetes, mycorrhizae, insects, nematodes, stones, and many other agents that may serve to help introduce noninfective frankiae into the plant tissue. It should be remembered that most actinomycetes capable of causing infections are, in fact, opportunists, gaining access to tissue via wounds or other alterations induced by other agents. A *Pseudomonas* strain was reported to be a "helper" in increasing the infectivity of *Frankia* strains (Knowlton *et al.*, 1980; Knowlton and Dawson, 1983). Also, certain frankiae are infective only when the plants are grown on a solid substratum rather than in hydroponic culture (Zhang and Torrey, 1985).

Co-infection of nodules may also occur. Under controlled conditions of growth, surface-sterilized nodule inocula may introduce more than one *Frankia* strain into the plant tissue (e.g., Zhang *et al.*, 1984). Independent infections or co-infections may be involved. The infective strain could pave the way for the noninfective strain, which could then penetrate the host root in an opportunistic manner. In fact, using a combined inoculum composed of a noninfective and an infective strain, we have been able to recover the noninfective strain from the resulting nodules (M. P. Lechevalier, unpublished observations).

Another puzzling finding is the seeming nonspecificity (in botanical terms) of some isolates (see Chapter 4). Some isolates from *Elaeagnus* will also infect plants from the unrelated genus *Alnus* (Lalonde *et al.*, 1981) and isolates from *Casuarina* may also infect species of *Myrica*. An extensive study by Baker (1987) has found that when some 50 *Frankia* strains from 10 host plant genera were studied, four distinct groups could be discerned: (1) strains nodulating *Casuarina* and *Myrica;* (2) strains nodulating *Alnus* and *Myrica;* (3) strains nodulating Elaeagnaceae and *Myrica;* and (4) strains nodulating only the Elaeagnaceae. For taxonomic evaluation, additional criteria will be needed to refine these divisions.

A start in this direction has been made using whole cell sugar analysis by means of gas–liquid chromagraphy and statistics. Using data on 79 *Frankia* strains isolated from 6 plant genera, St. Laurent *et al.* (1987) were able to classify most of their strains in either the "*Alnus*" or "*Elaeagnus*" specificity groups. A certain number of *Myrica gale* isolates fell between the two groups. Also, Gardes and Lalonde (1987) found that gel patterns of soluble proteins yielded host compatibility groupings in the "*Alnus*" and "*Elaeagnus*" isolates. Again, there were a few misfits. Examination of the isozyme patterns of various frankiae from *Alnus* and *Elaeagnus* permitted the resolution of three large, heterogeneous groups. Only a few *Alnus* isolates showed good homologous patterns (Gardes *et al.*, 1987). In contrast, Puppo *et al.* (1985) were able to show considerable electrophoretic homogeneity in the various superoxide dismutases among isolates from the Casuarinaceae. In other studies, serological analysis permitted Sun *et al.* (1987) to divide 30 *Frankia* strains into three host compatibility subgroups: (1) *Alnus–Comptonia–Myrica*, (2) *Hippophae*, and (3) *Casuarina*.

In the final analysis, host compatibility may not have taxonomic significance, since very unlike strains infect the same host plant; thus, strains in the same compatibility group may share only a small number of characteristics such as the external polysaccharide, which may play a role in the host–endophyte recognition, and possibly certain enzymatic batteries. This is underlined by the striking phenotypic diversity of isolates from the same host (Normand and Lalonde, 1982; Benson and Hanna, 1983; Lechevalier *et al.*, 1983; Benson *et al.*, 1984; Burggraaf and Valstar, 1984; Lechevalier and Ruan, 1984; Zhang *et al.*, 1984; Wu and Ding, 1987; J.-S. Ruan and Y.-L. Shi, personal communication), which has perplexed many researchers. Strains of frankiae also can vary spontaneously in culture (Burggraaf and Valstar, 1984; An *et al.*, 1985b; Parson *et al.*, 1985; Lechevalier *et al.*, 1987) and this may be an explanation of the diversity observed. In any case, such variability will have to be taken into consideration in any taxonomic scheme developed in the future.

2. Current Status of *Frankia* Taxonomy

a. Speciation within the Genus Frankia. Because of the lack of a clear-cut species criterion such as the interbreeding capability used for the species assignment in eukaroytes, criteria for assignment to a bacterial species have rested heavily on cell morphology, staining reactions, ecology, and physiological tests. It is a credit to the early taxonomists' talents that, despite the lack of what we now consider fundamental biological knowledge, many of their early groupings have often hung together. Nevertheless, one must now ask whether really basic criteria, such as DNA homology, should not be regarded as the prime foundation on which to base a species definition for prokaryotes, including *Frankia*. This relatively lengthy and often tedious procedure cannot be easily carried out by every laboratory, so that once relationships have been established in this way, phenotypic and cell chemical criteria will then have to be developed to enable all workers to do their own taxonomy. In any case, DNA homology studies such as those of An *et al.* (1985a, 1987) give us reason to hope that species groups in *Frankia* may yet be defined; however, at this time no recommendations can be made other than the previously expressed view of workers in the field (Lechevalier, 1984) that species names in *Frankia* be avoided until a clearer picture emerges.

Recently, Lalonde *et al.* (1988) have considered that enough was known about some of the frankiae to recognize the species *alni* and *elaeagni* and within the species *alni,* two subspecies. The rules of bacterial nomenclature dictate that one must honor previously validly published species names, even though their definition must be completely revised. In the case of *Frankia* it is indeed unfortunate and certainly confusing that the species *alni* has a proposed type strain isolated from a species of *Comptonia*. Moreover, this specific epithet will apply to some strains isolated from *Myrica* and will never apply to all strains from *Alnus*. However, it is necessary to make a beginning, which Lalonde and co-workers have commendably done. Whether their proposals will be supported by other workers' data remains to be seen.

b. Relation of Frankia *and* Geodermatophilus. Catalogs of 16*S* RNA tell us what the phylogenetic position of a given group is likely to be and give a clearer picture of its close relatives. This is not to say that other characteristics, especially cell chemistry (Stakebrandt, 1986), may not also give us results congruent with those of the RNA and DNA findings. In consequence of this belief, Hahn *et al.* (1989a) have proposed a new definition of the family Frankiaceae based on RNA catalogs and reverse transcriptase sequencing. Formerly, only the genus *Frankia* was contained within the family; now it will be redefined to include the closely allied genus *Geodermatophilus*.

II. Isolation of *Frankia*

More than a century passed between the time the presence of endophytes in root nodules of actinorhizal plants was suspected and their successful isolation. In general, the reason for this long delay was due more to the slow rate of growth of frankiae, which may have a doubling time of up to 5 days, than to any very esoteric growth requirements. Present evidence indicates that it is easier to isolate some frankiae than others and that they are easier to isolate from some plants, such as alders, than from others. To achieve isolation, Pommer used glucose–asparagine agar, a simple defined medium that was used by S. A. Waksman for the maintenance and growth of actinomycetes. Torrey and co-workers (Callaham *et al.*, 1978), whose isolate CpI1 really started all this, used a more complicated medium.

A. Sources

Frankiae have been isolated mainly from nodules of plants with which they are associated. Their presence as free-living soil forms is more suspected than proven since they have only been successfully isolated from soil from the rhizosphere of actinorhizal plants.

1. Nodules

Greenhouse-grown and -inoculated plants yield nodules from which the endophyte can be isolated with less interference from contaminating microorganisms than if the nodules had been collected in the field. The latter, however, have the advantage of authenticity and the strains obtained from such nodules do not carry the stigma of perhaps being the fruit of an association that is only a laboratory curiosity.

Collection of nodules in the field should be done with care to maximize the chances of cultivation of the endophyte. Young nodules should be collected in preference to old ones and they should be kept moist and cool until the laboratory is reached. If the nodules cannot be worked on immediately, the endophytes can be kept viable for months by storage of the nodules in the freezer at −20°C. It is also possible to store a small volume of nodule lobes on silica-gel desiccant. The dry nodules are stable for some time at room temperature but not enough is known about their viability over a long period of time to make a recommendation for long-term use of this technique.

Every investigator will want to keep a record of information about the nodules collected. This should include the name of the plant, the date and location of the collection, and so on. The information most usually needed is included on the "Data form for the registration of *Frankia* strains," which can be obtained from the authors.

2. Soil

Strains of *Frankia* are assumed to survive in soil, and soil is assumed to be the source of the root inoculum in nature. However, isolation from soil has rarely been achieved. Baker and O'Keefe (1984), using a modified sucrose fractionation procedure, have reported the isolation of a strain from the rhizosphere of an actinorhizal plant.

B. Methods

A number of authors have recently reviewed the methods currently used in the isolation of frankiae (Carpenter and Robertson, 1983; Stowers, 1987; Diem and Dommergues, 1988; Baker, 1989). Here, we will simply outline the most commonly used techniques and give details about the methods currently used in our laboratory. None of the procedures is always successful and no meaningful comparison between methods has been performed.

1. Nodule Workup

The nodules are first washed in a stream of water and the lobes separated from each other. This is followed by shaking in a dilute detergent solution to remove soil and other extraneous debris. The lobe surface is treated with any of a number of disinfectants such as phenol, bromine, sodium or calcium hypochlorite, mercuric chloride, hydrogen peroxide, chloramine T, glutaraldehyde, or osmium tetroxide. The last compound, although very useful, is also very dangerous and should be used only under a chemical fume hood. The action of the disinfectant can be potentiated with the addition of a wetting agent, or by placing the nodule under reduced pressure in a side-arm flask attached to a water pump to displace air pockets.

After surface sterilization, the nodules are rinsed thoroughly in sterile distilled water, buffer, or sodium bicarbonate solution. The disinfection is usefully repeated following overnight standing at room temperature. This permits many spore formers to grow out into a more susceptible vegetative stage. Even with the greatest care, some contaminating organisms, both bacteria and fungi, usually escape and surface sterilization is not complete.

a. Microdissection. In microdissection the nodules are cut into small fragments and pieces of mature nodule cortex taken near the lobe apex are either placed in tubes of liquid medium or imbedded in a layer of a nutrient agar. Prolonged incubation (several weeks) is needed before the characteristic growth of the frankiae can be recognized. The apparition of profuse growth after a few days is diagnostic of contaminated cultures.

Variations of the microdissection method have yielded many isolates of *Frankia,* but failures have often been attributed to the inhibitory action of plant phenolics present in the pieces of tissue. A slight modification of the method consists in adding activated carbon to the medium in the hope that it absorbs any inhibitory substances (Diem and Dommergues, 1988). Sephadex or polyvinylpyrrolidone (PVP) have also been used to remove the phenolics.

b. Differential Filtration. A comparatively simple method of separating the cells of frankiae from the tissue fragments containing the plant phenolics was proposed by Benson in 1982. Differential filtration consists of crushing surface-sterilized nodules to produce an aqueous suspension of plant tissue and actinomycetic fragments. This suspension is forced through a succession of two nylon mesh filters; the first, with pores of 50 μm, retains most of the plant debris, and the second, with pores of 20 μm, traps most of the *Frankia* vesicle clusters. After washing the actinomycetic concentrate, plates or tubes of suitable media can be inoculated.

c. Sucrose-Density Centrifugation. In sucrose-density centrifugation, the *Frankia* cells are selectively concentrated in one specific layer of a sucrose-density gradient (Baker and O'Keefe, 1984). Three concentrations of sucrose are used (2.5, 1.6, and 1.0 molar) in a step gradient. The crushed nodule or soil suspension is prepared in 0.7% phenol and layered on the gradient, which is centrifuged in a swinging bucket rotor until equilibrium is reached. The amount of time required depends on the speed of the centrifuge but high speed is not required. The greatest concentration of cells of *Frankia* is found at the interface between the 2.5 and 1.6 *M* sucrose layers from which the inoculum for plating out or serial dilution can be collected.

It was subsequently observed (Quispel and Burggraaf, 1981) that the discontinuous gradient was unnecessary: it is sufficient to layer the nodule fragments on a tube containing 60% sucrose and collect the sediment at the bottom when the centrifuge run is finished. The authors hypothesized that many of the competing organisms died during the procedure.

In our own laboratory, we prefer either full-strength sodium hypochlorite (commercial solution of 5.25%) or osmium tetroxide (1–3%) as sterilizing agents. We soak the nodules in the hypochlorite with the addition of 0.1% Tween 80 for 30 min while shaking at 200 rpm. This is followed by thorough rinsing in sterile water and soaking in 5% sodium bicarbonate for 10 min. After rinsing, the wet nodules are left at room temperature overnight and then resterilized as before. At this point, if necessary the sterile nodule lobes can be stored at −20°C.

For inoculation, the nodule lobes are placed in a Petri dish containing 0.5 ml of sterile broth or water and crushed with a heavy stirring rod. The suspension is pipetted into a liquid medium or incorporated into an agar medium and incubated at 22–33°C. We have also used the sucrose-density technique and the differential filtration method with success.

2. Isolation Media

Five considerations govern the selection of media and growth conditions for strains of *Frankia:* (1) most strains are microaerophilic and as such grow best in static or slowly shaking liquid media or in Petri dishes, embedded in an agar medium; (2) they grow very slowly compared to other actinomycetes and some strains more slowly than others; (3) they may not all have the same nutritional requirements; (4) the optimum temperature for their growth varies between 25 and 33°C; and (5) they are not sensitive to the action of purely antifungal antibiotics, which can be incorporated in the isolation media. Isolation failures are not unusual and the use of several different media when attempting isolations from nodules is helpful.

The media used for isolation should be relatively poor, nutritionally speaking, to keep down growth of competing bacteria. We find it best to use semisolid agars (0.8–1.0%) amended with cycloheximide (300 μg/ml) and nystatin (10 μg/ml) to reduce the growth of fungi. We have found the following media to be the most useful in our own laboratory:

Tap Water Agar

Crude agar (e.g., Sigma A7002)	10.0 g
Tap water	1.0 liter

No pH adjustment

Tap Water Agar/NZ

Above medium plus 0.2% NZ Amine A (casein hydrolysate; Sheffield Farms, Norwich, N.Y.

Bennett's Agar

Yeast extract	1.0 g
Beef extract	1.0 g
NZ Amine A	2.0 g
Glucose	10.0 g
Agar	10.0 g
Tap water	1.0 liter

pH 7.3 before sterilization

Bennett's medium may be used at half strength to one-sixteenth strength.

S Medium

NZ Amine A	4.0 g
Glucose	10.0 g
K_2HPO_4	0.5 g
$MgSO_4 \cdot 7H_2O$	0.2 g
$CaCl_2 \cdot 2H_2O$	0.1 g
Ferric citrate	1.0 ml
(Citric acid : ferric citrate, 1% aqueous)	
Q Mod minor salts (see below)	1.0 ml
Agar	10.0 g
Distilled water	1.0 liter

pH 6.8

Benson's *Frankia* Medium

K_2HPO_4	3.00 g	
KH_2PO_4	2.00 g	
$MgSO_4 \cdot 7H_2O$	0.20 g	
NaCl	0.30 g	
FeNa EDTA	0.16 g	
Trace elements	1.00 ml	
Vitamin solution	1.00 ml	
Casamino acids	3.00 g	
$CaCO_3$	0.05 g	(broth only)
Agar	10.0 g	
Distilled water	900.0 ml	

pH 6.9 prior to autoclaving

Sodium pyruvate	3.00 g

Distilled water	100.00 ml

Filter-sterilize and add to above autoclaved solution.

Trace Elements for preceding medium

H_3BO_3	1.50 mg
$ZnSO_4 \cdot 7H_2O$	1.50 mg
$MnSO_4 \cdot H_2O$	4.50 mg
$Na_2MoO_4 \cdot 2H_2O$	0.25 mg
$CuSO_4 \cdot 5H_2O$	0.04 mg
Distilled water	1.00 liter

Vitamin Solution for preceding medium

Thiamin · HCl	10.00 mg
Nicotinic acid	50.00 mg
Pyridoxine · HCl	50.00 mg
Distilled water	1.00 liter

HIML 3 Medium

L-Histidine	0.1 g
L-Methionine	0.1 g
L-Isoleucine	0.1 g
L-Leucine	0.1 g
Glucose	10.0 g
K_2HPO_4	0.5 g
$CaCl_2 \cdot 2H_2O$	0.1 g
$MgSO_4 \cdot 7H_2O$	0.2 g
NaFe EDTA	0.01 g
Distilled water	1.0 liter
Agar	10.0 g

pH 6.4

Q Mod Medium

K_2HPO_4	300 mg
NaH_2PO_4	200 mg
$MgSO_4 \cdot 7H_2O$	200 mg
KCl	200 mg
Yeast extract	500 mg
Bacto peptone	5 g
Glucose	10 g
Ferric citrate	1 ml
Minor salts	1 ml
Deionized water	1 liter

Adjust pH to 6.8–7.0 with NaOH or HCl then add:

$CaCO_3$	100 mg
Lipid supplement	5 mg
Agar	15 g

Minor Salts (g/liter): H_3BO_3, 1.5; $MnSO_4 \cdot 7H_2O$, 0.8; $ZnSO_4 \cdot 7H_2O$, 0.6; $CuSO_4 \cdot 7H_2O$, 0.1; $(NH_4)Mo_7O_{24} \cdot 4H_2O$, 0.2; $CoSO_4 \cdot 7H_2O$, 0.01.

Lipid Supplement: Dissolve 500.00 mg of L-a-lecithin (Sigma P-5638) in 50 ml of absolute ethanol, then add 50 ml distilled water (the result is not limpid).

Although we do not use the Q Mod medium of Lalonde and Calvert (1979), we have included it because it is widely used.

In our judgment, the most useful technique for isolation employs Petri dishes with agar media since this provides discrete colonies that can be monitored microscopically through the back of the plate by means of a long working distance objective and condenser. Alternately, whole lobes can be placed in liquid culture and if after a suitable period they appear to be sterile, they may be crushed or ground and added to fresh broth or plated out to obtain colonies.

3. Incubation

Plates and/or tubes of liquid media may be incubated at 25–35°C in the dark. In general, plates should be protected from desiccation by placing them in a moist environment. We use disposable plastic Petri dishes contained in the plastic sleeve in which they are supplied. We have kept agar plates (most are prepared with agar layers 8–10 mm thick) for over one year. Tubes of liquid media can be rehydrated as needed with sterile distilled water or amended with other solutions if outgrowth fails.

Frankia colonies may appear as soon as one week to 10 days after plating, but this is rare. More commonly, colonies are big enough to be identified morphologically after 2–4 months. We have had colonies that began to grow only after one year.

4. Identification of Isolated Colonies

Because frankiae form characteristic morphological structures (see Section I), identification of an actinomycete colony as a *Frankia* often relies on microscopic verification of the presence of the characteristic sporangia and/or vesicles in addition to branching hyphae and lack of aerial mycelium. Some strains form sporangia only under exceptional conditions and such morphologically sterile colonies should be isolated and grown to check their morphology on a different medium and, if need be, their cell chemistry.

5. Colony Workup

A colony of interest should be cut out of the agar with a sterile spatulate needle and ground in sterile water or broth using a small (2–5 ml) Ten Broeck tissue grinder. The ground cells are transferred to fresh agar plates or to tubes of various nutrient media and incubated as before. Outgrowth is often quite slow, but in tubes of broth it may be facilitated by gentle shaking every 3–7 days. As discussed in the following for subculturing, the use of several quite different media in parallel is often the best strategy to assure viability. It is not uncommon that a strain isolated on a given medium may not grow out again on that medium following transfer. This is most likely due to exhaustion of a critical trace nutrient. It may also reflect a selective process for adaptation to life *in vitro*.

C. Discussion

1. Utility of Methods

Methods of isolation such as those outlined earlier have permitted researchers to isolate endophytes from 15 of the 23 genera of actinorhizal plants in a little over 10 years. The large majority of isolates are from the genus *Alnus*, and it seems likely that strains of this host compatibility group may be the easiest to isolate and grow. The sucrose-density method may select for strains that form propagules that are inert to, or at least not killed by, dehydration from their passage in such high-sucrose concentrations. The differential filtration method was designed to free the vesicles from the accompanying plant tissue and its associated toxic compounds, notably phenolics. Yet using the pour-plate technique, we have observed that in some nodules there are endophytes that consistently grow out in association with nodule tissue (usually these are the slowest to grow) and some that produce colonies independent of nodular tissue. Thus, the differential filtration method may be selective for the latter type. It seems clear that no one method may be ideal, but the pour-plate method may be less selective than the others.

We are still confronted with the fact that the endophytes from at least eight actinorhizal genera have not yielded to the blandishments and complicated cocktails offered by expectant frankiologists. Do we need other methods? Probably not, but we may have to modify those we already have to accommodate the requirements of those frankiae that remain to be grown in culture.

2. Unsuccessful Isolations

There are many unsuccessful isolations. Like the fish that got away after it was hooked, some frankiae grow out once but do not seem to be able to be subcultured. Another frustration is the visible, healthy-looking vesicle clusters that never grow into colonies even after long periods of incubation. Recently, Anton Quispel of Leiden University, one of the earliest workers in the field who is still active, has successfully proven his long-held theory that root extracts contain substances that enhance the growth of frankiae (Quispel, 1960). He and his colleagues (Quispel *et al.*, 1989) have purified the principal *Frankia*-stimulating factor in the root extracts of *Alnus glutinosa* and characterized it as dipterocarpol (hydroxydammarenone II). The presence of this substance in isolation media increases the incidence of outgrowth from nodule fragments. Once the endophyte is isolated, it does not appear to increase growth in maintenance media. Obviously, not all frankiae require these special substances to grow *in vitro,* but it is not at all unlikely that some endophytes, those more committed to symbiosis and thus more metabolically constrained, may indeed require some special substance to stimulate enzyme production or modify external membranes or wall to permit transport and/or uptake of nutrients.

III. Culture of *Frankia*

A. Decontamination

Isolation plates (or tubes) or subsequent cultures may be contaminated by microorganisms. When this occurs, growth of fungi can be slowed or eliminated through the use of antifungal antibiotics such as cycloheximide (Actidione) (500 μg/ml) and a polyene, such as nystatin (10 μg/ml). However, frankiae being bacteria, it is difficult to eliminate bacterial contaminants. The antibacterial substance, nalidixic acid at about 10 μg/ml, can be used to good effect in some cases, since its principal activity is against gram-negative bacteria. If this fails, strenuous washing and grinding of colonies followed by growth on a nitrogen-free medium such as DPM reduces the growth of contaminants to the point where plating out to obtain isolated colonies can yield a pure culture.

B. Maintenance of Cultures

Once a *Frankia* strain is in pure culture, the question becomes: What is the best medium for maintenance and growth of the organism isolated? In general, nitrogen sources (other than atmospheric nitrogen) include ammonia, nitrate, urea, amino acids, and amides. Casein hydrolysate is a widely used source of amino acids; yeast extract is toxic to some strains. The best simple carbon sources are organic acids such as propionate, pyruvate, and acetate; some strains also dissimilate succinate and malate. Other excellent carbon sources for some strains are Tweens such as Tween 80 or Tween 20. Monosaccharides and disaccharides such as glucose, arabinose, mannose, xylose, sucrose, and trehalose are variously utilized, mostly by Type A strains (Lechevalier and Lechevalier, 1989). The effect of salts is complex (Lechevalier *et al.*, 1987) and may be highly individual. If nondistilled water is used, this could be a critical factor in the maintenance of strains; some waters support the growth of frankiae better than others. Most strains do not require vitamins.

In the following we list nine broth media that are used in our laboratory for the routine maintenance of strains.

(1) L/2 Medium

Basal:	Glycerol, 10% aqueous	1.0 ml
	NH_4Cl, 25% aqueous	1.0 ml
	Na_2HPO_4	1.0 g
	KH_2PO_4	0.3 g
	NaCl	1.0 g
	Thiamin, 0.5% aqueous	1.0 ml
	Distilled water	997.0 ml
	Adjust to pH 7.4; autoclave	
Supplement:	Bovine serum albumin (V)	20.0 g
	$CaCl_2 \cdot H_2O$, 1% soln.	2.0 ml
	$MgCl_2 \cdot 6H_2O$, 1% soln.	2.0 ml
	$ZnSO_4 \cdot 7H_2O$, 0.4% soln.	2.0 ml
	$CuSO_4 \cdot 5H_2O$, 0.3% soln.	2.0 ml
	$FeSO_4 \cdot 7H_2O$, 0.5% soln.	20.0 ml
	B_{12}, 0.2% soln.	2.0 ml
	Tween 80, 16% soln.	25.0 ml

Sterile distilled water to 200.00 ml
Seitz-filter and mix Basal : Supplement 9 : 1.
Test sterility before use.

(2) Defined Propionate Medium (DPM/N) (Baker and O'Keefe, 1984)

KH_2PO_4	1.0 g
$MgSO_4 \cdot 7H_2O$	0.1 g
$CaCl_2 \cdot 2H_2O$	0.01 g
Sodium propionate	1.2 g
NH_4Cl	0.43 g
Hoagland's micronutrients	1.0 ml
Hoagland's iron	1.8 ml

pH 6.8

Hoagland's Micronutrients (for the above)

	(gm/liter)
H_3BO_3	2.86
$MnCl_2 \cdot 4H_2O$	1.81
$ZnSO_4 \cdot 7H_2O$	0.22
$CuSO_4 \cdot 5H_2O$	0.08
Na_2MoO_4	0.025
$CoCl_2 \cdot 6H_2O$	0.025

Hoagland's Iron (for the above)

	(g/liter)
$FeSO_4 \cdot 7H_2O$	5.56
Na_2 EDTA	7.45

(3) DPM Medium

Omit the ammonium chloride from DPM/N.

(4) DPM/2PF

Substitute 2.4 g of sodium pyruvate for the sodium propionate of DPM. Filter-sterilize pyruvate separately and add to basal medium.

(5) S Medium

See Section II, B, 2.

(6) S + TW Medium

Add 2 ml/liter of Tween 80 to S medium

(7) Bennett's Half-Strength (B/2)

See Section II, B, 2.

(8) Yeast–Czapeks Medium (YCz)

$NaNO_3$	2.0 g
K_2HPO_4	1.0 g
$MgSO_4 \cdot 7H_2O$	0.5 g
KCl	0.5 g
$FeSO_4 \cdot 7H_2O$	0.01 g
Sucrose	30.0 g
Yeast extract	2.0 g
Distilled water	1.0 liter

pH 6.6

(9) Yeast Malt (YM)

Yeast extract	4.0 g
Malt extract	10.0 g
Glucose	4.0 g
Tap water	1.0 liter

pH 7.3

It is best to attempt to maintain a new isolate in at least four of these: for example, L/2, B/2, YCz, DPM, or DPM/2PF. In this way, following four or five transfers, the organism itself "picks" the medium or media in which it grows best. Although Lalonde's Q Mod (see the previous section) is used widely by many persons, in our hands it has never been a satisfactory isolation medium. It has also proved to be a poor medium for maintenance *except for those strains that were initially isolated on that medium*. This observation underlines two of our many unresolved problems: (1) Is the isolation medium one uses selective for certain segments of an endophyte population? and (2) Do the endophytes adapt to their new laboratory environment in ways we still know little about? The answer is that probably both situations occur.

Frankiae grow best at $>25°C$ to $<37°C$, and the frequency with which cultures should be transferred ranges from 2 to 6 months, depending on many factors. New isolates are often slower to grow out than older ones, and initially it is wise not to transfer a new isolate too frequently. Growth occurs at the bottom or along the sides (or even at times at the top) of the liquid medium, depending on the relative tendency of the strain to microaerophily and the composition of the medium. Some frankiae of the A type are sufficiently aerobic to be maintained on slants. Storage is best at room temperature as some strains will die at 4°C.

C. Long-Term Preservation

Preservation of *Frankia* spp. is possible by lyophilization (freeze-drying). Although most actinomycete cells are conveniently lyophilized as a skim milk suspension, this is not possible with frankiae; something in the milk appears to be inhibitory to outgrowth. We lyophilize these organisms in the spent medium in which they were grown. Horse serum can also be used. We have attempted to use storage in sterile blood-amended soil but this has proved less satisfactory than lyophilization in maintaining viability. Outgrowth from lyophile storage is usually reasonably rapid; however, some strains may take up to 6 months to grow. Fontaine *et al.* (1986) compared the preservation of a single strain from *Alnus rubra*. Of the three methods used, lyophilization was considered best. Another convenient method of storage is by freezing (−70°C) cells in growth medium with or without added (20%) glycerol (C. R. Schwintzer, personal communication).

D. Deposit of Strains in Recognized Culture Collections

Strains of frankiae isolated by scientists in North America may be sent to the USDA's Northern Regional Research Center in Peoria, Illinois (1815 University Street; Dr. D. P. Labeda, Curator) for accession in their famous actinomycete culture collection. The IFO collection in Japan has also accessioned a certain number of strains, as has the American Type Culture Collection (12301 Parklawn Drive, Rockville, Md. 20852).

E. Registry of Strain Numbers

In 1980, it was decided to create an international registry that would provide data on the previous designations, publications, origin, infectivity, effectiveness, cell chemistry, and availability of strains about which data had been or would be published (Lechevalier, 1983). Two editions of a catalog listing strains have been issued (Lechevalier, 1986). In an effort to eliminate many of the confusing and often cumbersome "lab" names in the literature, strains are now referred to in publications by an acronym of three letters (to be chosen by the individual scientists to represent their culture collection), followed by up to 10 numbers, the first four of which represent the plant host genus and species from which an isolate was obtained (Lechevalier, 1983).

References

An, C. S., Wills, J. W., Riggsby, W. S., and Mullin, B. C. (1983). Deoxyribonucleic acid base composition of 12 *Frankia* isolates. *Can. J. Bot.* **61**, 2859–2862.

An, C. S., Riggsby, W. S., and Mullin, B. C. (1985a). Relationships of *Frankia* isolates based on deoxyribonucleic acid homology studies. *Int. J. Syst. Bacteriol.* **35**, 140–146.

An, C. S., Riggsby, W. S., and Mullin, B. C. (1985b). Restriction pattern analysis of genomic DNA of *Frankia* isolates. *Plant Soil* **87,** 43–48.

An, C. S., Riggsby, W. S., and Mullin, B. C. (1987). DNA relatedness of *Frankia* isolates ArI4 and EuI1 to other actinomycetes of cell wall type III. *Actinomycetes* **20,** 50–59.

Anders, T., Schultz, N. A., Benson, D. R., Steele, D. B., and White, D. C. (1989). Differences in composition between vegetative cells and nitrogen-fixing vesicles of *Frankia* species strain CpI1. *Proc. Natl. Acad. Sci. U.S.A.* **86,** 3399–3403.

Baker, D. D. (1987). Relationships among pure cultured strains of *Frankia* based on host specificity. *Physiol. Plant.* **70,** 245–248.

Baker, D. (1989). Methods for the isolation, culture and characterization of the Frankiaceae, soil actinomycetes and symbionts of actinorhizal plants. *In* "Isolation of Biotechnical Organisms from Nature" (D. P. Labeda, ed.), pp. 213–236. McGraw-Hill, New York.

Baker, D., and O'Keefe, D. (1984). A modified sucrose fractionation procedure for the isolation of frankiae from actinorhizal root nodules and soil samples. *Plant Soil* **78,** 23–28.

Baker, D., Pengelly, W. L., and Torrey, J. G. (1981). Immunological analysis of relationships among the isolated frankiae (Actinomycetales). *Int. J. Syst. Bacteriol.* **31,** 148–151.

Becking, J. H. (1970). *Frankiaceae* fam. nov. (Actinomycetales) with one new combination and six new species of the genus *Frankia* Brunchorst 1886, 174. *Int. J. Syst. Bacteriol.* **20,** 201–220.

Becking, J. H. (1974). Family III. Frankiaceae Becking 1970, 201. *In* "Bergey's Manual of Determinative Bacteriology" (R. E. Buchanan and N. E. Gibbons, eds.), 8th ed., pp. 701–706. Williams & Wilkins, Baltimore, Maryland.

Benson, D. R. (1982). Isolation of *Frankia* strains from alder actinorhizal root nodules. *Appl. Environ. Microbiol.* **44,** 461–465.

Benson, D. R., and Hanna, D. (1983). *Frankia* diversity in an alder stand as estimated by sodium dodecyl sulfate–polyacrylamide gel electrophoresis of whole-cell protein. *Can. J. Bot.* **61,** 2919–2923.

Benson, D. R., Bucholz, S. E., and Hanna, D. G. (1984). Identification of *Frankia* strains by two-dimensional polyacrylamide gel electrophoresis. *Appl. Environ. Microbiol.* **47,** 489–494.

Blom, J., Roelofsen, W., and Akkermans, A. D. L. (1980). Growth of *Frankia* AvcI1 on media containing Tween 80 as C-source. *FEMS Microbiol. Lett.* **9,** 131–135.

Bloom, R. A., Lechevalier, M. P., and Tate, R. L., III (1989a). Physiological, chemical, morphological and plant-infectivity characteristics of *Frankia* isolates from *Myrica pensylvanica*: Correlation to DNA restriction patterns. *Appl. Environ. Microbiol.* **55,** 2161–2166.

Bloom, R. A., Mullin, B. C., and Tate, R. L., III (1989b). DNA restriction patterns and solution hybridization studies of *Frankia* isolates from *Myrica pensylvanica* (bayberry). *Appl. Environ. Microbiol.* **55,** 2155–2160.

Burggraaf, A. J. P., and Valstar, J. (1984). Heterogeneity within *Frankia* sp. LDAgp1 studied among clones and reisolates. *Plant Soil* **78,** 29–43.

Callaham, D., Del Tredici, P., and Torrey, J. G. (1978). Isolation and cultivation *in vitro* of the actinomycete causing root nodulation in *Comptonia*. *Science* **199,** 899–902.

Carpenter, C. V., and Robertson, L. R. (1983). Isolation and culture of nitrogen-fixing organisms. *In* "Biological Nitrogen Fixation in Forest Ecosystems: Foundations and Applications" (J. C. Gordon and C. T. Wheeler, eds.), pp. 89–106. Nijhoff/Dr. W. Junk, The Hague.

Crameri, R., Hintermann, G., and Hutter, R. (1983). Deoxyribonucleic acid restriction endonuclease fingerprint characterization of actinomycete strains. *Int. J. Syst. Bacteriol.* **33,** 652–655.

Diem, H. G., and Dommergues, Y. R. (1988). Isolation, characterization and cultivation of

Frankia. *In* "Biological Nitrogen Fixation, Recent Developments" (N. S. Subba Rao, ed.), pp. 227–254. Gordon and Breach Science Publishers, New York.

Dillon, J. T., and Baker, D. (1982). Variations in nitrogen fixing efficiency among pure-cultured *Frankia* strains tested on actinorhizal plants as an indication of symbiotic compatibility. *New Phytol.* **92,** 215–219.

Dobritsa, S. V. (1985). Restriction analysis of the *Frankia* spp. genome. *FEMS Microbiol. Lett.* **29,** 123–128.

Fontaine, M. S., Young, P. H., and Torrey, J. G. (1986). Effects of long term preservation of *Frankia* strains on infectivity, effectivity and *in vitro* nitrogenase activity. *Appl. Environ. Microbiol.* **51,** 694–698.

Gardes, M., and Lalonde, M. (1987). Identification and subgrouping of *Frankia* strains using sodium dodecyl sulfate polyacrylamide gel electrophoresis. *Physiol. Plant.* **70,** 237–244.

Gardes, M., Bousquet, J., and Lalonde, M. (1987). Isozyme variation among 40 *Frankia* strains. *Appl. Environ. Microbiol.* **53,** 1596–1603.

Gerber, N. N., and Lechevalier, M. P. (1984). Novel benzo(*a*)naphthacene quinones from an actinomycete, *Frankia* G-2 (ORS 020604). *Can. J. Chem.* **62,** 2818–2821.

Goodfellow, M., Jones, E., and Priest, F. G. (1985). "Computer-Assisted Bacterial Systematics." Academic Press, Orlando, Florida.

Hahn, D., Lechevalier, M. P., Fischer, A., and Stackebrandt, E. (1989a). Evidence for a close phylogenetic relationship between members of the genera *Frankia, Geodermatophilus,* and "*Blastococcus*" and emendation of the family Frankiaceae. *Syst. Appl. Microbiol.* **11,** 236–242.

Hahn, D., Dorsch, M., Stackebrandt, E., and Akkermans, A. D. L. (1989b). Synthetic nucleotide probes for identification of *Frankia* strains. *Plant Soil,* **118,** 211–219.

Holman, R. M., and Schwintzer, C. R. (1987). Distribution of spore-positive and spore-negative nodules of *Alnus incana* ssp. *rugosa* in Maine, USA. *Plant Soil* **104,** 103–111.

Horrière, F. (1984). *In vitro* physiological approach to classification of *Frankia* isolates of the "*Alnus* group," based on urease, protease and β-glucosidase activities. *Plant Soil* **78,** 7–13.

Knowlton, S., and Dawson, J. O. (1983). Effects of *Pseudomonas cepacia* and cultural factors on the nodulation of *Alnus rubra* roots by *Frankia*. *Can. J. Bot.* **61,** 2877–2882.

Knowlton, S., Berry, A., and Torrey, J. G. (1980). Evidence that associated soil bacteria may influence root hair infection of actinorhizal plants by *Frankia*. *Can. J. Microbiol.* **26,** 971–977.

Lalonde, M. (1978). Confirmation of the infectivity of a free-living actinomycete isolated from *Comptonia peregrina* root nodules by immunological and ultrastructural studies. *Can. J. Bot.* **56,** 2621–2635.

Lalonde, M., and Calvert, H. E. (1979). Production of *Frankia* hyphae and spores as an infective inoculant for *Alnus* species. *In* "Symbiotic Nitrogen Fixation in the Management of Temperate Forests" (J. C. Gordon, C. T. Wheeler, and D. A. Perry, eds.), pp. 95–110. For. Res. Lab., Oregon State Univ., Corvallis, Oregon.

Lalonde, M., Calvert, H. E., and Pine, S. (1981). Isolation and use of *Frankia* strains in actinorhizae formation. *In* "Current Perspectives in Nitrogen Fixation in the Management of Temperate Forests" (A. H. Gibson and W. E. Newton, eds.), pp. 296–299. Aust. Acad. Sci., Canberra, Australia.

Lalonde, M., Simon, L., Bousquet, J., and Séguin, A. (1988). Advances in the taxonomy of *Frankia*: Recognition of species *alni* and *elaeagni* and novel subspecies *pommerii* and *vandijkii*. *In* "Nitrogen Fixation: Hundred Years After" (H. Bothe, F. J. de Bruijn, and W. E. Newton, eds.), pp. 671–680. Fischer, Stuttgart.

Lechevalier, H. A. (1989). A practical guide to generic identification of actinomycetes. *In* "Bergey's Manual of Systematic Bacteriology" (S. T. Williams, M. E. Sharpe, and J. G. Holt, eds.) Vol. 4, pp. 2344–2347. Williams & Wilkins, Baltimore, Maryland.

Lechevalier, H. A., and Lechevalier, M. P. (1981). Introduction to the order Actinomycetales. *In* "The Prokaryotes" (M. P. Starr, H. Stolp, H. G. Truper, A. Balows, and H. G. Schlegel, eds.), pp. 1915–1922. Springer-Verlag, Berlin and New York.

Lechevalier, M. P. (1981). Ecological associations involving actinomycetes. *Zentralbl. Bakteriol., Suppl.* **11,** 159–166.

Lechevalier, M. P. (1983). Cataloging *Frankia* strains. *Can. J. Bot.* **61,** 2964–2967.

Lechevalier, M. P. (1984). The taxonomy of the genus *Frankia*. *Plant Soil* **78,** 1–6.

Lechevalier, M. P. (1986). Catalog of *Frankia* strains. *Actinomycetes* **19,** 131–162.

Lechevalier, M. P. (1988). Actinomycetes in agriculture and forestry. *In* "Actinomycetes in Biotechnology" (M. Goodfellow, S. T. Williams, and M. Mordarski, eds.), pp. 327–358. Academic Press, San Diego, California.

Lechevalier, M. P. (1989). Actinomycetes with multilocular sporangia. *In* "Bergey's Manual of Systematic Bacteriology" (S. T. Williams, M. E. Sharpe, and J. G. Holt, eds.), Vol. 4, pp. 2405–2406. Williams & Wilkins, Baltimore, Maryland.

Lechevalier, M. P., and Lechevalier, H. A. (1980). The chemotaxonomy of actinomycetes. *In* "Actinomycete Taxonomy" (A. Dietz and D. W. Thayer, eds.), pp. 227–291. Soc. Ind. Microbiol., Arlington, Virginia.

Lechevalier, M. P., and Lechevalier, H. A. (1989). Genus *Frankia* Brunchorst 1886, 174[AL]. *In* "Bergey's Manual of Systematic Bacteriology" (S. T. Williams, M. E. Sharpe, and J. G. Holt, eds.), Vol. 4, pp. 2410–2417. Williams and Wilkins, Baltimore, Maryland.

Lechevalier, M. P., and Ruan J.-S. (1984). Physiology and chemical diversity of *Frankia* spp. isolated from nodules of *Comptonia peregrina* (L.) Coult. and *Ceanothus americanus* L. *Plant Soil* **78,** 15–22.

Lechevalier, M. P., Horrière, F., and Lechevalier, H. (1982). The biology of *Frankia* and related organisms. *Dev. Ind. Microbiol.* **23,** 51–60.

Lechevalier, M. P., Baker, D., and Horrière, F. (1983). Physiology, chemistry, serology and infectivity of two *Frankia* isolates from *Alnus incana* subsp. *rugosa*. *Can. J. Bot.* **61,** 2826–2833.

Lechevalier, M. P., Labeda, D. P., and Ruan, J. S. (1987). Studies on *Frankia* sp. LLR 02022 from *Casuarina cunninghamiana* and its mutant LLR 02023. *Physiol. Plant.* **70,** 249–254.

Lopez, M. F., Whaling, C. S., and Torrey, J. G. (1983). The polar lipids and free sugars of *Frankia* in culture. *Can. J. Bot.* **61,** 2834–2842.

McBride, M. J., and Ensign, J. C. (1987). Effects of intracellular trehalose content on *Streptomyces griseus* spores. *J. Bacteriol.* **169,** 4995–5001.

Minnikin, D. E., and O'Donnell, A. G. (1983). Actinomycete envelope lipid and peptidoglycan composition. *In* "The Biology of the Actinomycetes" (M. Goodfellow, M. Mordarski, and S. T. Williams, eds.), pp. 337–388. Academic Press, New York.

Mort, A., Normand, P., and Lalonde, M. (1983). 2-O-Methyl D-mannose, a key sugar in the taxonomy of *Frankia*. *Can. J. Microbiol.* **29,** 993–1002.

Normand, P., and Lalonde, M. (1982). Evaluation of *Frankia* strains isolated from provenances of two *Alnus* species. *Can. J. Microbiol.* **28,** 1133–1142.

Parson, W. L., Robertson, L. R., and Carpenter, C. V. (1985). Characterization and infectivity of a spontaneous variant isolated from *Frankia* sp. WEY 0131391. *Plant Soil* **87,** 31–42.

Pommer, E. (1959). Über die Isolierung des Endophyten aus den Wurzelknöllchen *Alnus glutinosa* Gaertn. und über Erfolgreiche Reinfektionsversuche. *Ber. Dtsch. Bot. Ges.* **72,** 138–150.

Puppo, A., Dimitrijevic, L., Diem, H. G., and Dommergues, Y. R. (1985). Homogeneity of superoxide dismutase patterns in *Frankia* strains from Casuarinaceae. *FEMS Microbiol. Lett.* **30,** 43–46.

Quispel, A. (1960). Symbiotic nitrogen fixation in non-leguminous plants. V. The growth requirements of the endophyte of *Alnus glutinosa*. *Act. Bot. Neerl.* **9,** 380–396.

Quispel, A., and Burggraaf, A. J. P. (1981). *Frankia*, the diazotrophic endophyte from actinorhizas. *In* "Current Perspectives in Nitrogen Fixation" (A. Gibson and W. Newton, eds.), p. 229. Aust. Acad. Sci., Canberra, Australia.

Quispel, A., Svendsen, A. B., Schripsema, J., Baas, W. J., Erkelens, C., and Lugtenburg, J. (1989). Identification of dipterocarpol as isolation factor for the induction of primary isolation of *Frankia* from root nodules of *Alnus glutinosa* (L.) Gaertner. *Mol. Plant–Microbe Interact.* **2,** 107–112.

Rickards, R. W. (1989). Revision of the structures of the benzo(*a*)naphthacene quinone metabolites G-2N and G-2A from bacteria of the genus *Frankia*. *J. Antibiot.* **42,** 336–339.

Ruan, J.-S., Zhang, X, Liu, W., Wang, C., Liu, Z., and Zhang, Y. (1986). Taxonomic studies of *Frankia* spp. *In* "Biological, Biochemical and Biomedical Aspects of Actinomycetes" (G. Szabo, S. Biro, and M. Goodfellow, eds.), pp. 587–590. Akademiai Kiado, Budapest, Hungary.

St. Laurent, L., Bousquet, J., Simon, L., and Lalonde, M. (1987). Separation of various *Frankia* strains in the *Alnus* and *Elaeagnus* host specificity groups using sugar analysis. *Can. J. Microbiol.* **33,** 764–772.

Stackebrandt, E. (1986). The significance of "wall types" in phylogenetically based taxonomic studies on actinomycetes. *In* "Biological, Biochemical and Biomedical Aspects of Actinomycetes" (G. Szabo, S. Biro, and M. Goodfellow, eds.), Part B, pp. 497–506. Akademiai Kiado, Budapest, Hungary.

Stowers, M. D. (1987). Collection, isolation, cultivation and maintenance of *Frankia*. *In* "Symbiotic Nitrogen Fixation Technology" (G. H. Elkan, ed.), pp. 29–53. Dekker, New York.

Streeter, J. G. (1985). Accumulation of α,α-trehalose by *Rhizobium* bacteria and bacteroids. *J. Bacteriol.* **164,** 78–84.

Sun, H., Ding, J., Wu, Y., Xu, Q., Huang, Y., Su, F., Zhang, Z., and Li, W. (1987). The serogroups of *Frankia*. *J. Microbiol. (People's Republic of China)* **7,** 51–56.

Torrey, J. G. (1987). Endophyte sporulation in root nodules of actinorhizal plants. *Physiol. Plant.* **70,** 279–288.

van Dijk, C. (1978). Spore formation and endophyte diversity in root nodules of *Alnus glutinosa* (L.) Vill. *New Phytol.* **81,** 601–615.

Wu, Y., and Ding, J. (1987). Studies on biological characteristics of *Frankia* sp. from *Hippophae rhamnoides*. *Acta Microbiol. Sin.* (*Engl. Transl.*) **27,** 227–232.

Zhang, Z., and Torrey, J. G. (1985). Studies of an effective strain of *Frankia* from *Allocasuarina lehmanniana* of the Casuarinaceae. *Plant Soil* **87,** 1–16.

Zhang, Z., Lopez, M. F., and Torrey, J. G. (1984). A comparison of cultural characteristics and infectivity of *Frankia* isolates from root nodules of *Casuarina* species. *Plant Soil* **78,** 79–90.

4

The Infection Process and Nodule Development

Alison M. Berry and Leslie A. Sunell
Department of Environmental Horticulture
University of California
Davis, California, U.S.A.

Actinorhizal root nodules are established through a series of interactions between *Frankia* hyphae and host root cells. The nodule is a system of structurally and developmentally modified lateral roots. The structural changes observed in the development of nodules reflect molecular events in which normal patterns of plant development are altered. These modifications permit microsymbiotic invasion and the establishment of nitrogen-fixing root nodule tissue.

I. Infection Process

A. Conditions for Infection

The infection process begins in the rhizosphere. Conditions necessary for colonization and initial binding of *Frankia* to the host root have not been entirely elucidated. *Frankia* hyphae are embedded within a mucilage layer at the root hair or epidermal surface. This mucilage layer is evidently secreted by the root hairs (Berry *et al.*, 1983; Prin and Rougier, 1986), although the root cap also produces mucilage. Elements of the host-derived matrix may participate in attachment phenomena (see Section IV,A).

In general, as long as conditions favorable for root growth are present, environmental factors such as pH or nutrient availability do not appear to limit infection and nodulation, at least for *Frankia* strains isolated to date. However, Bond (1951) showed that nodulation of *Myrica gale* was poor at pH 3.3, even though plants supplied with nitrate grew vigorously at this pH. "Helper" microorganisms, for example, *Pseudomonas* sp., improve nodulation of both *Alnus* and *Casuarina* under controlled growth conditions (Knowlton *et al.*, 1980; S. R. Mansour and J. G. Torrey, unpublished observations). The effect of these organisms appears to be twofold, first in inducing root hair deformation, which may involve altered cell wall deposition, and second in increasing the production of root hair extracellular matrix (Berry and Torrey, 1983).

Even under favorable rhizosphere conditions, infection is not assured. Compatibility factors, about which little is known, influence the potential for successful infection and nodulation. This topic is explored in Chapter 4.

B. Infection Pathways

There are two known pathways by which *Frankia* initially infects compatible host roots. For some hosts, the endosymbiont penetrates a deformed root hair, while for other hosts, early colonization of root tissues by *Frankia* is entirely intercellular. Despite these differences in the early stages of cellular interactions between host and microsymbiont, eventually *Frankia* penetrates host cell walls in developing nodule lobes and differentiates within the cells. The same *Frankia* strain can, in some instances, infect different hosts by different early pathways (Miller and Baker, 1985; Racette and Torrey, 1989). These observations indicate a common genetic basis for infection.

1. Root Hair Infection

a. Root Hair Deformation. Root hair infection has been described in *Myrica, Comptonia, Alnus,* and *Casuarina* (Torrey, 1976; Callaham and Torrey, 1977; Callaham *et al.*, 1979; Berry *et al.*, 1986). This process involves extensive changes in host cells beginning with wall deformation of expanding root hairs. Growing root hairs are sensitive to perturbation by rhizosphere bacteria, especially by *Frankia*. Responding to unknown signals, growing root hairs initiate multiple tips, and tip growth is modified so that branches may bulge and wrap around each other (Fig. 1). Branching root hairs may be few in number, as in *Comptonia* (Callaham and Torrey, 1977), or extensive, as in *Alnus* (Berry, 1983). Cell walls of deformed root hairs fluoresce yellow-orange after staining with acridine orange, indicating a polyanionic component not evident in normal root

Figure 1 Scanning electron micrograph of the root epidermal surface of *Alnus rubra* inoculated with *Frankia*. Root hairs bulge at tips, curl, and branch. Adjacent root hairs or lobes may be closely appressed. Hyphae (arrow) adhere to root hair surfaces. Bar is 10 μm.

hairs. Other changes include increased secretion of extracellular matrix of unknown composition.

Even though all root hairs may become deformed in response to the presence of *Frankia* in the rhizosphere, only one or, in some hosts, a few root hairs become infected (Fig. 2). *Frankia* hyphae associated with deformed root hairs penetrate at the site of folding of the root hair cell wall. Where convolutions of the growing root hair juxtapose portions of the cell wall, *Frankia* filaments are embedded within a thick, densely staining accumulation of extracellular matrix.

At the site of wall penetration, the infecting hyphae are continuous with *Frankia* in the rhizosphere. Characteristic changes occur in patterns of host wall deposition, as observed in *Alnus rubra* (Berry *et al.*, 1983). Inner wall layers at the site of infection resemble transfer cell walls (Berry *et al.*, 1986). Pronounced deposition of wall material is also associated with microsymbiont invasion in *Myrica gale* and *Comptonia* (Callaham *et al.*, 1979). Within the infected hair, the microsymbiont may branch early

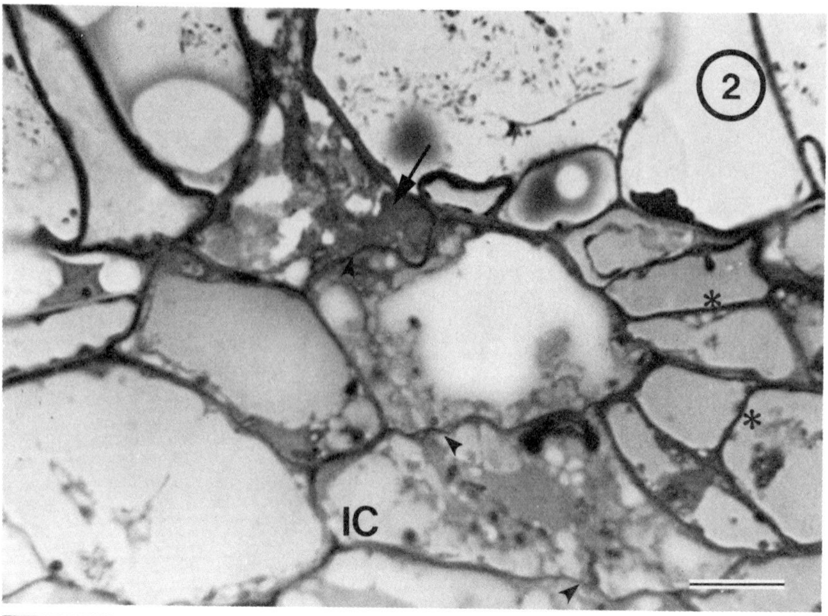
2
IC

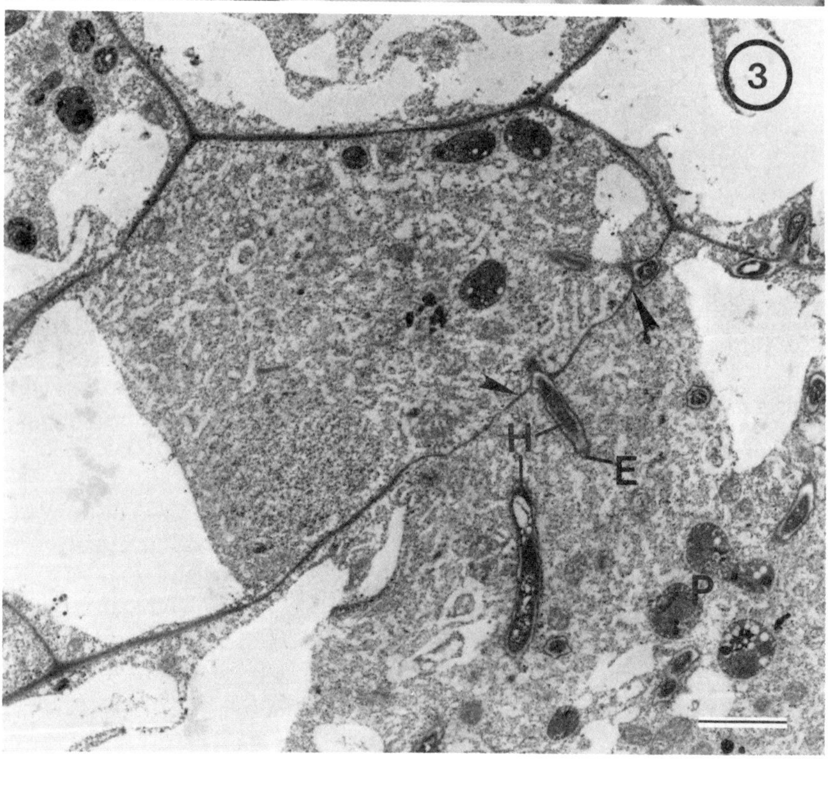
3
H
E
P

in infection. Whether branched or unbranched, the microsymbiont becomes surrounded by host plasmalemma and by host-derived wall material that is continuous with the wall accumulation at the site of penetration. The encapsulating wall material contains a polyanionic constituent (Lalonde and Knowles, 1975). Callose is not detectable in the wall accumulations at sites of successful *Frankia* infection (A. M. Berry and M. E. McCully, in press). In *Alnus*, multilamellar secondary wall deposition is observed only in infected root hairs. Toluidine blue O staining of the secondary wall layers in infected root hairs suggests a degree of phenolic permeation or binding, unlike that found in other root hairs (Callaham *et al.*, 1979; Berry *et al.*, 1986).

After *Frankia* infection, root hairs give the appearance of high metabolic activity, while uninfected hairs degenerate. Infected hairs retain cytoplasmic contents, more so at the base (Fig. 2), with numerous ribosomes and organelles, including a nucleus, Golgi bodies, endoplasmic reticulum, and large plastids. Vacuoles are relatively small. Mitochondria are often numerous, particularly in the vicinity of *Frankia* hyphae within the cell. Microtubules are also observed in infected root hairs in cytoplasmic arrays surrounding the encapsulated hyphae and in arrays oriented parallel to the longitudinal axis of the root hair (Berry *et al.*, 1986).

In some cases, penetration of *Frankia* into root hairs is aborted very early. Localized callose-containing deposits are associated with these aborted infections. The cytoplasm in root hairs containing such incomplete infections degenerates (Berry *et al.*, 1986; Berry and McCully, in press).

b. Prenodule Growth and Infection. With the ingression of the microsymbiont, increased mitotic activity occurs in the root cortex in proximity to root hair infections. Sustained cell division and expansion beneath the infected hair give rise to a region called the prenodule (Figs. 3 and 4). Externally, the prenodule is recognizable as a slight swelling of the host root, apparent before nodule lobe primordia have emerged.

Figure 2 Root hair infection and prenodule growth in *Alnus rubra*. Light micrograph showing root hair that is densely cytoplasmic at the base (arrow). Infected tip not shown. Cortical cell divisions (*) subjacent to the infected root hair and hypertrophy of infected cortical cells (IC) generate the prenodule region. The pathway of *Frankia* infection is traceable in the cytoplasm from the base of the hair through thin walls of the hypertrophied cortical cells (small arrows). Bar is 10 μm.

Figure 3 Prenodule growth in *Alnus rubra*. Transmission electron micrograph of prenodule cells similar to those in Fig. 2. *Frankia* hyphae (H), encapsulated in host wall material (E), traverse the prenodule through thin-walled regions (arrows) of hypertrophied cells. Plastids (P) in prenodule cells do not contain starch. Bar is 1 μm.

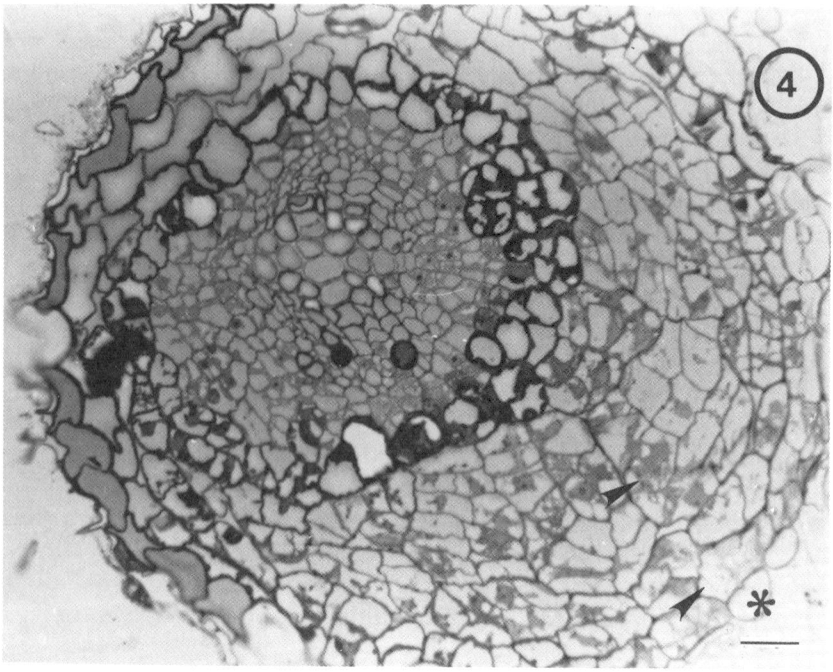
4
*

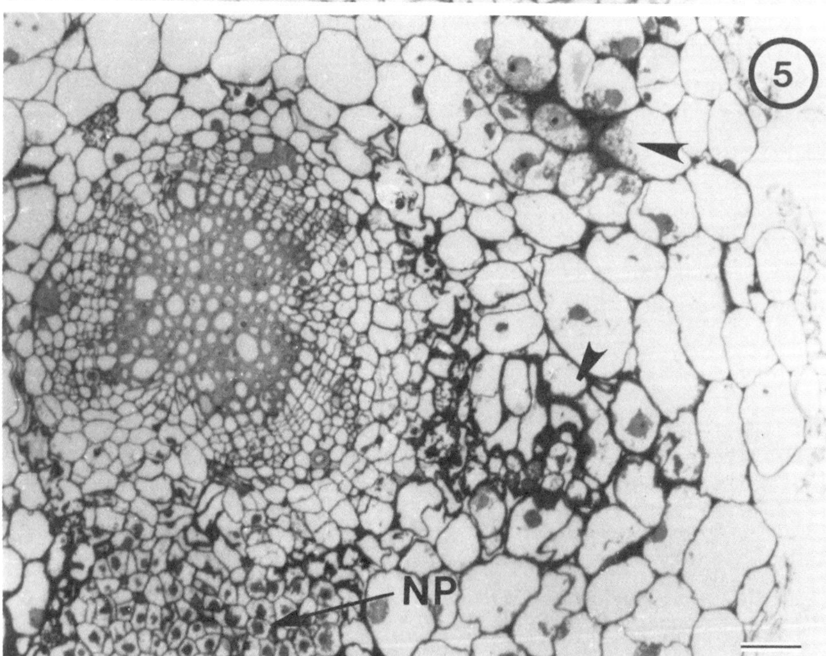
5
NP

Encapsulated hyphae within the root hair advance into the prenodule by penetrating the cell wall at the base of the root hair and continued cell-to-cell passage. *Frankia* further proliferates in groups of expanding cells in the prenodule (Fig. 3; see also Callaham and Torrey, 1977). These cells undergo considerable hypertrophy and contain dense cytoplasm with numerous small vacuoles, Golgi bodies, mitochondria, and plastids lacking starch grains. Portions of the cell walls of these recently expanded cells are relatively thin. *Frankia* appears to penetrate through these thin-walled regions; cell-to-cell passage does not appear to take place via plasmodesmata.

Cytological differences distinguish infected and uninfected prenodule cells. Uninfected cells contain numerous starch grains and may exhibit staining patterns consistent with the accumulation of tannins. In infected prenodule cells, *Frankia* generally persists in hyphal form; vesicles are occasionally formed within the prenodule.

In species where multiple root hair infections occur (e.g., *Myrica* and *Comptonia;* see Callaham *et al.*, 1977), several prenodules can be initiated. There is insufficient evidence to date to determine whether several incipient prenodules can subsequently contribute to the formation of a single nodule. It is clear that some prenodules do not contribute to nodule formation.

2. Intercellular Colonization

Early infection events in some species, as first described in *Elaeagnus* by Miller and Baker (1985), are distinctly different from the root hair deformation pathway described earlier. Root hairs are not deformed or invaded by *Frankia*. Instead, the microorganism grows between adjacent epidermal cells, penetrating the middle lamella. The microsymbiont, not encapsulated by host-derived wall substances, progresses through the root cortex in intercellular spaces entirely. Multiple sites of infection are

Figure 4 Histology of early stages in host root infection by *Frankia*. Light micrograph of prenodule on a root of *Alnus rubra* infected via root hair penetration. Cortical cell divisions have taken place within a broad sector of the host root subjacent to the base of the infected root hair (*). Several cortical cells, infected with *Frankia*, have expanded greatly in volume (arrows). Bar is 20 μm.

Figure 5 Histology of early stages in host root infection by *Frankia*. Light micrograph demonstrating characteristic features of intercellular colonization of *Ceanothus arboreus* root (courtesy of Q. Liu). Note groups of cortical cells surrounded by densely staining material deposited intercellularly (arrows). There is evidence of dense cytoplasm and some mitotic activity. Penetration of cells by *Frankia* and hypertrophy of cortical cells are not evident. A portion of a young nodule lobe primordium (NP) may be seen. Bar is 20 μm.

frequently observed during intercellular colonization. Epidermal and cortical cells slough or secrete material into the intercellular space. This material, present near sites of infection in the nodule lobe (Fig. 9), stains intensely with methylene blue and is electron dense in the transmission electron microscope.

The initial intercellular colonization by *Frankia* occurs in mature root cortical tissues, while at the same time a nodule lobe primordium is initiated in the pericycle. Neither cell division nor hypertrophy of root cortical cell characteristic of typical prenodules is observed in *Elaeagnus*. Ultimately, hyphae penetrate and invade host cells in the developing nodule lobe (Fig. 10).

In *Ceanothus* (Rhamnaceae), an intermediate situation exists in which initial invasion of host tissue is via intercellular colonization, as in *Elaeagnus*. A densely staining intercellular matrix in which *Frankia* hyphae are embedded surrounds the root cortical cells in very early stages of root infection (Fig. 5). In these zones of *Frankia* colonization, groups of cortical cells divide secondarily, forming what resembles an early prenodule. However, *Frankia* hyphae have not been observed within these cells, in contrast to prenodules derived from root hair infections (Q. Liu, unpublished observations). A third genus, *Cercocarpus* (Rosaceae), appears in preliminary studies (L. A. Sunell, unpublished observations) to be infected by an intercellular pathway (Figs. 9 and 10).

II. Nodule Development

A. Nodule Morphology

Actinorhizal root nodules are perennial, coralloid structures consisting of multiple nodule lobes. Field-collected nodules 3–5 cm in diameter are not uncommon. Nodules may consist of discrete lobes, as in *Myrica* or *Ceanothus* (Fig. 6), or the lobes may be densely packed, as in *Alnus* (Fig. 7). Each nodule lobe is a modified lateral root. The nodule meristem gives rise to mature tissues, including a vascular cylinder, cortical tissue, and a superficial periderm. In most cases, nodule lobe growth is determinate. *Frankia* infects only cells of the nodule cortex (Fig. 8).

In several actinorhizal hosts, a more indeterminate pattern of growth is reflected in the formation of nodule roots from the apex of mature nodule lobes. In *Myrica* and *Casuarina*, nodule roots, which are negatively geotropic, develop from most nodule lobes. Nodule roots are devoid of *Frankia* hyphae and vesicles, and so do not contribute to nitrogen fixation. They appear to function in increasing oxygen diffusion to the infected tissue (Tjepkema, 1978; Chapter 8). Nodule roots often con-

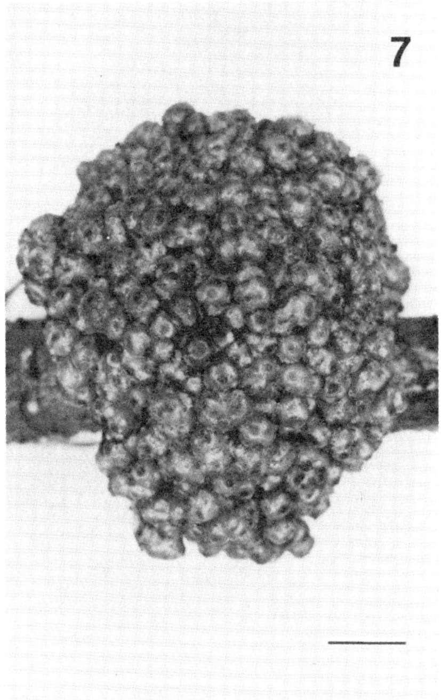

Figure 6 Actinorhizal root nodule morphology. Root nodule on *Ceanothus americanus* is composed of discrete branched lobes. Bar is 1 cm.

Figure 7 Actinorhizal root nodule morphology. Root nodule on *Alnus* sp. is densely compact. Bar is 1 cm.

tain large air spaces, an anatomical feature that facilitates oxygen diffusion.

B. Nodule Lobe Development

The nodule lobe meristem is initiated in the pericycle. Angulo Carmona (1974) showed that the number of lateral root primordia initiated on roots of *Alnus* following inoculation with *Frankia* is greater than that on uninoculated control roots, indicating that *Frankia* can stimulate lateral root initiation. One or more of the primordia developing within the infected cortical zone may be invaded by the microsymbiont. In cases reported to date, *Frankia* appears to transit from the root cortex into the developing nodule lobe primordium through cortical cells near the base of the lobe primordium. Further progress of the hyphae within the nodule lobe occurs either intercellularly or intracellularly depending on the

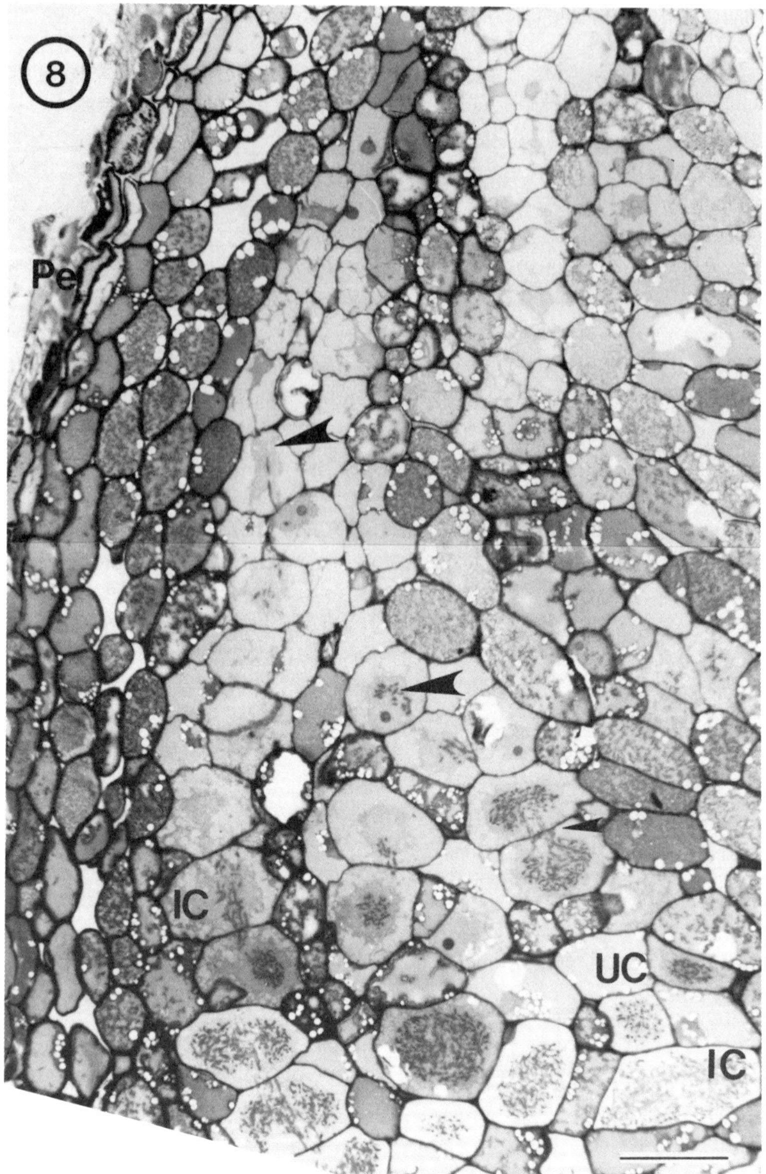

Figure 8 Light micrograph of a developing nodule lobe of *Myrica gale* in longitudinal section. The tissues of the nodule lobe include the outer periderm (Pe), bands of infected and uninfected cortical cells (IC, UC), and a vascular cylinder (not shown). *Frankia* progresses acropetally through the developing lobe cortex by penetration of expanding meristematic derivatives (arrows), giving rise to a series of infected cells. As *Frankia* hyphae proliferate within, cortical cells enlarge greatly (IC). Uninfected cells contain amyloplasts and vacuolar phenolic aggregates. Bar is 20 μm.

mode of infection. In *Ceanothus,* for the most part, hyphae do not pass from cell to cell directly (Strand and Laetsch, 1977; Q. Liu, unpublished observations), while in *Alnus, Myrica,* and *Casuarina,* hyphae are clearly continuous within several cells in a file or group.

Invasion of the developing nodule lobe by *Frankia* appears to be coordinated with expansion of postmeristematic cells. Stages of *Frankia* proliferation within infected cells correlate structurally with expansion and hypertrophy of the host cells (Fig. 8). Fully differentiated cells infected by *Frankia* in either infection mode are considerably larger than uninfected cortical cells, in the nodule lobe as in the prenodule, implying that *Frankia* has a considerable effect on cell growth. In both infection pathways, *Frankia* grows more or less acropetally through the nodule cortical tissue. In some hosts, conspicuous files of hypertrophic cells infected with *Frankia* result.

Once within a nodule cortical cell, *Frankia* hyphae branch and proliferate (Fig. 11). Encapsulating wall material, as well as plasma membrane, is deposited by the host cell around the hyphae as they grow and differentiate. The encapsulation may be utilized by *Frankia* as a metabolic substrate; *Frankia* has been reported to possess pectinolytic and cellulolytic activities (Safo-Sampah and Torrey, 1988; Séguin and Lalonde, 1989). In all of the actinorhizal genera examined to date, *Frankia* hyphae and vesicles are always surrounded by the host membrane and some encapsulating wall material.

C. Nitrogen-Fixing Nodule Tissue

In nitrogen-fixing nodule tissue, infected host cells are living. Cytoplasm surrounding the *Frankia* vesicle clusters occupies a large cell volume and is dense with ribosomes, mitochondria, and profiles of rough endoplasmic reticulum, indicating a high level of metabolic activity (Fig. 12). Nuclei in these cells are enlarged, with prominent nucleoli, and may be extensively lobed (Fig. 9). While plastids are present in nitrogen-fixing cells, these are rarely, if ever, amyloplasts. In most cases, the vacuole is peripheral, but in *Datisca* and *Coriaria,* the vacuole is central and cytoplasm containing *Frankia* vesicles is peripheral (Calvert *et al.,* 1979; Newcomb and Pankhurst, 1982).

Uninfected cells in the same tissue are smaller than infected cells, are more highly vacuolate, and contain little cytoplasm (Fig. 9). Numerous amyloplasts are present within the cytoplasm. Vacuoles are often filled with tannins or other phenolic compounds.

In all the actinorhizal genera except *Casuarina* and *Allocasuarina,* the onset of nitrogen fixation is associated with the appearance of *Frankia* vesicles in the hypertrophied host cortical cells. Vesicles are terminal

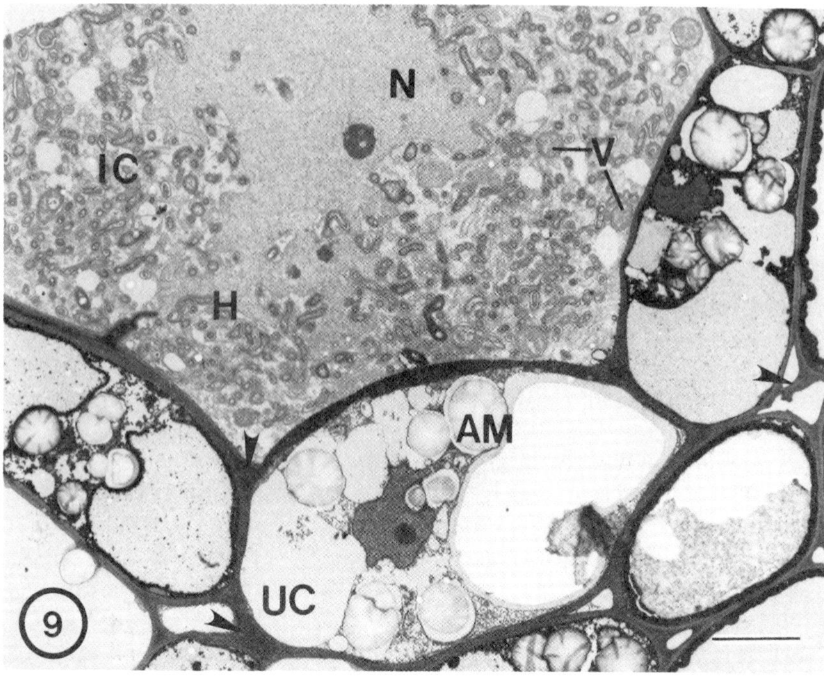
N
IC
V
H
AM
UC
9

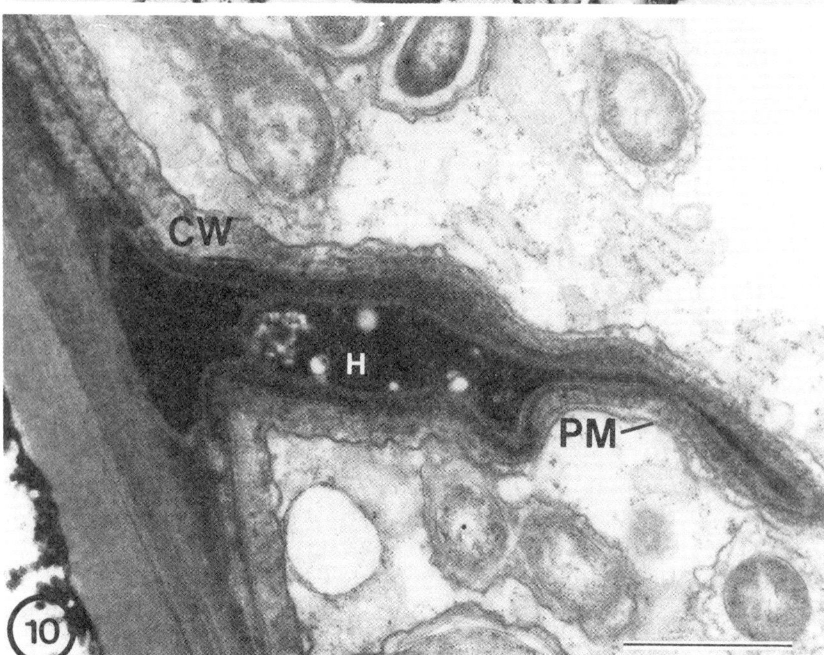
CW
H
PM
10

swellings that differentiate from short branch hyphae (Fig. 12). The shape and spatial organization of *Frankia* vesicles in the nodule are influenced by the host. Symbiotic vesicles may be spherical, club-shaped, elliptical, or filamentous (Newcomb and Wood, 1987). Internally, vesicles may be septate, consisting of multiple irregular segments, or nonseptate. *Frankia* vesicles have conspicuously thick, lamellar walls, with properties suggestive of a glycolipid (Torrey and Callaham, 1982; Lamont *et al.*, 1988; Chapter 6). The host plasmalemma and host-derived encapsulation material surrounding *Frankia* further compartmentalize the *Frankia* vesicles within the host cytoplasm.

The vesicle wall probably functions as a structural barrier to oxygen diffusion, thereby protecting nitrogenase within the vesicle from oxygen damage (see Chapter 8). In *Casuarina,* nitrogen fixation occurs in the absence of *Frankia* vesicles. A specialized host cell-wall layer has been described in *Casuarina* with hydrophobic properties, which could function as an oxygen diffusion barrier (Berg and McDowell, 1988). Other structural and metabolic regulations of oxygen tension have been examined in actinorhizal nodules, and these are discussed in detail in Chapter 8.

Frankia sporangia are observed in the cortical tissue of some actinorhizal nodules (Fig. 13). Sporangia in the nodule structurally resemble those formed in free-living cultures, and are located both within infected cortical cells and in intercellular spaces of some hosts. In *Alnus, Myrica,* and *Comptonia,* differentiation of sporangia coincides with vesicle senescence (Suetin *et al.*, 1981; VandenBosch and Torrey, 1984). Host cytoplasm degenerates in cells where sporangia are present. More information on spores and sporangia is given in Chapter 9.

Frankia symbioses may be characterized as either effective, that is, nitrogen-fixing, or ineffective. There is a basis in nodule structure for these distinctions. In ineffective nodules, fewer cells contain *Frankia*

Figure 9 Symbiotic nodule tissue in *Cercocarpus betuloides.* Transmission electron micrograph of infected and uninfected nodule lobe cortical cells. The large *Frankia*-infected cell (IC) remains cytoplasmic and contains small vacuoles, plastids, a large lobed nucleus (N), and other organelles. *Frankia* hyphae (H) and vesicles (V) are numerous. Uninfected cells (UC) contain amyloplasts (AM) and accumulate tannins. An electron-dense material is apparent in intercellular spaces (arrows). Bar is 5 μm.

Figure 10 Symbiotic nodule tissue in *Cerocarpus betuloides.* Transmission electron micrograph of the same infected cell shown in Fig. 9, but of a different section, showing mode of penetration of the cortical cell by *Frankia* following intercellular colonization. A single hyphal branch (H) projects into the cell from an origin between adjacent cortical cell walls. Layers of host cell-wall material (CW) and the host plasma membrane (PM) separate the *Frankia* hypha from infected cell cytoplasm. Bar is 1 μm.

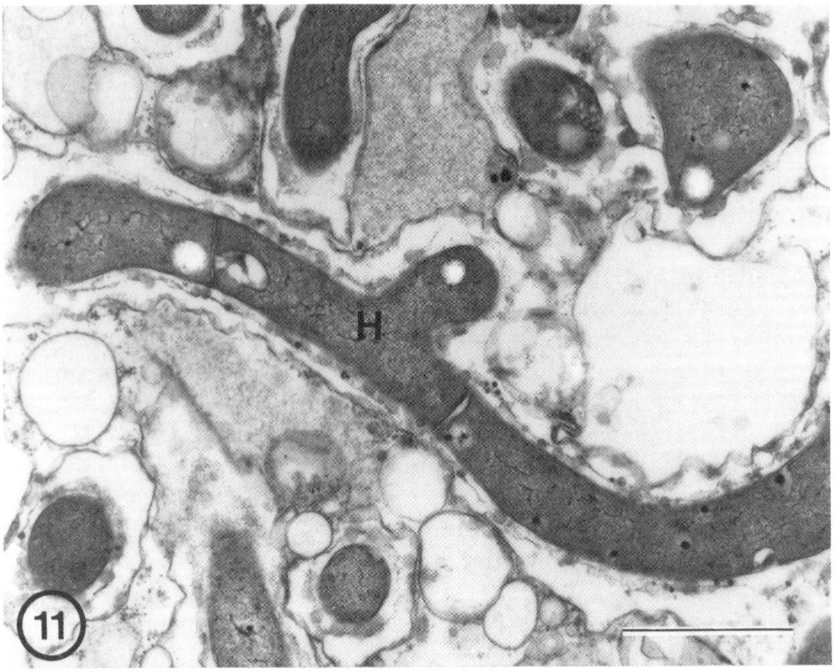
H
11

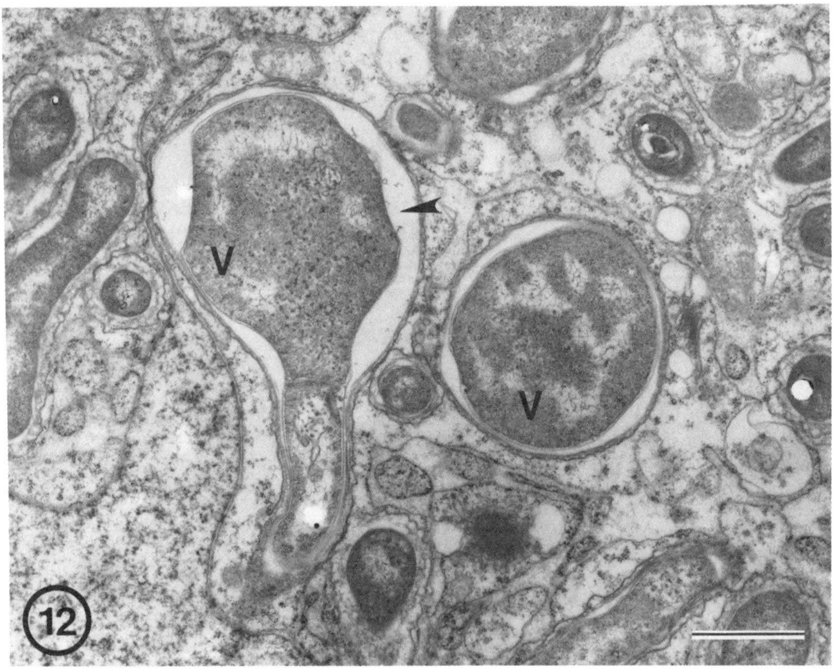
V
V
12

hyphae, and within infected cells, varying degrees of hyphal proliferation occur. Host cytoplasm may degenerate. Generally, vesicles do not form in ineffective nodules (Hahn *et al.*, 1988), although vesiclelike structures have been reported in certain ineffective *Alnus* nodules (Weber *et al.*, 1987). In some ineffective host—microsymbiont combinations, sporangia develop from the proliferating hyphae (VandenBosch and Torrey, 1983).

D. Senescence

In older portions of the root nodule, infected cells degenerate (Fig. 13). *Frankia* vesicles may exhibit increased electron density in the electron microscope, a possible early indicator of senescence (Newcomb and Wood, 1987). Degradation of host cytoplasm as well as that of the microsymbiont may be observed. There is an apparent seasonal correlation between the observed course of infected cell senescence and the decline of nitrogenase activity in symbiotic tissue (Schwintzer *et al.*, 1982).

E. Seasonal Patterns of Nodule Lobe Development

There have been few studies of nodule growth and activity in relation to annual cycles of plant growth or seasonal changes in environmental conditions. In the winter-deciduous species, *Myrica gale*, the onset of nodule lobe growth in the spring occurs from preformed nodule lobe primordia, and the timing correlates with budbreak and the onset of shoot growth (Schwintzer *et al.*, 1982). Nodule development continues until about midsummer, a time when leaf area is at a maximum and shoot growth has stopped. Maximum nitrogenase activity, at least in *M. gale*, occurs during the midsummer period. Nitrogenase activity drops off rapidly, however, in a pattern that does not correlate with changes in

Figure 11 Growth and differentiation of *Frankia* in nodules of *Cercocarpus betuloides*. Transmission electron micrograph of *Frankia* hyphae (H). Proliferating hyphae branch within host cells. Hyphae are surrounded by host plasma membrane. Host-derived encapsulating material is minimally apparent in this section. Bar is 1 μm.

Figure 12 Growth and differentiation of *Frankia* in nodules of *Cercocarpus betuloides*. Transmission electron micrograph of *Frankia* vesicles (V), which form from short branch hyphae in nitrogen-fixing nodule tissue. Vesicles in this species lack internal septae, except between the vesicle and the vesicle stalk. Void space between the *Frankia* vesicle and the host plasma membrane (arrow) may represent loss of lamellar vesicle wall material during fixation. A thin layer of host-derived encapsulation separates the vesicles and hyphae from the host cytoplasm. Host cytoplasm contains Golgi bodies, ribosomes, endoplasmic reticulum profiles, and numerous mitochondria. Bar is 1 μm.

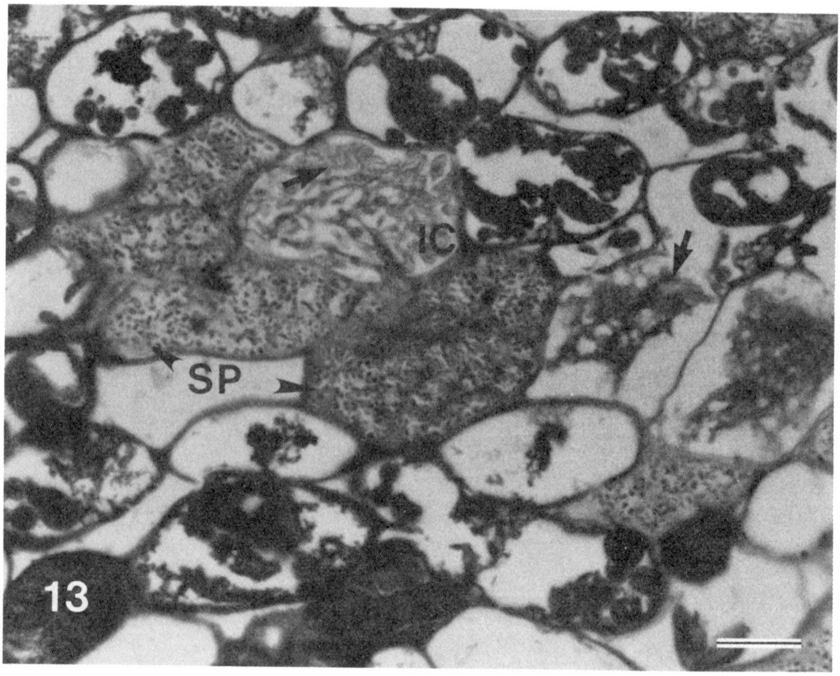

Figure 13 Light micrograph of an older region of a nodule lobe of *Myrica gale*. *Frankia* hyphae within some infected cells have collapsed or appear hollow (arrows). Other cells contain *Frankia* sporangia (SP). Note darkly staining deposits, presumably tannins, in surrounding uninfected cells. Bar is 10 μm.

shoot phenology (e.g., leaf drop) or with obvious changes in nodule cytology (e.g., large-scale vesicle senescence).

The next year's first nodule lobe primordia are formed during the active growing season, but remain uninfected until the following spring. By late fall, the current season's nodule cortex is completely senescent, except for a few residual apical derivatives that remain in a state of early infection. These infected cells have thick walls and thick encapsulating material around the *Frankia* hyphae. They are located close to the overwintering nodule lobe primordia, and when growth recommences in the spring, the residual infected cells contribute colonizing hyphae for continued infection.

There are probably other annual cycles of nodule growth and activity, although none has yet been as carefully described. In *Ceanothus*, an evergreen shrub in California, nitrogenase activity was detected in October, January, and February (Delwiche *et al.*, 1965). Nitrogenase activity in *Ceanothus* seedlings in southern California ceased entirely in the summer months (Ellis and Kummerow, 1988). It may be that in Mediterranean climates, nodule growth and nitrogenase activity are linked closely with periods of adequate soil moisture, so long as leaves are present.

III. Regulation of Infection and Nodule Development

Although the processes of infection and nodule formation are complex, there are certain key molecular and cellular phenomena that are likely to be pivotal for regulation. We presently know very little about possible molecular levels of regulation. Cellular aspects of *Frankia* symbioses that may determine developmental patterns include the initial contact between *Frankia* and host root cells, cell divisions in the host cortex, expansion of host cells, and deposition of modified cell walls.

A. Regulation of Early Interactions

Root exudates are one category of signaling molecule generated by the host that could regulate *Frankia* genes concerned with nodulation. Such an interaction has been demonstrated conclusively to occur in legume–*Rhizobium* symbioses (Zaat *et al.*, 1988), where specific phenolic compounds secreted in host root exudates regulate the promoters of bacterial nodulation genes. Specific phenolics, or some other regulatory signal, might also mediate compatibility between *Frankia* and actinorhizal hosts. Prin and Rougier (1987) reported that while culture filtrates of *Frankia* had no effect on growth of axenic roots of alder, filtrates from cultures incubated with host root exudates caused extensive root hair deformation. A host exudate factor is probably involved in regulating aspects of early nodulation.

Extracellular matrices such as mucilage, or the host-derived encapsulation layer, may be important in regulating infection. Mucilage secreted by the host, both on the epidermal surface and within intercellular spaces, may facilitate bacterial colonization; *Frankia* and other microorganisms have been shown to adhere to mucilage at the root hair surface (Berry *et al.*, 1983). Binding of *Frankia* to the host cell exterior could directly trigger host responses, initially at the cell surface and secondarily within the host cell. Additionally, the polyanionic nature of these matrices may provide a medium for flux of cations, for example, calcium, which can affect a variety of cellular phenomena.

Extracellular polysaccharides of host origin may provide a substrate for *Frankia* enzyme activity. Oligosaccharides derived from plant cell wall degradation are known to function as signaling molecules in certain pathogenic infections (Darvill and Albersheim, 1984). Séguin and Lalonde (1989) detected pectinolytic activity in culture and reported positive hybridization between *Frankia* DNA and genes coding for pectate lyase. These results suggest that *Frankia* might degrade pectic substances within nodule tissues. Resulting oligosaccharides might mediate compatibility in actinorhizal associations.

B. Control of Cell Division Centers and Cell Expansion

An early consequence of *Frankia* inoculation is a localized renewal of mitotic activity in cortical tissues near sites of *Frankia* ingress. Concomitantly, lateral root primordia are initiated *de novo* in the infected zone. Such stimulation of meristematic centers during early infection suggests a regulatory role for plant growth substances. Elevated levels of cytokinins, auxins, and gibberellins in root nodules have been reported (Wheeler *et al.*, 1979) providing further circumstantial evidence for the involvement of plant growth regulators in nodule development. Auxins and a cytokinin have been shown to be secreted by *Frankia* (Stevens and Berry, 1988; Berry *et al.*, 1989); ethylene is not produced by *Frankia in vitro* (A. M. Berry, unpublished observations). Auxins and cytokinins may also be involved in the hypertrophy of infected cells. Another possibility is that localized cell divisions, or cell hypertrophy, result indirectly from a signal from *Frankia* that triggers a host-derived plant growth regulator.

C. The Host Cell Wall and *Frankia* Penetration

The host cell wall is a physical barrier to microbial invasion. Penetration of the host wall by *Frankia* may involve wall loosening, changes in wall structure, or wall degradation. Wall-degrading enzymes produced by *Frankia* could soften or dissolve host walls. As mentioned earlier, *Frankia* appears to have pectate lyase gene sequences (Seguin and Lalonde, 1989). Cellulase activity has been demonstrated in *Frankia* as well (Safo-Sampah and Torrey, 1988). Endogenous pectinolytic activity is important in fruit ripening and other developmental processes and could also be involved in wall hydrolysis during nodule development. Alternatively, wall loosening may be effected by high levels of IAA or changes in external pH (Cleland, 1987). Whether fluxes in pH occur at the host cell wall in response to inoculation by *Frankia* has yet to be investigated.

Several structural observations suggest that new wall deposition is closely linked to successful infection. *Frankia* infects host cells during cell expansion, when primary wall is deposited, both in tip-growing root hairs and in the postmeristematic derivatives in the nodule lobe. Wall encapsulation material is deposited around the invading and proliferating hyphae. Many of the distinctive structural features of newly infected cells, including hypertrophy, the presence of numerous Golgi bodies, Golgi-derived vesicles, and microtubular arrays, indicate that cell-wall synthesis and deposition are primary activities at this stage.

IV. Summary

Symbiotic infection of actinorhizal plants by *Frankia* and the formation of nitrogen-fixing root nodules involve uniquely altered patterns of plant development. Presumably, the mechanisms regulating these changes are mediated by common physiological or molecular controls. A fuller understanding of the cellular phenomena associated with nodulation and the mechanisms governing these processes may contribute to a better understanding of patterns and regulation of root development. It may also be possible to extend the capacity to nodulate to other currently nonnodulating species of economic importance.

References

Angulo Carmona, A. F. (1974). La formation des nodules fixateurs d'azote chez *Alnus glutinosa* (L.) Vill. *Acta Bot. Neerl.* **23,** 257–303.

Berg, R. H., and McDowell, L. (1988). Cytochemistry of the wall of infected cells in *Casuarina* actinorhizae. *Can. J. Bot.* **66,** 2038–2047.

Berry, A. M. (1983). "The Development of the Actinorhizal Association between *Frankia* and *Alnus rubra* Bong," Ph.D. dissertation. Univ. of Massachusetts, Amherst, Massachusetts.

Berry, A. M., and Torrey, J. G. (1983). Root hair deformation in the infection process of *Alnus rubra*. *Can. J. Bot.* **61,** 2863–2876.

Berry, A. M., Torrey, J. G., and McCully, M. E. (1983). The fine structure of the root hair wall and surface mucilage in the actinorhizal host, *Alnus rubra*. *In* "Plant Molecular Biology" (R. Goldberg, ed.), pp. 319–327. Liss, New York.

Berry, A. M., McIntyre, L., and McCully, M. E. (1986). Fine structure of root hair infection leading to nodulation in the *Frankia–Alnus* symbiosis. *Can. J. Bot.* **64,** 292–305.

Berry, A. M., Kahn, R. K. S., and Booth, M. C. (1989). Identification of indole compounds secreted by *Frankia* HFPAr13 in defined culture medium. *Plant and Soil* **118,** 205–209.

Berry, A. M., and McCully, M. E. (1990). Callose-containing deposits in relation to root hair infections of *Alnus rubra* Bong. by *Frankia*. *Can. J. Bot.*, in press.

Bond, G. (1951). The fixation of nitrogen associated with the root nodules of *Myrica gale* L., with special reference to its pH relation and ecological significance. *Ann. Bot. (London)* **15,** 447–459.

Callaham, D., and Torrey, J. G. (1977). Prenodule formation and primary nodule development in roots of *Comptonia* (Myricaceae). *Can. J. Bot.* **51,** 2306–2318.

Callaham, D., Newcomb, W., Torrey, J. G., and Peterson, R. L. (1979). Root hair infection in actinomycete-induced root nodule initiation in *Casuarina, Myrica,* and *Comptonia. Bot. Gaz. (Chicago), Suppl.* **140,** S1–S9.

Calvert, H. E., Chaudhary, A. H., and Lalonde, M. (1979). Structure of an unusual root nodule symbiosis in a non-leguminous herbaceous dicotyledon. *In* "Symbiotic Nitrogen Fixation in the Management of Temperate Forests" (J. E. Gordon, C. T. Wheeler, and D. A. Perry, eds.), pp. 474–475. For. Res. Lab., Oregon State Univ. Press, Corvallis, Oregon.

Cleland, R. E. (1987). The mechanism of wall loosening and wall extension. *In* "Physiology of Cell Expansion During Plant Growth" (D. J. Cosgrove and D. P. Knievel, eds.), pp. 18–27. Amer. Soc. Plant Physiol., Rockville, Maryland.

Darvill, A. G., and Albersheim, P. (1984). Phytoalexins and their elicitors—a defense against microbial infection in plants. *Ann. Rev. Plant Physiol.* **35,** 243–275.

Delwiche, C. C., Zinke, P. J., and Johnson, C. M. (1965). Nitrogen fixation by *Ceanothus. Plant Physiol.* **40**(6), 1045–1047.

Ellis, B. A., and Kummerow, J. (1988). N_2 (C_2H_2–C_2H_4) fixation in two species of *Ceanothus* seedlings in second year postfire chaparral. *Plant Soil* **109,** 207–213.

Hahn, D., Starrenburg, M. J. C., and Akkermans, A. D. L. (1988). Variable compatibility of cloned *Alnus glutinosa* ecotypes against ineffective *Frankia* strains. *Plant Soil* **107,** 233–243.

Knowlton, S., Berry, A. M., and Torrey, J. G. (1980). Evidence that associated soil bacteria may influence root hair infection of actinorhizal plants by *Frankia. Can. J. Microbiol.* **26,** 971–977.

Lalonde, M., and Knowles, R. (1975). Ultrastructure, composition and biogenesis of the encapsulation material surrounding the endophyte in *Alnus crispa* var. *mollis* root nodules. *Can. J. Bot.* **53,** 1951–1971.

Lamont, H. C., Silvester, W. B., and Torrey, J. G. (1988). Nile red fluorescence demonstrates lipid in the envelope of vesicles from N_2-fixing cultures of *Frankia. Can. J. Microbiol.* **34,** 656–660.

Miller, I. M., and Baker, D. D. (1985). The initiation, development and structure of root nodules in *Elaeagnus angustifolia* L. (Elaeagnaceae). *Protoplasma* **128,** 107–119.

Newcomb, W., and Pankhurst, C. E. (1982). Fine structure of actinorhizal root nodules of *Coriaria arborea* (Coriariaceae). *N.Z. J. Bot.* **20,** 93–102.

Newcomb, W., and Wood, S. M. (1987). Morphogenesis and fine structure of *Frankia* (Actinomycetales): The microsymbiont of nitrogen-fixing actinorhizal root nodules. *Int. Rev. Cytol.* **109,** 1–88.

Prin, Y., and Rougier, M. (1986). Cytological and histochemical characteristics of the axenic root surface of *Alnus glutinosa. Can. J. Bot.* **64,** 2216–2226.

Prin, Y., and Rougier, M. (1987). Preinfection events in the establishment of *Alnus–Frankia* symbiosis: Study of the root hair deformation step. *Plant Physiol. (Life Sci. Adv.)* **6,** 99–106.

Racette, S., and Torrey, J. G. (1989). The isolation, culture, and ineffectivity of a *Frankia* strain from *Gymnostoma papuanum* (Casuarinaceae). *Plant and Soil* **118,** 165–170.

Safo-Sampah, S., and Torrey, J. G. (1988). Polysaccharide-hydrolyzing enzymes of *Frankia* (Actinomycetales). *Plant Soil* **112,** 89–97.

Schwintzer, C. R., Berry, A. M., and Disney, L. D. (1982). Seasonal patterns of root nodule growth, endophyte morphology, nitrogenase activity and shoot development in *Myrica gale. Can. J. Bot.* **60,** 746–757.

Séguin, A., and Lalonde, M. (1989). Detection of pectolytic activity and *pel* homologous sequences in *Frankia. Plant Soil* **118,** 221–229.

Stevens, G. A., and Berry, A. M. (1988). Cytokinin secretion by *Frankia* sp. HFPArI3 in defined medium. *Plant Physiol.* **87,** 15–16.

Strand, R., and Laetsch, W. M. (1977). Cell and endophyte structure of the nitrogen-fixing root nodules of *Ceanothus integerrimus* H. and A. II. Progress of the endophyte into young cells of the growing nodule. *Protoplasma* **93,** 179–190.

Suetin, S. O., Pariiskaya, A. N., and Kalakutskii, L. V. (1981). Electron-microscopic study of the growth cycle of an actinomycete microsymbiont in nitrogen-fixing nodules on *Alnus glutinosa* roots. *Microbiology (Engl. Transl.)* **49,** 524–528.

Tjepkema, J. D. (1978). The role of oxygen diffusion from the shoots and nodule roots in nitrogen fixation by root nodules of *Myrica gale* L. *Can. J. Bot.* **56,** 1365–1371.

Torrey, J. G. (1976). Initiation and development of root nodules of *Casuarina* (Casuarinaceae). *Am. J. Bot.* **63,** 335–344.

Torrey, J. G., and Callaham, D. (1982). Structural features of the vesicle of *Frankia* sp. CpI1 in culture. *Can. J. Microbiol.* **28,** 749–757.

VandenBosch, K. A., and Torrey, J. G. (1983). Host–endophyte interactions in effective and ineffective nodules induced by the endophyte of *Myrica gale*. *Can. J. Bot.* **61,** 2898–2909.

VandenBosch, K. A., and Torrey, J. G. (1984). Consequences of sporangial development for nodule function in root nodules of *Comptonia peregrina* and *Myrica gale*. *Plant Physiol.* **76,** 556–560.

Weber, A., Nurmiaho-Lassila, E.-L., and Sundman, V. (1987). Features of the intragenic *Alnus–Frankia* specificity. *Physiol. Plant.* **70,** 289–296.

Wheeler, C. T., Henson, I. E., and McLaughlin, M. E. (1979). Hormones in plants bearing actinomycete nodules. *Bot. Gaz. (Chicago), Suppl.* **140,** S52–S57.

Zaat, S. A. J., Wuffelman, C. A., Mulders, I. H. M., van Brussel, A. A. N., and Lugtenberg, B. J. J. (1988). Root exudates of various host plants of *Rhizobium leguminosarum* contain different sets of inducers of *Rhizobium* nodulation genes. *Plant Physiol.* **86,** 1298–1303.

5 Cross-Inoculation Groups within *Frankia* and Host–Endosymbiont Associations

John G. Torrey
Harvard Forest
Harvard University
Petersham, Massachusetts, U.S.A.

I. Introduction

In this chapter I shall focus on two major topics. The first section will be concerned with methods of testing host–microsymbiont interactions, which are crucial to successful cross-inoculum tests. The second section reviews the evidence accumulated to the present time on the actual results of cross-inoculation trials using either crushed nodule suspensions as inoculum or pure-cultured cells of *Frankia*. The successes and failures of such trials serve as the basis for defining host specificity in the *Frankia* infection process and, in so doing, help define the genus *Frankia* and its different strains as well as delimiting the hosts limits among actinorhizal plants.

II. Methods of Testing Host–Microsymbiont Associations

A. The Microbial Inoculum

Before 1978 when pure cultures of *Frankia* first became available for manipulation (Callaham *et al.*, 1978), one used complex sources of bacterial populations for inoculating seedling plants. Usually, the richest and most infective source of *Frankia* inoculum was root nodule material, freshly collected either from plants growing in the field or from the greenhouse. Young healthy nodule lobes from the current year's nodules served best. Such nodule pieces were washed free of soil, weighed, surface sterilized with hypochlorite solution or hydrogen peroxide, rinsed, and then pulverized in a sterilized mortar and pestle or glass homogenizer to form a slurry of nodule pieces and filamentous bacteria and, if present, spores. The nodule suspension was diluted with sterile distilled water. Bond's formula was 1 g fresh weight of nodule to 20 ml of water (e.g., MacIntosh and Bond, 1970). Lalonde (1979) washed twice with centrifugation and resuspended in 1% NaCl solution or physiological saline. Seedling roots cleaned of substrate were dipped in this suspension or, if placed in water culture, the suspension was brushed onto the seedling root surface or applied dropwise to the root surfaces. In some cases, addition of the crushed nodule suspension was made directly into the water culture solution surrounding the seedling root or added dropwise to the soil or substrate surrounding the seedling root. Nodulation would occur within 2–3 weeks in the best cases. Sometimes nodule initiation would be delayed, depending on the state of the seedling.

Crushed nodule suspensions contain *Frankia* in homogenized form but also the broken cells of the nodule that may contain a complex mixture of phenolic compounds, tannins, and/or oxidizing enzymes deleterious to the microorganism and possibly to the seedling root to which the inoculum is applied. In some cases the bacterial component is irreversibly damaged or inactivated and no infection is achieved. Efforts to protect against oxidative damage have led to adding soluble polyvinylpyrrolidone (PVP) (Lalonde *et al.*, 1981) or ascorbic acid or other antioxidants or adsorbants to nodule suspensions. In our experience, these efforts have not proved very successful. Nodule suspensions from *Casuarina* nodules turn gray and then blacken rapidly from polyphenoloxidase activities and phenolic release. In *Alnus* spp. crushed nodule suspensions may turn bright yellow from released nodule pigments and then rapidly darken. Dilution of the suspension and rapid application to the seedlings may be the simplest effective way to achieve a viable inoculum. Baker *et al.* (1979) showed that separation of *Frankia* from nodule suspensions using Sephadex fractionation substantially improved nodulation in *Elaeagnus* and in *Myrica*.

In their studies of *Alnus glutinosa* inoculation, Quispel and Burggraaf (1981) have found that a lipid-soluble fraction extracted from roots of *A. glutinosa* provided some sort of protection, leading to a more infective inoculum (and also the possibility of achieving successful isolation of the *Frankia* strain) that is somehow protected by the extract. The active principle in the lipid-soluble fraction has been identified recently by Quispel *et al.* (1989) as a triterpenoid compound, dipterocarpol. How it acts remains to be understood.

Nodules from spore(+) plants have consistently proved to be most infective sources of inoculum. In a series of studies made on *A. glutinosa*, Akkermans and van Dijk (1976) and van Dijk (1984) demonstrated that weight for weight, nodule suspensions from spore(+) nodules were up to 1000 times more infective when applied to seedlings than crushed nodule suspensions from spore(−) plants. It is presumed that *Frankia* spores are more resistant to damage from chemical oxidants, from desiccation, or from other deleterious factors and that therefore more "infective particles" survive in the inoculum to act in the infective process.

This idea has been confirmed recently in our laboratory by the unpublished studies of S. Burleigh on inoculum preparation for *Casuarina* spp. from the cultured *Frankia* strain HFPCcI3. He showed that *Frankia* hyphal filaments grown in culture, washed, and air-dried lost infective capacity when applied to seedling roots. Spore suspensions separated from *Frankia* cultures, washed, and air-dried retained their infective capacity. Presumably the infective agent was a newly germinated spore population.

Perradin *et al.* (1983) tested the effects of phenolic compounds on pure-cultured *Frankia* strains by studying their influence on growth and on host infection. Their studies confirmed that specific phenolic compounds in relatively low concentrations were deleterious. If present as natural products released by nodule suspensions, they could be expected to interfere with the infection process.

Other sources of bacterial inoculum when pure-cultured *Frankia* strains are not available include soil taken from around an effectively nodulated plant in the field or greenhouse. In some cases, such soil preparations prove to be more infective than crushed nodule suspensions (cf. Bond, 1974). In studies with *Ceanothus integerrimus*, Strand and Laetsch (1977) reported they were unable to infect seedlings with prepared nodule suspensions, but rather soil from around the roots of nodulated plants proved to be the best inoculum.

When available, pure cultures of *Frankia* provide the most convenient and efficient source of inoculum. It must be stated at the outset that, for a large proportion of the 24 genera of currently known actinorhizal plants, pure-cultured cells of *Frankia* are not available since successful isolations and culture have not been achieved. *Frankia* from the genera in the families Rosaceae and Rhamnaceae have proved particularly difficult to bring into pure culture despite many efforts in many laboratories. The reasons for these failures are not known. On the other hand, several of the 24 known actinorhizal genera have hardly been tried and, until serious efforts have been made, should not be placed in the group of recalcitrant genera. In the cases where pure cultures are not available, no other option exists except to use nodule suspensions or ambient soil as inoculum.

Pure-cultured *Frankia* exists in different morphological forms, that is, hyphae, sporangia, spores, and vesicles. In most studies over the past ten years in which pure-cultured *Frankia* inoculum has been used, a complex mixture of these morphological forms has been used as inoculum. The common practice is to grow *Frankia* in liquid standing or shaken culture in defined or complex media, whichever is best for bacterial growth, for several weeks and then harvest the mycelial mat by centrifugation. After several washings with distilled water to remove medium residues, the bacterial cell mass is homogenized in a glass homogenizer, if on a small scale. The homogenate is again centrifuged and the bacterial homogenate measured in a calibrated conical tube to determine packed cell volume (pcv). Time and speed of centrifugation should be standardized to achieve consistency. Then the homogenate is diluted to a standard concentration and applied as liquid inoculum. Some standards have been worked out. Our laboratory uses 0.01 ml pcv per seedling for inoculation in water culture (Fontaine *et al.*, 1986). In their study in *Alnus*, Lalonde and Calvert (1979) reported the effectiveness of inoculum after serial dilutions of a standard inoculum based on optical density.

It is probably true that, under appropriate nutrient conditions, all pure cultures of *Frankia* form sporangia and spores (Torrey, 1987). The evidence is strong that nutrient medium conditions influence sporulation with suppression of spore formation occurring in complex media with high levels of organic nitrogen components and sporulation enhancement occurring in defined medium low in organic nitrogen or reduced nitrogen compounds. Other nutrient deficiencies in the medium may also stimulate sporulation. Thus, using routine cultures of *Frankia* with media that facilitate growth of the organism, the older the culture before autolysis sets in, the more likely sporulation will occur. So most cultures selected for inocula at 3–4 weeks contain sporangia and possibly released spores. Homogenization further facilitates spore release.

Careful studies using pure hyphal preparations of *Frankia* as inocula and pure spore suspensions have not been published. Efforts in this direction have been made in our laboratory but with equivocal results. In early studies using liquid cultures (A. M. Berry, S. A. Lancelle, M. F. Lopez, P. H. Young, and J. G. Torrey, unpublished observations), we were unable to demonstrate any difference in infectivity between highly enriched hyphal inocula versus highly enriched spore suspensions derived from freshly harvested liquid cultures of *Frankia*. Only recently in experiments using desiccated inocula have we observed the superiority of spore-containing inocula over hyphal inocula (S. Burleigh and J. G. Torrey, unpublished observations).

Spores may be infective particles only upon germination with the formation of nascent hyphal growth. The process of infection involving spores in the early stage of germination remains to be demonstrated. This area represents an important field for future studies.

B. Host Plants for Testing Infectivity

Equally important in studying questions of host–microbial specificity in symbioses are the status of the host plant, its origins, and how it is grown. Information on plant propagation can be found in Chapters 11 and 14.

The choice of plant culture method depends on the inherent genetic traits of the host plant, the purposes of the cross-inoculation trials, and the facilities available. Many actinorhizal plants started from seed or from rooted cuttings grow well in some form of water culture or hydroponics.

1. Water Culture Methods

In the water culture method, young, well-developed seedlings, usually advanced to the first leaf stage, are removed from the germinating substrate, and the roots are washed gently and transferred to nutrient solution containers and supported in place. Most water culture containers allow one to examine the roots of the plants at regular intervals, a great advantage when following the onset of nodulation. Seedlings are inoculated and placed in a greenhouse or growth chamber. Aeration of the nutrient solution may be essential or desirable. This can be arranged by using aquarium pumps to bubble air into each jar (Fig. 1) or in continuous-flow systems. In some cases, aeroponics boxes are useful (Zobel *et al.*, 1976) for studying populations of plants. Temperatures compatible with good growth of the seedlings are essential and may be important for optimum nodulation. Tropical species prefer raised temperatures day and night. Adequate light is important and usually day lengths matching full summer regimes are most effective for nodulation. In temperate zones supplementary greenhouse lighting may be needed to extend day lengths for favorable growth. There is no firm evidence of the

Figure 1 Water culture jars in a cross-inoculation trial using seedlings of *Comptonia peregrina* placed on a greenhouse bench. Aeration of each jar is achieved by small plastic tubes connected to two electric aquarium pumps in the center of the picture.

importance of photoperiod in controlling or influencing nodulation other than through the health and vigor of the plant.

The choice of the nutrient solution is important and different labs have used different conditions. In Bond's lab modified Crone's solution was used (Bond, 1950). In our lab modified Hoagland's solution was selected (Hoagland and Arnon, 1950). A detailed recipe can be found in the book by Machlis and Torrey (1957). Usually these solutions provide complete nutrients but in different balanced states. Attention must be given to micronutrient requirements. Molybdenum and cobalt play special roles in symbiotic nitrogen fixation (cf. discussion by Becking, 1977). Crone's solution even at one-fourth strength contains insoluble components; Hoagland's solution at one-fourth strength is fully dissolved.

Two types of solutions are used. For seedling development and control plants "complete" solutions are used in that they contain nitrogen components, either as nitrate or as ammonium ions or both. For inoculation, seedlings are transferred to the solution lacking nitrogen. To achieve this N-deficiency, one may have to modify other salts in the nutrient solution to retain ionic balance. In the early stages after inoculation of actinorhizal plants it is common practice to start seedlings in the -N solution to which a small amount (<10 ppm N) of N is added. This nitrogen level usually does not inhibit nodulation and sustains the plants until nodules develop and dinitrogen fixation commences (Bond *et al.*, 1954). In ineffective combinations this arrangement allows nodules to be initiated before the plants begin to fail.

Careful control of the pH of the nutrient solution must be maintained and should be monitored at the outset and at regular intervals. Changing nutrient solutions in standing cultures on a regular basis is advisable to avoid specific elemental deficiencies unrelated to nitrogen fixation. In between changes water levels are maintained by distilled water additions. Nutrient solution containers should be opaque to discourage growth of photosynthetic microorganisms, especially cyanobacteria, that succeed in N-deficient media by fixing atmospheric nitrogen. Aluminum-foil-wrapped canning jars serve very well.

In some circumstances the above described arrangements fail to allow optimum growth of the plants or effective nodulation to occur. Cultivation of the seedlings in sand with the same nutrient solutions may provide better aeration to the roots or a more favorable environment, allowing plants to grow and nodulate better (for an example, see Zhang and Torrey, 1985). The disadvantage is that following the time course of nodulation by visual observation is interfered with, although examination of the roots at regular intervals may still be managed with some stress to the root systems.

Control treatments are essential in studies of nodulation and cross-inoculations. In any inoculation trial one should always have a number of plants grown in parallel but without inoculation. These plants, failing to nodulate and growing in -N solutions, soon show nitrogen deficiency symptoms and will die. An alternative control would be to grow plants in nutrient solution containing limiting amounts of fixed nitrogen. Such plants will survive longer and perhaps be susceptible to infection longer than plants grown in solutions that completely lack nitrogen.

Contamination may sometimes be due to *Frankia* spores. The spores are very small (~1 μm) and survive desiccation. In relatively closed systems with positive airflow such as growth chambers, airborne spores may infect plants in water culture and cause false results. Care in handling plants and especially in examining root systems must be practiced to minimize contamination by droplet as well.

The use of nutrient solutions combined with substrates other than sand adds complications to the system. Vermiculite can be autoclaved and used as a substrate but some sources inhibit plant growth. Furthermore, roots are difficult to separate from vermiculite if weight measurements are to be made. Sterilized soil presents the same problems with the added difficulty that autoclaving soil tends to change its chemical and physical character.

2. The Growth Pouch Method

A modified form of water culture is provided by growing seedlings in growth pouches—a method advocated by Lalonde and co-workers (Lalonde, 1979). Growth pouches are polypropylene envelopes containing a paper-towel-like insert with a flap to hold the seed or seedling in place. These pouches offer a number of advantages. Visibility is excellent and one can monitor progress of infection and nodule development easily. They provide economy of space as many pouches can be placed in parallel in a carrier and easily moved from place to place. They must be monitored regularly for moisture level. However, they are not axenic and are not easily maintained free of other microorganisms. They offer conditions similar to aerated water cultures.

3. Axenic Seedling Culture

Under a strict interpretation of fulfilling Koch's postulates in the demonstration of the causal organism for *Frankia*-induced nodulation, the reinfection of the root by the cultured organism should be performed axenically with only *Frankia* present in the association. Following this rule considerably complicates the cross-inoculation test. In our experience the cultivation of seedlings of actinorhizal plants aseptically can be difficult. In a study using dilute Hoagland's solution solidified with agar to produce agar slants in test tubes to facilitate handling aseptically germinated seedlings, Knowlton *et al.* (1980) found that seedlings of *Alnus rubra* formed nodules infrequently or not at all when inoculated with homogenized cultures of a homologous strain of *Frankia* (i.e., a strain isolated from the same host). Adding a suspension of another bacterium, for example, *Pseudomonas* or *Rhizobium*, increased the infection rate dramatically, usually to 100% of all seedlings tested. The "helper" organism itself caused root hair deformation and facilitated infection. Recently we have gone through this same experience with *Casuarina glauca* seedlings inoculated with a homologous strain of *Frankia* (unpublished research of S. Mansour).

This interaction in the infection process of *Frankia*, the host plant, and a "helper" organism has been contested by Lalonde *et al.* (1981), who demonstrated infection by *Frankia* strains in culture of axenically grown micropropagated host plants. Their test medium was a modified Crone's solution. It is conceivable that differences in the inorganic nutrient status of the test host plants played a role in these studies. Further work is needed to explain these differences in experiences. Suffice it to say that demonstration of infectivity in *Frankia*–actinorhizal plant associations in a water culture series in which uninoculated plants fail to nodulate while inoculated plants form nodules in a reasonable time, usually about 3 weeks, is generally accepted as proof of the causal organism.

4. Scoring for Infectivity

Some years before pure cultures of *Frankia* were available for study, Bond (1967) listed the criteria that should be used for demonstration of the infectivity of an actinorhizal plant leading to nodulation by an introduced microbial strain. His first requirement was that nodules develop in a reasonable time. He suggested 2–3 weeks for the appearance of macroscopically visible nodules on seedling roots after inoculation. However, in some of our experiments, the time to first nodule appearance was prolonged (e.g., *Allocasuarina* ssp. inoculated with HFPA11I1), requiring 1–2 months for first nodule appearance, yet the evidence supported the specificity of this association (Zhang and Torrey, 1985).

Nodule initiation with the formation of a prenodule, if formed, and the development of the first nodule lobe, while constituting infection, does not assure the development of effective nodules. Ineffective nodules may be formed that fail to lead to the development of the enzyme nitrogenase or to fixation of dinitrogen (cf. VandenBosch and Torrey, 1983). The event of vesicle development and/or nitrogenase production are separate events in nodule formation. Effectivity may be measured in the crudest way by successful growth of the plant that turns healthy green and shows good shoot elongation. This growth is ultimately measurable as growth increment in increased fresh and dry weight and incremental increase in protein or total nitrogen.

Alternatively, rapid assessment of effectivity can be made with nondestructive acetylene reduction assays of whole root or whole plant systems. Such a demonstration early in nodule development is a useful guide to the effective versus the ineffective nodule association. It is very important in scoring for host specificity derived from cross-inoculation trials to record whether the nodules formed by the successful infection are effective or ineffective in dinitrogen fixation.

III. Evidence Concerning Host Specificity of *Frankia* Strains and the Phenomenon of Cross-Inoculation Infectivity

Having outlined the various methods of testing host–microsymbiont associations and the pitfalls in making these assessments, one can attempt to review or summarize the evidence. Reviews of cross-inoculation studies include those by MacIntosh and Bond (1970), Rodriguez-Barrueco and Bond (1976), Rodriguez-Barrueco and Miguel (1979), Dillon and Baker (1982), VandenBosch and Torrey (1983), Huang *et al.* (1985), Baker (1987), Rodriguez-Barrueco and Subramaniam (1988), and Torrey and Racette (1989). The number of genera among the actinorhizal plants that have been studied in any extensive way is quite limited. The major efforts have been made among members of the Betulaceae, the Myricaceae, the Casuarinaceae, and the Elaeagnaceae. This selectivity reflects the fact that genera in other families have been difficult to nodulate experimentally in any case and that, for the most part, isolations of infective *Frankia* have not been made in members of genera in other families, including the Coriariaceae, Datiscaceae, Rhamnaceae, and Rosaceae.

A. Problems of Inoculum Source

1. Nodule Suspensions

Nodule suspensions offer difficulties when one uses them as inoculum in cross-inoculation studies, as has been pointed out by VandenBosch and Torrey (1983). In the first place, removing contaminating microorganisms from root nodule surfaces is difficult and offers a source of possible confusion. Methods of surface sterilization are discussed elsewhere in this volume in relation to isolation procedures for *Frankia*. This problem is well illustrated by the experiments reported by Rodriguez-Barrueco and Bond (1976) on cross-inoculation in *Alnus* and *Myrica* using nodule suspensions. Reddell and Bowen (1985) made special efforts to prepare their nodule suspensions used as inocula for *Casuarina* species. Surface-sterilized nodules were prepared as inoculum of axenically grown seedlings and these nodulated plants were used in turn as source for the nodule suspensions (of "test-tube-synthesized nodules") used as inoculum in cross-inoculation trials.

A second problem is that grinding up root nodules releases many secondary plant products that may be toxic to the microbial symbiont, as discussed previously.

A third difficulty with nodule suspensions as inoculum is that individual nodule lobes may contain more than one strain of *Frankia,* as has been shown in studies by Benson and Hanna (1983). Thus one cannot with certainty identify a single *Frankia* strain as component in a nodule suspension.

2. Pure-Cultured *Frankia* Isolates

With the availability of pure-cultured isolates of *Frankia* from many actinorhizal host plants, somewhat more definitive experiments in cross-inoculation trials can be made. Even with isolates as inoculum, there are difficulties in characterizing the microbial component. For example, it is likely (cf. Benson and Hanna, 1983; Normand and Lalonde, 1982; and other studies) that more than one strain of *Frankia* occurs in any isolated cultured organism. This problem is discussed elsewhere in this volume (e.g., cf. isolation of *Frankia* and genetics of *Frankia*). Identification of multiple strains in a single cultured inoculum is not a simple matter.

One possible resolution of this problem may be to start each culture to be used as inoculum from a single germinated spore. Spore germination has proven feasible (Tzean and Torrey, 1989) in several strains and the establishment of single-spore isolate strains of *Frankia* may become commonplace.

If one is to depend on cross-inoculation studies made only with pure-cultured isolates of *Frankia,* then all host groups in which no *Frankia* isolations have been made are excluded and will be until we can work out the difficulties of isolation and culture of *Frankia* in these groups. At the present time, indications of similarities and differences among host specificities must be gotten from nodule suspension studies.

B. Cross-Inoculation Studies Using Nodule Suspensions

In their review of cross-inoculation studies in the *Alnus–Myrica* host group, VandenBosch and Torrey (1983) divided the results of cross-inoculation studies into two groups, viz., those performed with nodule suspensions and those performed with pure cultures of *Frankia* isolates (see VandenBosch and Torrey, 1983, p. 2900, Table I).

Examination of the results summarized in their table shows some of the complex interactions in the *Alnus–Myrica* relationship. Nodule suspensions from *Alnus glutinosa* have been reported to produce effective nodules on *A. glutinosa* itself, and on *A. incana* ssp. *rugosa, A. rubra,* and *A. sieboldiana,* but ineffective nodules on *A. jorulensis*. What is surprising is that nodule suspensions of *A. glutinosa* effectively nodulate *Myrica faya* and *M. gale*. Here is a cross-inoculation that is usually satisfactory among species of the genus but can also extend to several species of a genus in a different host family. *Alnus jorulensis* nodule suspensions are less generally accepted by *Alnus* host plants as source of nodule inoculum, producing effective nodules in *A. incana* ssp. *rugosa*, *A. multinervis*, and *A. rubra* but producing ineffective nodules in *A. glutinosa*, *A. sieboldiana*, and in its own host, *A. jorulensis*. Inoculation of *Myrica* species by *A. jorulensis* was not reported.

Myrica gale nodule suspensions produce variable responses in the *Alnus–Myrica* group (VandenBosch and Torrey, 1983, Table I). Thus nodule suspensions of *M. gale* effectively nodulate *M. gale*, *M. faya*, and *Alnus rubra*, but produce ineffective nodules on *M. cerifera*, *M. cordifolia*, and *M. faya* as well as *Alnus incana* ssp. *rugosa* and *Comptonia peregrina*. *Myrica gale* nodule suspensions fail to nodulate *A. glutinosa*. On the basis of these types of responses, *Alnus* and *Myrica* have been placed in a common group as *Frankia* source (group 1 of Baker, 1987) but it is clear that there are some exceptions in which no nodulation occurs and, further, that ineffective nodulation is not uncommon, suggesting that the first processes of infection and early nodule development may proceed normally but that a later step in host–microsymbiont interaction fails.

It should be noted that although *Alnus* and *Myrica* are placed together in receptivity to a common *Frankia* group, no explanation exists as to why these two different host families possess in common a susceptibility to certain specific *Frankia* strains.

Table I summarizes additional results of experiments variously reported using nodule suspensions as inoculum. *Alnus glutinosa* is nodulated effectively by species in groups other than *Alnus* and *Myrica*. Thus, nodule suspensions from *Elaeagnus* produce nodules on *A. glutinosa* (Miguel *et al.*, 1978). In *Alnus nitida*, nodulation is caused by nodule suspensions from *Datisca cannabina* (Datiscaceae) (Hafeez *et al.*, 1984).

Coriaria myrtifolia (Coriariaceae) has been studied, especially by Miguel *et al.* (1978), Rodriguez-Barrueco and Miguel (1979), and Cervantes *et al.* (1988). *Coriaria myrtifolia* is nodulated with nodule suspensions from itself and other *Coriaria* species tested and from root nodules of several species of *Myrica* (in the family Myricaceae). It is not nodulated by nodule suspensions from *Elaeagnus* or from *Casuarina*. *Alnus glutinosa* nodule suspensions also fail to nodulate *Coriaria myrtifolia* with one interesting and novel exception. Cervantes *et al.* (1988) have shown that a filtered nodule extract from *C. myrtifolia*, if added together with a nodule suspension from *A. glutinosa*, causes infection and effective nodulation of *Coriaria*. In this case, chemical extracts from root nodules of the host plant facilitate infection by an otherwise noninfective nodule suspension.

Table I Cross-Inoculation Trials Using Nodule Suspensions

Host species	Inoculum from nodule suspensions	Response	Reference[a]
Allocasuarina torulosa	*A. torulosa*	+	1
	Gymnostoma sp.	−	1
Alnus glutinosa	*Alnus glutinosa*	+	2
	Coriaria myrtifolia	−	3
	Elaeagnus angustifolia	±	3
	Hippophaë rhamnoides	−	3
	Myrica faya	±	3
	Myrica gale	+	3
Alnus nitida	*Alnus nitida*	+	4
	Datisca cannabina	+	4
Casuarina cristata	*Casuarina cristata*	+	5
Casuarina cunninghamiana	*Allocasuarina torulosa*	+	6
	C. cunninghamiana	+	7
	C. equisetifolia	+	7
	C. timorensis	+	7
	C. glauca	+	5
	Gymnostoma	−	6
Casuarina equisetifolia	*C. equisetifolia*	+	6
	C. cunninghamiana	+	6
	C. junghuhniana	+	8
	C. timorensis	+	7
Casuarina glauca	*C. glauca*	+	8
	C. cunninghamiana	+	5
Casuarina junghuhniana	*C. glauca*	+	8
Coriaria myrtifolia	*Alnus glutinosa*	−	2
	Alnus glutinosa + filtered nodule extr. of *C. myrtifolia*	+	9
	Casuarina cunninghamiana	−	9
	Coriaria myrtifolia	+	
	Coriaria nepalensis	+	3
	Elaeagnus angustifolia	−	3
	Hippophaë rhamnoides	+	3
	Myrica faya	+	3
	Myrica gale	+	3
Coriaria nepalensis	*Coriaria myrtifolia*	+	2
Datisca cannabina	*Alnus nitida*	−	4
	Coriaria nepalensis	+	10
	Datisca cannabina	+	4
	Datisca glomerata	+	4
Elaeagnus angustifolia	*Alnus glutinosa*	±	2
	Hippophaë rhamnoides	+	3
Gymnostoma vitiensis	*Allocasuarina* sp.	−	1
	Casuarina sp.	−	1
	G. vitiensis	+	1

(*Continued*)

Table I (*Continued*)

Host species	Inoculum from nodule suspensions	Response	Reference[a]
Hippophaë rhamnoides	*Alnus glutinosa*	−	2
	Coriaria myrtifolia	−	3
	Elaeagnus angustifolia	+	3
	Myrica faya	−	3
	Myrica gale	−	3
Myrica faya	*Alnus glutinosa*	+	3
	Coriaria myrtifolia	+	3
	Hippophaë rhamnoides	+	3
	Myrica gale	+	3
Myrica gale	*Alnus glutinosa*	+	3

[a]References: (1) Reddell (1986); (2) Miguel *et al.* (1978); (3) Rodriguez-Barrueco and Miguel (1979); (4) Hafeez *et al.* (1984); (5) Coyne (1983); (6) Reddell and Bowen (1986); (7) Reddell and Bowen (1985); (8) Mowry (1934); (9) Cervantes *et al.* (1988); (10) Chaudhary and Mirza (1987).

Datisca cannabina of the family Datiscaceae is nodulated by nodule suspensions from other families, for example, Coriaraceae (*Coriaria nepalensis*), as well as other species of *Datisca* (*D. glomerata*). In our laboratory, L. J. Winship (unpublished observations) found that *Datisca cannabina* was most easily nodulated in water culture with soil from around plant roots of *Ceanothus americanus* growing in the field.

Elaeagnus angustifolia is reported to be nodulated by its own nodule suspensions and those from its close relative *Hippophaë rhamnoides*, but also, although poorly by nodule suspensions from *Alnus glutinosa* (Miguel *et al.*, 1978). *Hippophaë rhamnoides* is not a very susceptible host plant, being nodulated by nodule suspensions of *Elaeagnus angustifolia* but not from *Alnus glutinosa*, *Coriaria myrtifolia*, *Myrica faya*, or *M. gale*. *Myrica faya*, referred to by some as a promiscuous host, may be easily nodulated by nodule suspensions from different genera but frequently the nodules are ineffective and show arrested development (Rodriguez-Barrueco and Miguel, 1979).

The *Casuarina* group was made subject of a special study, first by Mowry (1934), then by Coyne (1983), and most recently by Reddell (1986) and by Reddell and Bowen (1986). In Table I these results are summarized. Within the genus *Casuarina* cross-inoculations using nodule suspensions are generally successful. Application of nodule suspensions of *Casuarina* spp. to *Gymnostoma* and *Allocasuarina* often may not succeed. *Gymnostoma* tends to be nodulated best by nodule suspensions from *Gymnostoma*. The same is true of *Allocasuarina*. Altogether the group reflects considerable specificity, at least among genera.

C. Cross-Inoculation Studies Using Pure Isolated Cultures as Inoculum

Host specificity studies of the past decade based on cross-inoculation of seedlings grown in water culture under carefully controlled conditions, inoculated with pure-cultured strains of *Frankia*, permit more precise analysis of host–microsymbiont interactions than has been previously possible. Data from experiments with pure isolated culture inoculations are summarized in Tables II and III.

1. Cross-Inoculation Groups

The most extensive study of host specificity of pure-cultured *Frankia* strains was published by Baker (1987) and served as the basis for his Fig. 1 (Baker, 1987, p. 247). In this figure Baker attempted to summarize the evidence concerning differences among *Frankia* strains with respect to their capacity to nodulate seedlings of genera from among four families. This summary is based on the results of 173 cross-inoculation trials using 50 different pure-cultured strains of *Frankia*. Table III lists in a form comparable to other tables presented here some of the many cross-inoculation tests described by Baker (1987).

On the basis of this study, Baker (1987) divided *Frankia* into four host specificity groups, viz., strains that nodulate (1) *Alnus* (Betulaceae) and *Myrica* (Myricaceae), (2) *Casuarina* (Casuarinaceae) and *Myrica*, (3) *Myrica* and *Elaeagnus* (Elaeagnaceae), and (4) only members of the Elaeagnaceae (*Elaeagnus, Hippophaë, Shepherdia*). As pointed out by Baker, additional groupings may be expected when we have had more success in isolation and testing of strains from root nodules of other families that are excluded from this summary (Coriariaceae, Datiscaceae, Rhamnaceae, and Rosaceae).

It is clear at the outset that host specificity, in itself, is not a basis for classification among *Frankia* strains as was attempted by Becking (1970), who designated species of *Frankia* based on the name of the host plant they would nodulate. The difficulty in the use of this character led researchers working on *Frankia* isolations and culture to agree not to make species designations based on any one character and, for convenience, to use a numerical cataloging system to identify strains (Lechevalier, 1983, 1984, 1985) until a large data bank was developed to designate more explicitly the taxa within the genus *Frankia*. That system has proven to be very useful and allows careful systematic studies to be made without applying confusing species designations that may require changing as we learn more about the nature and characters of *Frankia*.

Table II Cross-Inoculation Studies Involving Pure-Cultured Strains of *Frankia* Isolated from Root Nodules of Members of the Casuarinaceae[a]

Trivial designation	Catalog number	Host of origin	*C. cunn.*	*C. equis.*	*C. glauca*	*Gym. pap.*	*E. ang.*	*E. umb.*	*Hip. rh.*	*Shep. ar.*	*M. gale*	*M. cerf.*	Reference
	Allocasuarina lehmannia												
AllI1	HFP022801	+	+	+	+	±	−	−		−	+		Zhang and Torrey (1985)
	Casuarina cunninghamiana												
CcI2	HFP020202	−	−	−	−	+	+	+		+	+		Zhang *et al.* (1984)
CcI3	HFP020203	+	+	+	+	±	−	−		−	+	+	Zhang *et al.* (1984)
R43	LLR02022	−	−	−			+	+	+			+	Lechevalier *et al.* (1987)
	Casuarina equisetifolia												
CeI2	DDB020210	−		−			+					+	Baker (1987)
	DDB020510	−		−			+					+	Baker (1987)
CeI5	UFG		−		±	+	+	±			+	?	R. H. Berg (unpublished observations)
D1	ORS020601	−		−									Diem *et al.* (1982)
D11	ORS020602	−		−			+		+				Gauthier *et al.* (1981)
G2	ORS020604	−		−			+		+				Gauthier *et al.* (1983)
Ce D	ORS020606	+		+									
Ce F	ORS020607	+	+	+	+								
JCT287	—	+	+	+	+								Rosbrook and Bowen (1987)

CeC14	—	±		±									Huang *et al.* (1985)
	Casuarina glauca												
CAQP1	—	+			+								Chaudhary and Mirza (1987)
CAQP2	—	+			+								Chaudhary and Mirza (1987)
CAQP3	—	−			−								Chaudhary and Mirza (1987)
CAQP4	—	−			−								Chaudhary and Mirza (1987)
CAQP5	—	−			−								Chaudhary and Mirza (1987)
CgII	UFG	±	−		±	+	+				+	?	R. H. Berg (unpublished observations)
	Casuarina junghuhniana												
Cjl-82	ORS021001	+	+	+	+								Diem *et al.* (1983)
JCT295	—		+		+								Rosbrook and Bowen (1987)
	Gymnostoma papuanum												
Gp1	HFP021801	+	−	−	−	+	+	+		+	+		Racette and Torrey (1989)

[a]Host plants tested included seedlings from the three families Casuarinaceae, Elaeagnaceae, and Myricaceae. Catalog numbers are listed by Lechevalier (1985). Scores: + = all plants nodulated; ± = less than 50% nodulated; − = no nodulation; space = not tested. Reproduced with permission from Kluwer Academic Publishers.

Table III Cross-Inoculation Trials of Members of the Elaeagnaceae Using Pure-Cultured Isolates of *Frankia*

Host species	Designation of *Frankia* isolate (host)	Response
Elaeagnus angustifolia	DDB 54004 (*Alnus incana*)[a]	–
	LLR AirI1 (*Alnus incana* ssp. *rugosa*)[a]	–
	LLR R52 (*Alnus incana* ssp. *tenuifolia*)[a]	–
	DDB 54005 (*Alnus glutinosa*)[a]	–
	DDB ArhI2 (*Alnus rhombifolia*)[a]	+
	DDB ArI5 (*Alnus rubra*)[a]	–
	DDB AvcI1 (*Alnus viridis* ssp. *crispa*)[a]	–
	DDB AvsI6a (*Alnus viridis* ssp. *sinuata*)[a]	–
	DDB 53081 (*Casuarina equisetifolia*)[a]	+
	ORS Cjl-82 (*Casuarina junghuhniana*)[a]	–
	DDB 52065 (*Ceanothus americanus*)[a]	–
	LLR R2 (*Ceanothus americanus*)[a]	+
	DDB 53140 (*Cercocarpus ledifolius*)[a]	+
	WgCc1.17 (*Colletia cruciata*)[a]	+
	DDB CpI3 (*Comptonia peregrina*)[a]	–
	LLR R100 (*Comptonia peregrina*)[a]	+
	DDB 54010 (*Cowania mexicana*)[a]	–
	DDB EcI1a (*Elaeagnus commutata*)[a]	+
	DDB EuIlb (*Elaeagnus umbellata*)[a]	+
	DDB 53010 (*Myrica cerifera*)[a]	–
	DDB 53051 (*Myrica cerifera*)[a]	+
	DDB MgI5 (*Myrica gale*)[a]	+
	DDB MpI3 (*Myrica pensylvanica*)[a]	–
	DDB P+I1 (*Purshia tridentata*)[a]	–
Elaeagnus umbellata	HFP A11I1 (*Allocasuarina lehmanniana*)[b]	–
	HFP ArI3 (*Alnus rubra*)[b]	–
	HFP CcI3 (*Casuarina cunninghamiana*)[b]	–
	HFP CcI2 (*Casuarina cunninghamiana*)[b]	+
	LLR R43 (*Casuarina cunninghamiana*)[b]	+
	HFP CpI1 (*Comptonia peregrina*)[b]	–
	ULQ EaN1pec (*Elaeagnus angustifolia*)[b]	+
	HFPGpI1 (*Gymnostoma papuanum*)[b]	+
	HFPM+gI5 (*Myrica gale*)[b]	–
Hippophaë rhamnoides	DDB 54004 (*Alnus incana*)[a]	–
	DDB 54012 (*Alnus virids* ssp. *sinuata*)[a]	–
	DDB 53081 (*Casuarina equisetifolia*)[a]	+
	LLR R2 (*Ceanothus americanus*)[a]	+
	WgCc1.17 (*Colletia cruciata*)[a]	+
	DDB EuI1b (*Elaeagnus umbellata*)[a]	+
	DDB 53010 (*Myrica cerifera*)[a]	–
	DDB 53051 (*Myrica cerifera*)[a]	+
	DDB MgI5 (*Myrica gale*)[a]	+
	DDB MpI3 (*Myrica pensylvanica*)[a]	–
Shepherdia argentea	HFP A11I1 (*Allocasuarina lehmanniana*)[b]	–
	HFP ArI3 (*Alnus rubra*)[b]	–

(Continued)

Table III (*Continued*)

Host species	Designation of *Frankia* isolate (host)	Response
	HFP CcI2 (*Casuarina cunninghamiana*)[b]	+
	HFP CcI3 (*Casuarina cunninghamiana*)[b]	−
	HFP CpI1 (*Comptonia peregrina*)[b]	−
	ULQ EaN1pec (*Elaeagnus angustifolia*)[b]	+
	HFP GpI1 (*Gymnostoma papuanum*)[b]	+
	HFP M+gI5 (*Myrica gale*)[b]	−

[a]From Baker (1987).
[b]From J. G. Torrey and S. Racette (1989; unpublished observations).

Host specificity is just one of many characters that should prove useful and interesting. The capacity to nodulate effectively is of course one of the characters of greatest practical importance and is the basis for selecting and using strains for inoculation of host plants useful in commerce.

2. Degrees of Specificity

Having looked at the broad generalizations concerning host specificity as we currently view the subject, let us look at specific, well-studied cases to see how specific the *Frankia*–host plant relationship really is. At the host family level, the Elaeagnaceae present the clearest case. *Frankia* isolates made from root nodules of genera within the Elaeagnaceae, if infective at all, will usually nodulate the host of origin, other species in that genus, and other genera of the family. Thus, Baker (1987) reported that *Frankia* isolates from *E. umbellata* (EuI1b) effectively nodulated *E. angustifolia* and also *Hippophaë rhamnoides,* a genus in the same family (see Table III). Thus a *Frankia* isolate that nodulates one species in the family Elaeagnaceae may be expected to nodulate other species of that and other genera in the family.

Looking at the question of specificity from the host point of view rather than that of the microsymbiont, the situation is not so straightforward. Baker (1987) has shown that *Elaeagnus angustifolia* is nodulated effectively not only by *Frankia* isolates from *Elaeagnus* root nodules but also by isolates from *Alnus* (Betulaceae), *Casuarina* (Casuarinaceae), *Cercocarpus* (Rosaceae), *Myrica* (Myricaceae), *Ceanothus* (Rhamnaceae), *Comptonia* (Myricaceae), and *Colletia* (Rhamnaceae). Thus, *Elaeagnus angustifolia,* like other members of the Eleagnaceae, is promiscuous, that is, it will accept infection by *Frankia* strains from many other host species. Similarly, *Alnus glutinosa* is a promiscuous host by this definition.

From the evidence available now, *Myrica gale* and *Myrica cerifera* are likewise promiscuous. J. G. Torrey and S. Racette (1989; unpublished observations) have shown that *Myrica gale* is nodulated by a *Frankia* strain isolated from root nodules of *M. gale* (M^{+}gI5) (cf. Silvester *et al.*, 1988), but this host is also nodulated by *Frankia* strains from *Alnus, Casuarina, Comptonia, Cowania* (Rosaceae), *Elaeagnus, Myrica, Ceanothus,* and *Colletia*!

Studies by Torrey and Racette (1989; Racette and Torrey, 1989) with *Gymnostoma papuanum* suggest that this host plant is also promiscuous, being nodulated by pure-cultured isolates from *Gymnostoma* but also from *Alnus, Comptonia, Myrica, Elaeagnus,* and *Casuarina* (see Table II).

At the other extreme are actinorhizal host plants that fail to be nodulated by any of the pure-cultured isolates of *Frankia* available at the present time, including cultures derived from nodules of the host plant itself. Thus *Ceanothus americanus,* which has been studied extensively in several labs and from which *Frankia* isolations have been made (e.g., LLR R2 made by Ruan in Lechevalier's lab), fails to be nodulated by any pure-cultured isolate currently available. *Purshia tridentata* (Rosaceae) shows the same recalcitrance to infection by known *Frankia* isolates (S. Racette and J. G. Torrey, unpublished observations). The only interpretation that comes to mind is that there exist one or perhaps several classes of *Frankia* that we have failed to isolate successfully, perhaps through peculiar characters relating to nutrition, physiology, or biochemistry that we do not yet understand.

Between these two extremes of host specificity are *Frankia* strains with different degrees of infective capacity. Viewed from the host point of view, *Casuarina* and *Allocasuarina* species tend to be very restrictive, being nodulated only by *Casuarina* isolates such as HFPCcI3 (Torrey and Racette, 1989) or ORS Cj1-82 (Diem *et al.*, 1983). *Allocasuarina verticillata* is nodulated by both of these isolates and HFP A11I1 from root nodules of *A. lehmanniana* but by none of the other isolates tested (Torrey and Racette, 1989). Thus far, no *Frankia* isolates have been made that nodulate more than a few of the *Allocasuarina* species tested (Torrey and Racette, 1989).

A *Frankia* isolate from *Colletia spinosissima* of the Rhamnaceae (Gauthier *et al.*, 1984) nodulates not only its own host but also species of the Elaeagnaceae (*Hippophaë* and *Elaeagnus*). These authors suggested the existence of a cross-inoculation group for the Rhamnales but further evidence is needed.

In general, the groupings described by Baker (1987) hold up quite well with some uncertain exceptions. Judging from experiments both with pure-cultured strains of *Frankia* and with nodule suspensions, it seems clear that there must be additional groups that will include strains that infect members of the Rhamnaceae, the Rosaceae, and other families not yet adequately studied.

Acknowledgments

The author expresses his thanks to S. Racette and R. Lundquist for technical assistance over the years involving our own studies of host specificity of *Frankia*. My appreciation for secretarial support is expressed to B. Flye. For ongoing discussions of *Frankia* associations I am pleased to acknowledge the stimulation provided by D. Baker.

References

Akkermans, A. D. L., and van Dijk, C. (1976). The formation and nitrogen-fixing activity of root nodules of *Alnus glutinosa* under field conditions. *In* "Symbiotic Nitrogen Fixation in Plants" (P. S. Nutman, ed.), pp. 511–520. Cambridge Univ. Press, Cambridge, England.

Baker, D. D. (1987). Relationships among pure cultured strains of *Frankia* based on host specificity. *Physiol. Plant.* **70,** 245–248.

Baker, D., Kidd, G. H., and Torrey, J. G. (1979). Separation of actinomycete nodule endophytes from crushed nodule suspensions by Sephadex fractionation. *Bot. Gaz. (Chicago), Suppl.* **140,** S849–S851.

Becking, J. H. (1970). *Frankiaceae* fam. nov. (Actinomycetales) with one new combination and six new species of the genus *Frankia* Brunchorst 1886. *Int. J. Syst. Bact.* **20,** 201–220.

Becking, J. H. (1977). Dinitrogen-fixing associations in higher plants other than legumes. *In* "A Treatise on Dinitrogen Fixation: III. Biology" (R. W. F. Hardy and W. S. Silver, eds.), pp. 185–275. Wiley, New York.

Benson, D. R., and Hanna, D. (1983). *Frankia* diversity in an alder stand as estimated by sodium dodecyl sulfate–polyacrylamide gel electrophoresis of whole-cell proteins. *Can. J. Bot.* **61,** 2919–2923.

Bond, G. (1950). Symbiosis of leguminous plants and nodule bacteria. III. Observations on the growth of soya bean in water culture. *Ann. Bot. (London)* **14,** 245–261.

Bond, G. (1967). Fixation of nitrogen by higher plants other than legumes. *Annu. Rev. Plant Physiol.* **18,** 107–126.

Bond, G. (1974). Root-nodule symbioses with actinomycete-like organisms. *In* "The Biology of Nitrogen Fixation" (A. Quispel, ed.), pp. 342–378. Elsevier/North-Holland Publ., Amsterdam.

Bond, G., Fletcher, W. W., and Ferguson, T. P. (1954). The development and function of the root nodules of *Alnus, Myrica* and *Hippophaë. Plant Soil* **5,** 309–323.

Callaham, D., Del Tredici, P., and Torrey, J. G. (1978). Isolation and cultivation *in vitro* of the actinomycete causing root nodulation in *Comptonia. Science* **199,** 899–902.

Cervantes, E., Godoy, C., and Rodriguez-Barrueco, C. (1988). The ability of heterologous inocula to nodulate the actinorhizal plant *Coriaria myrtifolia* L. *In* "Nitrogen Fixation: Hundred Years After" (H. Bothe, F. J. de Bruijn, and W. E. Newton, eds.), pp. 687–691. Fischer, Stuttgart.

Chaudhary, A. H., and Mirza, M. S. (1987). Isolation and characterization of *Frankia* from nodules of actinorhizal plants of Pakistan. *Physiol. Plant.* **70,** 255–258.

Coyne, P. D. (1983). Specificity between *Casuarina* species and root nodule organisms. *In* "Casuarina Ecology, Management and Utilization" (S. J. Midgley, J. W. Turnbull, and R. D. Johnston, eds.), pp. 205–210. Commonw. Sci. Ind. Res. Org., Melbourne, Australia.

Diem, H. G., Gauthier, D., and Dommergues, Y. R. (1982). Isolation of *Frankia* from nodules of *Casuarina equisetifolia*. *Can. J. Microbiol.* **28,** 526–530.

Diem, H. G., Gauthier, D., and Dommergues, Y. R. (1983). An effective strain of *Frankia* from *Casuarina* sp. *Can. J. Bot.* **61,** 2815–2821.

Dillon, J. T., and Baker, D. (1982). Variations in nitrogen fixing efficiency among pure-cultured *Frankia* strains tested on actinorhizal plants as an indication of symbiotic compatibility. *New Phytol.* **92,** 215–219.

Fontaine, M. S., Young, P. H., and Torrey, J. G. (1986). Effects of long-term preservation of *Frankia* strains on infectivity, effectivity, and *in vitro* nitrogenase activity. *Appl. Environ. Microbiol.* **51,** 694–698.

Gauthier, D., Diem, H. G., and Dommergues, Y. (1981). *In vitro* nitrogen fixation by two actinomycete strains isolated from *Casuarina* nodules. *Appl. Environ. Microbiol.* **41,** 306–308.

Gauthier, D., Diem, H. G., and Dommergues, Y. (1983). Preliminary results of research on *Frankia* and endomycorrhizae associated with *Casuarina equisetifolia*. *In* "Casuarina Ecology, Management and Utilization" (S. J. Midgley, J. W. Turnbull, and R. D. Johnston, eds.), pp. 211–217. Commonw. Sci. Ind. Res. Org., Melbourne, Australia.

Gauthier, D., Frioni, L., Diem, H. G., and Dommergues, Y. (1984). The *Colletia spinosissima–Frankia* symbiosis. *Acta Oecol./Oecol. Plant.* **5,** 231–239.

Hafeez, F., Akkermans, A. D. L., and Chaudhary, A. H. (1984). Morphology, physiology and infectivity of two *Frankia* isolates An 1 and An 2 from root nodules of *Alnus nitida*. *Plant Soil* **78,** 45–59.

Hoagland, D. R., and Arnon, D. I. (1950). The water-culture method for growing plants without soil. *Circ.—Univ. Calif., Coll. Agric., Agric. Exp. Stn.* **347.**

Huang, J.-B., Zhao, Z.-Y., Chen, G.-X., and Liu, H.-C. (1985). Host range of *Frankia* endophytes. *Plant Soil* **87,** 61–65.

Knowlton, S., Berry, A., and Torrey, J. G. (1980). Evidence that associated soil bacteria may influence root hair infection of actinorhizal plants by *Frankia*. *Can. J. Microbiol.* **26,** 971–977.

Lalonde, M. (1979). Techniques and observations of the nitrogen-fixing *Alnus* root nodule symbiosis. *In* "Recent Advances in Biological Nitrogen Fixation" (N. S. Subba Rao, ed.), pp. 421–434. Oxford Univ. Press/IBH Publ., New Delhi, India.

Lalonde, M., and Calvert, H. E. (1979). Production of *Frankia* hyphae and spores as an infective inoculant for *Alnus* species. *In* "Symbiotic Nitrogen Fixation in the Management of Temperate Forests" (J. C. Gordon, C. T. Wheeler, and D. A. Perry, eds.), pp. 95–110. For. Res. Lab., Oregon State Univ., Corvallis, Oregon.

Lalonde, M., Calvert, H. E., and Pine, S. (1981). Isolation and use of *Frankia* strains in actinorhizal formation. *In* "Current Perspectives in Nitrogen Fixation" (A. H. Gibson and W. E. Newton, eds.), pp. 296–299. Aust. Acad. Sci., Canberra, Australia.

Lechevalier, M. P. (1983). Cataloging *Frankia* strains. *Can. J. Bot.* **61,** 2964–2967.

Lechevalier, M. P. (1984). The taxonomy of the genus *Frankia*. *Plant Soil* **78,** 1–6.

Lechevalier, M. P. (1985). Catalog of *Frankia* strains. *Actinomycetes* **19,** 131–162.

Lechevalier, M. P., Labeda, D. P., and Ruan, J.-S. (1987). Studies on *Frankia* sp. LLR 02022 from *Casuarina cunninghamiana* and its mutant LLR 02023. *Physiol. Plant.* **70,** 249–254.

Machlis, L., and Torrey, J. G. (1957). "Plants in Action." Freeman, San Francisco, California.

Mackintosh, A. H., and Bond, G. (1970). Diversity in the nodular endophytes of *Alnus* and *Myrica*. *Phyton (Buenos Aires)* **27,** 79–90.

Miguel, C., Cañizo, A., Costa, A., and Rodriguez-Barrueco, C. (1978). Some aspects of the *Alnus*-type root nodule symbiosis. *In* "Limitations and Potentials for Biological Nitrogen Fixation in the Tropics" (J. Dobereiner, R. H. Burris, and A. Hollaender, eds.), pp. 121–133. Plenum, New York.

Mowry, H. (1934). "Symbiotic Nitrogen Fixation in the Casuarinaceae," M.S. thesis. Univ. of Florida, Gainesville, Florida.

Normand, P., and Lalonde, M. (1982). Evaluation of *Frankia* strains isolated from provenances of two *Alnus* species. *Can. J. Microbiol.* **28,** 1133–1142.

Perradin, Y., Mottet, M. J., and Lalonde, M. (1983). Influence of phenolics on *in vitro* growth of *Frankia* strains. *Can. J. Bot.* **61,** 2807–2814.

Quispel, A., and Burggraaf, A. J. P. (1981). *Frankia,* the diazotrophic endophyte from actinorhizas. *In* "Current Perspectives in Nitrogen Fixation" (A. H. Gibson and W. E. Newton, eds.), pp. 229–236. Aust. Acad. Sci., Canberra, Australia.

Quispel, A., Svendsen, A. B., Schripsema, J., Baas, W. J., Erkelens, C., and Lugtenburg, J. (1989). Identification of dipterocarpol as isolation factor for the induction of primary isolation of *Frankia* from root nodules of *Alnus glutinosa* (L.) Gaertner. Molecular Plant–Microbe Interactions **2,** 107–112.

Racette, S., and Torrey, J. G. (1989). The isolation, culture and infectivity of a *Frankia* strain from *Gymnostoma papuanum* (Casuarinaceae). *Plant Soil* **118,** 165–170.

Reddell, P. (1986). "Soil and Plant Factors Affecting Nodulation and Nitrogen Fixation in Casuarinaceae–*Frankia* Symbioses," Ph.D. thesis. Univ. of Western Australia, Perth, Western Australia, Australia.

Reddell, P., and Bowen, G. D. (1985). *Frankia* source affects growth, nodulation and nitrogen fixation in *Casuarina* species. *New Phytol.* **100,** 115–122.

Reddell, P., and Bowen, G. D. (1986). Host–*Frankia* specificity within Casuarinaceae. *Plant Soil* **93,** 293–298.

Rodriguez-Barrueco, C., and Bond, G. (1976). A discussion of the results of cross-inoculation trials between *Alnus glutinosa* and *Myrica gale. In* "Symbiotic Nitrogen Fixation in Plants" (P. S. Nutman, ed.), pp. 561–565. Cambridge Univ. Press, Cambridge, England.

Rodriguez-Barrueco, C., and Miguel, C. (1979). Host plant–endophyte specificity in actinomycete-nodulated plants. *In* "Symbiotic Nitrogen Fixation in the Management of Temperate Forests" (J. C. Gordon, C. T. Wheeler, and D. A. Perry, eds.), pp. 143–159. For. Res. Lab., Oregon State Univ., Corvallis, Oregon.

Rodriguez-Barrueco, C., and Subramanian, P. (1988). Host–endophyte specificity in *Frankia* symbiosis. *In* "Biological Nitrogen Fixation—Recent Developments" (N. S. Subba Rao, ed.), pp. 283–310. Oxford Univ. Press/IBH Publ., New Delhi, India.

Rosbrook, P. A., and Bowen, G. D. (1987). The abilities of three *Frankia* isolates to nodulate and fix nitrogen with four species of *Casuarina. Physiol. Plant.* **70,** 373–377.

Silvester, W. B., Whitbeck, J., Silvester, J. K., and Torrey, J. C. (1988) Growth, nodule morphology, and nitrogenase activity of *Myrica gale* with roots grown at various oxygen levels. *Can. J. Bot.* **66,** 1762–1771.

Strand, R., and Laetsch, W. M. (1977). Cell and endophyte structure of the nitrogen-fixing root nodule of *Ceanothus integerrimus* H. and A. I. Fine structure of the nodule and its endosymbiont. *Protoplasma* **93,** 165–178.

Torrey, J. G. (1987). Endophyte sporulation in root nodules of actinorhizal plants. *Physiol. Plant.* **70,** 279–288.

Torrey, J. G., and Racette, S. (1989). Specificity among the Casuarinaceae in root nodulation by *Frankia. Plant Soil,* **118,** 157–164.

Tzean, S. S., and Torrey, J. G. (1989). Spore germination and the life cycle of *Frankia in vitro. Can. J. Microbiol.,* **35,** 801–806.

VandenBosch, K. A., and Torrey, J. G. (1983). Host–endophyte interactions in effective and ineffective nodules induced by the endophyte of *Myrica gale. Can. J. Bot.* **61,** 2898–2909.

van Dijk, C. (1984). "Ecological Aspects of Spore Formation in the *Frankia–Alnus* Symbiosis," Ph.D. thesis. State Univ., Leiden, The Netherlands.

Zhang, Z., and Torrey, J. G. (1985). Studies of an effective strain of *Frankia* from *Allocasuarina lehmanniana* of the Casuarinaceae. *Plant Soil* **87,** 1–16.

Zhang, Z., Lopez, M. F., and Torrey, J. G. (1984). A comparison of cultural characteristics and infectivity of *Frankia* isolates from root nodules of *Casuarina* species. *Plant Soil* **78,** 79–90.

Zobel, R. W., Del Tredici, P., and Torrey, J. G. (1976). Method for growing plants aeroponically. *Plant Physiol.* **57,** 344–346.

6

Physiology and Biochemistry of *Frankia* in Culture

David R. Benson and Nancy A. Schultz
Department of Molecular and Cell Biology
University of Connecticut
Storrs, Connecticut, U.S.A.

I. Introduction

Studies on the physiology and biochemistry of *Frankia* prior to 1978 necessarily used heterogeneous populations of symbiotic "vesicle clusters" from actinorhizal root nodules. Some progress was made in detecting bacteria-specific enzymes such as nitrogenase and hydrogenase but other aspects were simply not approachable before the first successful isolation of *Frankia* sp. strain CpI1 at Harvard Forest (Callaham *et al.*, 1978).

Since *Frankia* strains have become readily available, the most frequently encountered expletives used to describe their behavior in culture have concerned their slow growth rates and low amount of available biomass compared with "standard" laboratory bacteria. Growth periods of weeks to months have been commonly used to obtain sufficient biomass to do biochemical or physiological experiments, or for DNA isolations. Such long-term cultures generally contain sporangia, vesicles, and hyphae, and are to a considerable degree senescent. The lack of relative homogeneity makes interpretation of early physiological experiments problematical. More recent work has concentrated on relatively young cultures, from 3 to 5 days old, that have less diverse cell populations, although the problems inherent in dealing with the physiology of a filamentous organism are still present. The purpose of this chapter is to collate information concerning the physiology and growth of frankiae and to point out opportunities for future work.

The Biology of *Frankia* and Actinorhizal Plants

II. General Cultural Characteristics

A. Conclusions from Early Growth Studies

Most early physiological work concentrated on developing techniques for isolating and maintaining *Frankia* strains. Many media were developed empirically for this purpose and most were complex in composition. Defined liquid media for subcultivating frankiae were also developed and refinements continue today. The most commonly used media are discussed in Chapter 3, but the conclusions that can be drawn from the composition of routine cultivation media concerning the physiology of frankiae form the starting point for the discussion in this chapter.

The first conclusion one may draw from early media improvement studies is that frankiae are heterotrophic, aerobic, and sometimes microaerophilic. Of interest here is that most frankiae need no growth factors once established in laboratory culture although, in some instances, growth factors are necessary for isolation. Quispel *et al.* (1988) recently described a triterpene "isolation factor" extracted from alder roots that markedly improved the isolation efficiency of some frankiae from nodules. Once the strains were in culture, however, they no longer required the factor for growth. The function of the factor is unclear, although it might be necessary for maintaining the integrity of endophytic membranes during isolation (Quispel *et al.*, 1988). It is possible, even probable, that some strains will be found to have true growth requirements. Auxotrophy may explain why there are still plants from which no frankiae have been isolated, and why some strains from plants that normally yield isolates fail to grow. An example of the latter are the spore-positive strains from alders. However, most laboratory strains in common use today are prototrophs.

Another conclusion that may be drawn from early studies is that many frankiae are remarkably forgiving when neglected in cultures for months or even years. An effective way to maintain many strains in culture is simply to put them on the shelf in complex medium. No studies have been done on the survival mechanism used by frankiae in such a situation but such cultures are usually heavily sporulated. Survival is probably due to the longevity of the spores.

B. The Problem with Filamentous Organisms

Since *Frankia* spp. are filamentous bacteria, experiments on their physiology are somewhat more difficult to interpret than analogous experiments on unicellular organisms. The main difficulty is that actively growing cells in filamentous organisms are located at hyphal tips or at branch points along hyphae. Cells that are left behind are in a state that probably resembles the stationary phase in other bacterial cultures. Thus, for a *Frankia* culture that has been vigorously homogenized to provide numerous new growing points, "balanced growth" cannot be assumed past about one or two doublings. Balanced growth is where every "extensive property" (i.e., mass and volume) is increasing exponentially (Koch, 1981). The degree to which a *Frankia* growth curve resembles "exponential growth" will depend on the initial fragment size and the frequency of branching in culture. This problem is seen in a typical *Frankia* growth curve (done using well-buffered medium to moderate pH changes), which starts by resembling an exponential curve but quickly becomes linear or pseudolinear over time (Burggraaf and Shipton, 1983).

The effect of dealing with cultures that are not in balanced growth, aside from frustrating manuscript reviewers, is that the outcome of experiments depends to a great extent on the ratio of growing to stationary phase cells. When measuring an enzyme like glutamine synthetase, for example, the enzyme will be more abundant in cells that are growing rapidly and less abundant in those cells subsisting on maintenance energy. Extracts from such cultures represent the average levels of enzymes present at the time of harvest.

Another effect of hyphal growth is the formation of hyphal masses, the mycelia. Mycelia have nutrient gradients. Substrates like O_2, carbon sources, and nitrogen sources and other essential elements cannot diffuse readily into such structures. Conversely, waste products like CO_2 or organic acids exit slowly. This gradient makes accurate measurements of O_2 uptake rates, for example, problematical, and can lead to anaerobic microsites and regions of nutrient starvation, and stress amid apparent plenty. The presence of nutrient gradients probably accounts for the frequent observation of spores and vesicles being produced in the center of mycelial masses (Burggraaf and Shipton, 1983). The practical utility of growing cultures with large mycelial masses is limited.

With these problems in mind, it is still possible to obtain useful physiological information as long as the cultural conditions are carefully circumscribed to limit variability in cell type. Thus, only healthy cultures that have been homogenized in a tissue homogenizer, sonicator, or some other device, and then transferred to, and acclimated in, fresh medium, should be used for physiological studies.

III. Carbon Metabolism of *Frankia* in Culture

Relatively few studies have addressed the central metabolic pathways used by frankiae for cell growth and division. Of those available, most have concentrated on enzymes associated with major catabolic and anabolic pathways or with specific segments of ancillary pathways used for catabolizing individual compounds.

A. Carbohydrate Catabolism

Since a majority of frankiae grow best on media containing short-chain fatty acids like acetate or propionate, or TCA cycle intermediates like succinate, the importance of carbohydrate catabolism in culture, and in symbiosis, is difficult to assess. However, the presence of glycogen and trehalose as major storage compounds, even in strains that grow poorly on carbohydrates, implies the ability to catabolize carbohydrates (Benson and Eveleigh, 1979; Lopez *et al.*, 1983, 1984). Several frankiae degrade cellulose (J. G. Torrey, personal communication), and some have pectinase genes and activity (Séguin and Lalonde, 1989). Thus, carbohydrates are probably used to a greater extent then revealed by early nutritional studies.

Some workers have commented on the inability of certain frankiae to transport carbohydrates (Blom and Harkink, 1981; Lechevalier *et al.*, 1983) although comprehensive studies on the generality of this observation have not been done. Some *Frankia* subgroups, particularly those belonging to the heterogeneous Group A organisms as defined by Lechevalier (1984; see also Chapter 3), corresponding roughly to the *Frankia elaeagni* group as proposed by Lalonde *et al.* (1988), will grow on several carbohydrates (Lechevalier *et al.*, 1983; Tisa *et al.*, 1983). The sugars used include glucose, fructose, mannitol, trehalose, maltose, and sucrose (Burggraaf and Shipton, 1983; Lechevalier *et al.*, 1983; Tisa *et al.*, 1983; Lopez and Torrey, 1985a). Group B organisms, which generally have been isolated from *Alnus, Comptonia*, or *Myrica* root nodules, are less fond of carbohydrates and prefer short-chain fatty acids or TCA cycle intermediates for their carbon source, although some of these also use carbohydrates (Shipton and Burggraaf, 1982; Lopez and Torrey, 1985a).

Enzymes of the Embden–Meyerhof–Parnas (EMP) glycolytic sequence, including glucokinase, phosphofructokinase, and pyruvate kinase, have been detected in crude extracts from strain HFPArI3 grown on propionate and trehalose (Lopez and Torrey, 1985a). This work contrasts with earlier findings with strain AvcI1, where cells grown on Tween medium lacked hexokinase, pyruvate kinase, and pyruvate dehydrogenase activities (Blom and Harkink, 1981). The difference can probably be attributed to improvements in handling large cultures of frankiae in later work or to differences in culture media or enzyme assays.

The pentose phosphate pathway enzymes, glucose-6-phosphate dehydrogenase ($NADP^+$-linked) and 6-phosphogluconate dehydrogenase (NAD^+-linked), have also been demonstrated in HFPArI3 growing on propionate with and without ammonia, and on trehalose; similar enzymes were detected in HFPCcI2 growing on propionate but were not detected in the alder isolate NPI0136010 growing on propionate (Stowers *et al.*, 1986). Other enzymes of the pentose phosphate pathway have not been detected but can be assumed to be present since no requirements for pentoses or tetroses have been noted (Lopez and Torrey, 1985a). Entner–Douderoff (ED) pathway enzymes have not been detected (Lopez and Torrey, 1985a; Stowers *et al.*, 1986) but this is not surprising since gram-positive organisms like *Frankia* generally do not catabolize glucose via the ED pathway.

B. Metabolism of Storage Compounds

1. Glycogen

Glycogen and trehalose have been identified as major storage compounds in free-living and symbiotic frankiae (Benson and Eveleigh, 1979; Lopez *et al.*, 1983). Glycogen granules are visible in transmission electron micrographs of symbiotic frankiae hyphae and vesicles in root nodules of *Comptonia peregrina, Elaeagnus, Chamaebatia, Coriaria,* and *Myrica pensylvanica* but not in *Discaria* and *Dryas* (Newcomb and Wood, 1987). The presence or absence of glycogen appears to correlate with the metabolic status of the cell. Thus, in HFPArI3 cultures, immature (non-N_2-fixing) vesicles contain large amounts of glycogen deposits but fewer are seen in vesicles that are taken from N_2-fixing cultures (Fontaine *et al.*, 1984). The diminution of glycogen in vesicles that are presumably fixing N_2 suggests that carbohydrate catabolism can contribute to the supply of ATP and reducing power for nitrogenase. Glycogen can constitute about 10% of the dry weight of HFPArI3 cells growing on NH_4^+, and decreases to about 1% in cultures starved for carbon (Lopez *et al.*, 1984).

2. Trehalose

Trehalose (α-D-glucopyranosyl-α-D-glucopyranoside) has received attention in frankiae because of its abundance as a storage carbohydrate in frankial cells (Lopez *et al.*, 1983); the disaccharide may account for 10 to 20% of the dry weight of cells (Lopez *et al.*, 1984). The function of trehalose is not clear. In many organisms, including fungi (Thevelein, 1984), rhizobial bacteroids (Streeter, 1985), other actinomycetes (Hey-Ferguson *et al.*, 1973), cyanobacteria (MacKay *et al.*, 1984), and even *E. coli* (Rod *et al.*, 1988), trehalose is a storage carbohydrate in non-growing cells, or in spores. It has also been implicated as an osmoprotectant during conditions of osmotic stress (MacKay *et al.*, 1984; Rod *et al.*, 1988). Trehalose can be used as a carbon source by some frankiae when added to culture media (Lopez *et al.*, 1984). It is also metabolized internally in response to the metabolic state of the cell, as demonstrated by studies on *Frankia* sp. strain HFPArI3, in which stored trehalose was used during periods of rapid growth or during N_2 fixation (Lopez *et al.*, 1984).

Any direct role for trehalose, or glycogen, in N_2 fixation remains uncertain. Cultures that have been removed from propionate-containing medium and then resuspended in medium without a carbon source maintain between 20 and 50% of their nitrogenase (acetylene-reducing) activity (Lopez *et al.*, 1984). This observation suggests that endogenous energy reserves are being tapped to supply energy for N_2 fixation. Since N_2 fixation occurs in vesicles (Noridge and Benson, 1986), the metabolism of storage carbohydrate such as trehalose and glycogen in vesicles isolated from culture would be an interesting subject for future study. In addition, a study on trehalose levels as a function of culture osmolarity might suggest alternate functions for the disaccharide.

Trehalase is the enzyme that hydrolyzes trehalose to D-glucose in frankial cells (Lopez and Torrey, 1985b). Enzyme levels, measured as specific activity, drop during the period of most active N_2 fixation in HFPArI3 (Lopez and Torrey, 1985b), at a time when total trehalose content also decreases; thus, diminution of trehalose pools in N_2-fixing cells is probably due to lack of synthesis caused by a rerouting of carbon, rather than increased hydrolysis (Lopez and Torrey, 1985b). Specific activities of trehalase in symbiotic vesicle clusters of HFPArI3 were reportedly 6–7 times greater than in free-living N_2-fixing cells, and 26–33 times greater than in NH_4^+-grown cells (Lopez and Torrey, 1985b). It should be noted in this context that accurate protein determinations on pure vesicles and vesicle clusters taken from plants are difficult to obtain because of interference from plant-derived compounds and/or the relatively large quantities of lipid present in N_2-fixing cultures (D. R. Benson and N. A. Schultz, unpublished observations). However, the presence of trehalase in vesicle clusters may suggest that plant-derived or -stored trehalose is available for catabolism in the root nodule.

C. Pathways of Gluconeogenesis

Since frankiae grow on propionate, acetate, and in some cases succinate, a functional TCA cycle, glyoxylate cycle, and gluconeogenesis pathway may be assumed to operate. Early studies on *Frankia* sp. strain AvcI1, a typical alder isolate, demonstrated the presence of enzymes forming a complete TCA cycle in cells grown on Tween and QMOD/Tween media (Blom and Harkink, 1981). The glyoxylate cycle enzymes, isocitrate lyase and malate synthase, were also present during growth with Tween 80 (polyoxyethylene sorbitan monooleate), which is presumably catabolized as a fatty acid. The glyoxylate cycle enzymes were repressed by growth on propionate but induced by growth on acetate (Blom, 1982).

During growth on acetate or on substrates entering central metabolism as acetate, gluconeogenesis was proposed to occur from glyoxylate cycle-derived succinate through the actions of succinate dehydrogenase, fumarase, and malic enzyme to pyruvate and then to phosphoenolpyruvate via pyruvate orthophosphate dikinase (Blom and Harkink, 1981). An $NADP^+$-linked malic enzyme has been demonstrated in HFPArI3 grown on propionate; in agreement with its function in gluconeogenesis, the enzyme was present but repressed during growth on trehalose (Lopez and Torrey, 1985a). The conclusion to be drawn from current information is that gluconeogenic pathways commonly used by other bacteria also operate in frankiae.

A drawback with some enzyme studies that have been done is that little information has been given about optimizing the activities of the enzymes under study. It is difficult to draw conclusions about relative levels of catabolic or anabolic enzymes or even to conclude if the enzymes are present or absent without some indication of the conditions tried for assay.

D. Propionate Metabolism

Propionate comes close to being a universal carbon source for the growth of *Frankia*, since virtually all strains grow to some degree in propionate-containing medium; the sole reported exception is an isolate from *Elaeagnus*, EAN1pec (Tisa *et al.*, 1983). Its universality makes propionate metabolism of interest even though it may not provide the most rapid growth for individual strains.

Propionate transport into frankiae is inhibited by sodium azide, *p*-hydroxymercuribenzoate, *N*-ethylmaleimide, and CCCP (carbonyl cyanide *m*-chlorophenylhydrazone), suggesting that uptake is mediated by an active transport process and by a specific permease (Stowers *et al.*, 1986). In CpI1, propionate uptake and oxidation are probably constitutive activities since succinate-grown cells oxidize propionate at an appreciable rate (Tisa *et al.*, 1983). In contrast, succinate transport in CpI1 is probably inducible since propionate-grown cells oxidize succinate only slowly.

In bacteria, propionate can be catabolized by several pathways that direct carbon to central metabolism ultimately as either acetyl-CoA or succinate (Wegener *et al.*, 1968). Pathways leading to acetyl-CoA include oxidation via malonic semialdehyde, malonyl semialdehyde-CoA, or lactate. In each case, propionate is first activated as propionyl-CoA. The pathway leading to succinate begins with the biotin-dependent carboxylation of propionyl-CoA to yield methylmalonyl-CoA, which is subsequently racemized to succinyl-CoA. The carboxylase pathway is the one most frequently encountered in bacteria (Wegener *et al.*, 1968), and has been assumed to operate in *Frankia.*

Propionate assimilation has been studied in *Frankia* strain NPI0136010 isolated from *Alnus viridis* ssp. *sinuata* by using radiorespirometry with $^{14}C_1$-, $^{14}C_2$-, and $^{14}C_3$-labeled propionate (Stowers *et al.*, 1986). The pathway of assimilation was concluded to be via carboxylation to methylmalonyl-CoA with subsequent entry into central metabolism as succinyl-CoA. However, the radiorespirometric data ($C_1 > C_2 > C_3$) are also consistent with a pathway involving propionyl-CoA oxidation to acrylyl-CoA, hydration to lactyl-CoA, oxidation of lactate to pyruvate, and entry into central metabolism as acetate (Wegener *et al.*, 1968). The detection of glyoxylate cycle enzymes in NPI0136010 cells growing on propionate is also consistent with this interpretation (Stowers *et al.*, 1986).

Previous studies on AvcI1, also an alder strain, showed repression of glyoxylate enzymes during growth on propionate, stimulation of growth by CO_2, and the presence of propionyl-CoA carboxylase in cell-free extracts, although no carboxylase enzyme data were presented (Akkermans *et al.*, 1983). In addition, a variant of CpI1 has been reported that requires biotin for growth on propionate, and some reports have indicated that elevated CO_2 levels and biotin enhance growth and N_2 fixation (Shipton and Burggraaf, 1982; Tisa *et al.*, 1983; Murry *et al.*, 1984). These observations suggest the involvement of a carboxylation pathway since propionyl-CoA carboxylase is a biotin-dependent enzyme (Tisa *et al.*, 1983). However, a biotin stimulation does not preclude the presence of other pathways since other carboxylases essential for metabolism, including acetyl-CoA carboxylase and pyruvate carboxylase, use biotin as a coenzyme (Gottschalk, 1986). Clearly, additional work is necessary to clarify the pathways of propionate metabolism in frankiae. It is possible that not all frankiae are alike in their choice of pathways. Even in *E. coli,* there is evidence that different pathways are used depending on the availability of propionate; apparently, the CO_2 fixation pathway is used during growth on low propionate concentrations and the lactate pathway is used during growth on higher propionate concentrations (Wegener *et al.*, 1968).

IV. Nitrogen Metabolism

A. General Pathways of Nitrogen Assimilation

A vast literature now exists on nitrogen metabolism in bacteria, particularly in enterics like *E. coli*, *Salmonella typhimurium*, and *Klebsiella pneumoniae* and the obligate aerobes *Azotobacter* and *Rhizobium*. Using other systems as a guide, a general outline of nitrogen metabolism of *Frankia* has emerged over the past decade. The physiology and biochemistry of some strains have been addressed and molecular genetic studies have been initiated. Missing are the abilities to generate and complement nitrogen metabolism mutants. Transformation, transfection, and even mutational techniques have not been developed as of this writing.

In virtually all diazotrophs, including *Frankia*, N_2 assimilation primarily follows the nitrogenase–glutamine synthetase (GS)–glutamate synthase (GOGAT) path:

$$\text{Nitrogenase: } N_2 + 12\text{–}24\text{ATP} + 8e^- + 8H^+ \xrightarrow{Mg^{2+}} 2NH_3 + 12\text{–}24\text{ADP} + 12\text{–}24\text{Pi} + H_2$$

$$\text{Glutamine synthetase: } NH_3 + \text{Glu} + \text{ATP} \xrightarrow{Mg^{2+}} \text{Gln} + \text{ADP} + \text{Pi}$$

$$\text{Glutamate synthase: } \text{Gln} + \alpha\text{-ketoglutarate} + \text{NAD(P)H} + H^+ \longrightarrow 2\text{Glu} + \text{NAD(P)}^+$$

The energy requirements for the reduction and assimilation of 1 N_2 to 2 NH_4^+ are between 14 and 26 ATP and 2 NAD(P)H. Also required are a source of electrons and an anaerobic environment in which nitrogenase and associated electron carriers can be protected from O_2.

1. Nitrogenase

The biochemical characteristics of *Frankia* nitrogenases have been studied in very few systems to date. In common with other nitrogenases, nitrogenases from actinorhizal root nodules show ATP-dependent H_2 gas production when incubated under an argon atmosphere, are O_2 sensitive, and have an absolute requirement for MgATP and a source of electrons (Benson *et al.*, 1979). Again, like other nitrogenases, the MoFe protein and the Fe protein can form catalytically active complexes with the Fe and MoFe proteins, respectively, from *Azotobacter vinelandii* (Benson *et al.*, 1979). Several laboratories have demonstrated DNA sequence similarity of *Frankia nif* genes with those from other organisms (see Chapter 10). Therefore, there are probably no "major" differences in the biochemistry of *Frankia* nitrogenase as compared with other more conveniently studied systems. The most interesting aspects of *Frankia* nitrogenases are their location, their regulation, and their function in symbiosis.

Several lines of evidence indicate that vesicles contain nitrogenase when cultures are grown on N_2. Older studies implicated vesicles in N_2 fixation by correlating vesicle production in culture with the onset of nitrogenase activity (Tjepkema *et al.*, 1980a,b; Gauthier *et al.*, 1981; Fontaine *et al.*, 1984). Western blotting using antibodies raised against *Rhizobium leguminosarum* nitrogenase components on proteins from "vesicle-enriched" preparations and hyphal fractions showed that more nitrogenase cross-reactive material was present in vesicle-enriched fractions than in hyphal fractions (Meesters *et al.*, 1985). Noridge and Benson (1986) provided a direct demonstration of nitrogenase activity in pure preparations of vesicles isolated on sucrose gradients. Nitrogenase activity was found in vesicles prepared anaerobically, permeabilized with cetyltrimethyl ammonium bromide, and assayed with ATP and dithionite. Tisa and Ensign (1987b) isolated vesicles by selectively disrupting hyphae by passage through a French pressure cell followed by differential and isopycnic gradient centrifugation. Nitrogenase activity was not detected in the supernatant containing the vegetative cell lysate and was associated solely with the vesicles, suggesting that nitrogenase was located in the vesicles and not in the hyphae. A similar conclusion was reached by other workers using immunogold labeling of vesicles and hyphae with antibodies to nitrogenase from rhizobia; label was primarily associated with vesicles (Meesters *et al.*, 1987; Sasakawa *et al.*, 1988).

The weight of the evidence to date is that vesicles do indeed contain nitrogenase and that hyphae in aerobically growing cultures do not contain the enzyme complex. A caveat to the last statement is that cultures of HFPCcI3 induced for N_2 fixation while sparged with 0.3% O_2 do not form vesicles but are able to fix N_2 (Murry *et al.*, 1985). The nitrogenase formed under these conditions is sensitive to inactivation by O_2. This observation implies that nitrogenase can be synthesized in the hyphae and that vesicle formation is regulated by the availability of both nitrogen and oxygen.

A calcium requirement for vesicle development, nitrogen fixation, and cyanide-insensitive respiration has been reported for EAN1pec and CpI1 (Tisa and Ensign, 1987a). This study followed earlier reports that calcium as well as other common elements was necessary for nitrogen fixation (Murry *et al.*, 1984). The requirement for calcium is interesting since few reports of a calcium requirement by bacteria are available. However, the biochemical basis of the observation remains unclear.

2. Glutamine Synthetase

The fate of nitrogen in cells is governed by cellular needs for individual amino acids. In bacteria, glutamate is the central amino acid from which most others are made either directly or via transamination. Glutamine is used for protein synthesis and is involved in the biosynthesis of several metabolites, including asparagine, histidine, tryptophan, purines, pyrimidines, and amino sugars. As in most anabolic pathways, regulatory controls are exerted at the first step into the pathway. In the case of ammonia assimilation, primary regulation occurs on glutamine synthetase and the types of controls used by individual bacteria can differ.

To put the frankial system in perspective an understanding of the basic patterns of regulation of glutamine synthetase (GS) in other systems is essential. In enterics, GS is regulated transcriptionally and posttranslationally (Fig. 1). Enteric GS DNA is transcribed from two promoters, one for constitutive synthesis in a relatively high ammonia concentration and the other for inducible high-level synthesis in media with a low ammonia concentration. In addition, positive regulation at the constitutive promoter occurs through the mediation of the catabolite-activating protein (CAP) and cyclic AMP (Rietzer and Magasanik, 1987).

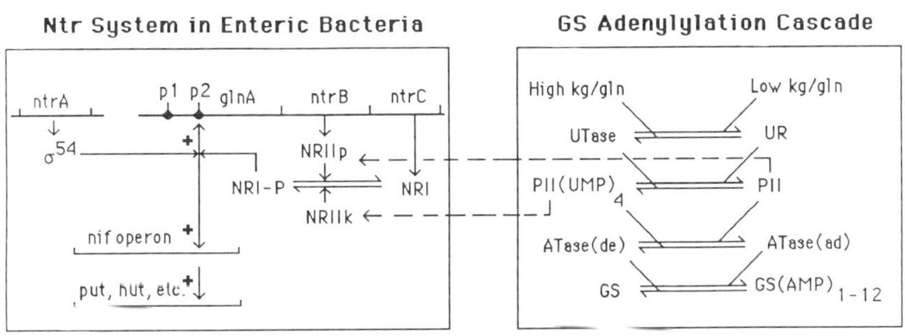

Figure 1 The regulation of glutamine synthetase and its relationship to the general nitrogen regulatory (ntr) system in enteric bacteria. Abbreviations: *ntrA*, gene for sigma 54; *ntrB*, gene for NRII; *ntrC*, gene for NRI; *nif* operon, nitrogen fixation genes (in *Klebsiella*); *put*, *hut* genes for proline and histidine utilization, respectively, also controlled by ntr; p1, constitutive promoter for glutamine synthetase expression; p2, inducible Ntr-like promoter for high-level glutamine synthetase expression; NRIIk, NRIIp, nitrogen regulatory sensor proteins in the kinase form and phosphatase forms, respectively; NRI and NRI-P, nitrogen regulatory transcriptional activator proteins in the inactive form and active phosphorylated forms, respectively; kg/gln, the ratio of α-ketoglutarate to glutamine; UTase, uridylyltransferase; UR, uridyl-removing enzyme; $PII(UMP)_4$, uridylylated PII protein; ATase(de) and ATase(ad), adenylyltransferase in the deadenylylating and adenylylating forms, respectively; GS and $GS(AMP)_{1-12}$, glutamine synthetase in unadenylylated and adenylylated forms, respectively.

At the level of enzyme activity, enteric GS is covalently modified with an AMP residue on each of twelve subunits through three proteins of an adenylylation cascade, including uridylyltransferase/uridylyl-removing enzyme, PII protein, and adenylyltransferase (Rietzer and Magasanik, 1987). As ammonia is depleted from the medium, the ratio of α-ketoglutarate/glutamine rises and stimulates the uridylyltransferase/uridylyl-removing enzyme to transfer UMP residues to the PII protein; $PII(UMP)_4$ activates adenylyltransferase to deadenylylate GS. In high concentrations of ammonia, the ratio of α-ketoglutarate/glutamine decreases and the uridylyltransferase/uridylyl-removing enzyme removes UMP from PII. Unmodified PII interacts with adenylyltransferase to adenylylate GS. Glutamine synthetase becomes increasingly sensitive to feedback inhibition by metabolites derived from glutamine as it is progressively adenylylated. The metabolic inhibitors include serine, glycine, alanine, histidine, tryptophan, glucosamine, AMP, CTP, and carbamyl-phosphate (Almassy *et al.*, 1986).

Transcriptional and posttranslational regulation are coupled through the PII protein; it interacts with both adenylyltransferase and the product of the *ntrB* gene, NRII (Fig. 1). NRII can act as either a kinase or a phosphatase on its substrate NRI, depending on the uridylylated state of PII. NRI is a transcriptional activator that, when phosphorylated, acts in conjunction with a novel sigma factor, σ^{54}, and RNA polymerase to initiate high-level transcription from the inducible promoter in front of the *glnA* gene. In *Klebsiella*, σ^{54} and NRI are also involved in transcribing the *nifLA* operon, which controls *nif* gene expression.

The enteric pattern of regulation seems to occur in many but not all gram-negative bacteria. Its essential features include the use of more than one promoter to control constitutive and inducible synthesis of GS, the coupling of regulation through the PII protein, the covalent adenylylation of the enzyme in response to the nitrogen status of the cell, and the ability to rapidly change the activity of GS in response to downstream metabolites.

In the few cyanobacteria studied, the pattern of GS regulation is similar to that in the enterics at the transcriptional level but differs at the level of the enzyme. In *Anabaena*, multiple promoters have been found that respond to the nitrogen status of the cell (Tumer *et al.*, 1983). However, the enzyme is not adenylylated either in cyanobacteria or when its gene is expressed in *E. coli*. Instead, short-term regulation of GS occurs by feedback inhibition by metabolites that originate from glutamine (Orr and Haselkorn, 1982). A similar pattern of regulation has also been described for the gram-positive *Bacillus* sp. (Deuel and Stadtman, 1970). In contrast, enzymes described from the gram-positive actinomycetes

Streptomyces cattleya and *Streptomyces hygroscopicus* are regulated by adenylylation in a manner reminiscent of that seen for enterics (Streicher and Tyler, 1981; Xia and Jiao, 1986).

In rhizobia, transcriptional regulation has apparently been split between two (or more) enzymes, called GSI and GSII, rather than between two promoters. This arrangement is found in most members of the *Rhizobiaceae,* including *Rhizobium, Bradyrhizobium, Agrobacterium,* and *Phyllobacterium,* all plant-associated bacteria. One enzyme, GSI, is similar to enteric GS in its regulation by adenylylation but is expressed from a single promoter at constitutive levels. The other enzyme, GSII, is synthesized during growth on poor nitrogen sources and is expressed from an *ntr*-like promoter at high levels (Martin *et al.,* 1988). A third locus involved in glutamine synthesis in rhizobia has been recently described (de Bruijn *et al.,* 1989).

In *Frankia* sp. strain CpI1, the pattern of GS regulation resembles that of members of the Rhizobiaceae rather than that of enterics, bacilli, cyanobacteria, or streptomycetes. Two GSs have been described, GSI and GSII (Edmands *et al.,* 1987). GSI is regulated by adenylylation and has a holoenzyme molecular weight consistent with a dodecameric structure (Edmands *et al.,* 1987). The second enzyme is synthesized in response to nitrogen starvation, or during growth on glutamate as the sole nitrogen source (Tsai and Benson, 1989). No evidence has been found for its adenylylation, and its holoenzyme molecular weight and subunit molecular weight are consistent with an octomeric structure (Y.-L. Tsai and D. R. Benson, unpublished observations). In common with rhizobial GSII, the frankial GSII is heat labile and is subject to feedback inhibition by glutamine (Tsai and Benson, 1989). In addition, the GSII gene, *glnII,* from *Bradyrhizobium* has been used as a heterologous probe to clone the *glnII* gene from a size-selected *Frankia* sp. strain CpI1 library (D. A. Rochefort and D. R. Benson, unpublished observations). Thus the regulation of ammonia assimilation in frankiae bears a striking resemblance to that in rhizobia.

The question arises, what does this observation say about the symbiosis? The details of GSI and GSII regulation may somehow predispose organisms with dual GSs to form N_2-fixing symbioses. For example, all symbiotic diazotrophs (frankiae, cyanobacteria, and rhizobia) have a GS in symbiosis that is not regulated by an adenylylation cascade. Could this be an energy-conserving mechanism to avoid using ATP and UTP in the adenylylation cascade, or does this arrangement make the enzyme more sensitive to its product rather than to the ratio of α-ketoglutarate/glutamine? Much work remains on this question but the similarity of *Frankia* nitrogen metabolism with other distantly related organisms provides a fertile field for investigations into the molecular genetics and biochemistry of nitrogen regulation in *Frankia.*

The physiology of GSI and GSII from CpI1 has been studied in culture (Tsai and Benson, 1989). In cells transferred from NH_4^+-containing medium to medium free of combined nitrogen, total GS activity more than doubled over 48 hr and the contribution of GSII rose from 0 to more than 90% of the total activity. GSI activity dropped gradually over the incubation period. In cells transferred from N_2-fixing culture to NH_4^+-containing medium, total GS activity dropped precipitously over 5 hr, and by 20 hr virtually all GS activity was contributed by GSI. When cells were transferred from NH_4^+-containing medium to glutamate medium, GSII activity became detectable at about 20 hr of incubation but contributed less than about 50% of the total GS activity after 48 hr. In long-term culture on glutamate, GSII contributed more than 80% of the total GS activity, vesicles were produced, and nitrogenase activity was detected. These results suggest that CpI1 synthesizes GSII in response to nitrogen starvation or during growth on a poor nitrogen source, such as glutamate.

When *Frankia* sp. strain CpI1 is grown on N_2, GSII is the predominant GS (>90%) in both vesicles and hyphae, with only a small amount of activity contributed by GSI (Schultz, 1989). The specific activity of GSII present in vesicles in such cultures is similar to that detected in hyphae, and the relative amounts of GSI and GSII parallel their staining on two-dimensional gels of whole cell proteins from vesicles and hyphae grown on N_2 (Schultz, 1989). For NH_4^+-grown cells, the two-dimensional gels show little or no GSII. Thus, although GSII is formed under N_2-fixing conditions, the enzyme is present in similar amounts in both the vesicles and hyphae.

3. Glutamate Synthase

The next enzyme in the sequence of N_2 assimilation is glutamate synthase (GOGAT). This enzyme forms glutamate for biosynthesis and for the continued functioning of GS. Whereas GS is generally highly regulated in most organisms, glutamate synthase is apparently primarily regulated in response to glutamate availability in those organisms that have been studied (Rietzer and Magasanik, 1987).

Glutamate synthase from *Frankia* sp. strain CpI1 has been studied with regard to its biochemical characteristics and its distribution between vesicles and hyphae. It appears to be similar to other prokaryotic GOGATs with the exception that it has a high K_m value for glutamine corresponding to approximately 5–6 m*M* glutamine (Schultz, 1989). In addition, GOGAT activity is present in NH_4^+-grown cells and in N_2-grown cells, but not in vesicles. The absence of the enzyme from vesicles was confirmed on polyacrylamide gels (Schultz, 1989). This finding, together with the presence of GSII in the vesicles, repeats the pattern established for nitrogenase-containing cyanobacterial heterocysts, which contain GS but not GOGAT.

4. Communication between Vesicles and Hyphae

Since vesicles are the sole sites of nitrogenase in aerobically cultured *Frankia,* they are also the only source of nitrogen for the vegetative hyphae. Vesicles have the potential to produce glutamine by GSII; such glutamine could conceivably be transferred to the vegetative hyphae through the constricted stem cell of the vesicle (Meesters, 1988). After transfer, glutamine would be converted by GOGAT to glutamate, with one of the resulting glutamates returning to the vesicle to act as an ammonia acceptor in subsequent reactions. Such cycling occurs in cyanobacterial heterocysts (Haselkorn, 1978).

Alternatively, it is possible, despite the presence of GSII in vesicles, that ammonia is not assimilated in the vesicle. Support for this proposal rests on the observations that GSII is present in the hyphae and that it is inhibited by glutamine and by a low energy charge as might prevail in actively N_2-fixing vesicles (Tsai and Benson, 1989). Physiologically inactive GS would lead to an accumulation of NH_4^+, which could diffuse out as NH_3. Any free NH_4^+ in the medium would be efficiently scavenged by the high-affinity ammonia permease present in nitrogen-starved hyphae (Mazzucco and Benson, 1984). Once transported into hyphae, NH_4^+ would be assimilated by the GS–GOGAT system. It should be pointed out that NH_4^+ is the compound assumed to be transported by *Frankia* in symbiosis, although direct evidence for this assumption is lacking. Rhizobial bacteroids and certain heterocystous cyanobacteria do excrete NH_4^+ when fixing N_2 symbiotically. It is possible that, for N_2-fixing frankiae in culture, the site of N_2 fixation is distinct from the site of NH_4^+ assimilation.

The situation in the root nodule is not clear at present (see Chapter 7). Studies on vesicle cluster extracts have failed to detect any enzymes of ammonia assimilation, including GS, GDH, and GOGAT (Akkermans *et al.,* 1983). It is not clear whether the lack of activities is due to enzyme absence or to the inactivation of the enzymes during preparation. GSII and GOGAT from CpI1 are both very heat sensitive and are readily inactivated (Schultz, 1989).

5. Other Enzymes of Ammonia Assimilation

Other enzymes less commonly used by bacteria to assimilate ammonia include glutamate dehydrogenase and alanine dehydrogenase. Both enzymes have been detected in CpI1 but the activities are low and essentially constant in ammonia-rich and combined nitrogen-deficient media, suggesting that they are not involved in the primary steps of ammonia assimilation (Schultz, 1989).

V. Protection of Nitrogenase from Oxygen in Culture

The location of nitrogenase within vesicles argues that vesicles protect the enzyme from exposure to O_2. Since protection of nitrogenase from O_2 is the subject of Chapter 8, we will limit our discussion here to the proposed composition of the vesicle envelope, which may form the first line of defense against oxidative damage.

Vesicle Envelopes

Nitrogenase from all N_2-fixing organisms, and from *Frankia,* is inactivated by brief exposure to O_2 (Benson *et al.*, 1979). Many lines of evidence point to the specialized architecture of vesicles as important in limiting O_2 diffusion (see Chapter 8). Vesicles have a capsule, or envelope, that appears laminated in electron micrographs; as many as 50 laminations have been reported (Parsons *et al.*, 1987). The laminations are lipid or glycolipid monolayers that stain with the lipid-specific fluorescent dye, Nile red (Lamont *et al.*, 1988), and the number of laminations increases with increasing O_2 levels (Parsons *et al.*, 1987), suggesting that envelope thickness is controlled in response to O_2.

The composition of *Frankia* vesicle envelopes is of obvious interest. The vesicle envelope has not been isolated from the vesicle proper but a total fatty acid analysis of NH_4^+-grown cells, N_2-grown cells, and vesicles isolated from N_2-grown cells of CpI1 is available (Tunlid *et al.*, 1989). Vesicles contain more (w/w) neutral lipids and glycolipids than either the ammonia-grown cells or vegetative hyphae from N_2-fixing cultures. Sixty percent (w/w) of the total fatty acids from NH_4^+-grown cells are in the polar (presumably phospholipid) fraction, whereas 26% (w/w) of the fatty acids from N_2-grown cells and 10% (w/w) of the fatty acids from isolated vesicles are in the polar fraction. In the vesicles, the neutral lipid fraction has 70% (w/w) of the total fatty acids. The neutral lipids are enriched with a series of long-chain fatty acids or alcohols with molecular ions in mass spectra at *m/z* 408 and *m/z* 410. The chain length of these compounds is estimated to be between 22 and 26 carbon atoms depending on the chemical structure (Tunlid *et al.*, 1989). The structure determination has not yet been reported.

The obvious interpretation of these results is that the long-chain compounds are components of the vesicle envelope, although confirmation of this hypothesis awaits isolation of the envelope free from other cellular components. Similar long-chain compounds have been found in lower amounts in vegetative hyphae from N_2- and NH_4^+-grown cultures. Thus, these fatty acids or alcohols may be normal cellular constituents that accumulate when vesicles differentiate (Tunlid *et al.*, 1989). Ultrastructurally similar laminae have been noted in electron micrographs of vegetative cells but never with the abundance seen in vesicles (Newcomb and Wood, 1987).

A parallel exists with the envelopes surrounding cyanobacterial heterocysts. The laminated layer of the heterocyst envelope is composed of glycolipid with long-chain (C26–C28) hydroxy fatty acids and polyhydroxy alcohols linked by ether or ester bonds to hexoses (Lambein and Wolk, 1973). The heterocyst glycolipid is poorly preserved with normal fixatives and solvents used for electron microscopy, and each layer measures around 3.5 to 4.0 nm (Winkenbach *et al.*, 1972). About 10 layers have been counted in the thinnest region of the heterocyst envelope. It will be interesting to compare the final structure of the fatty acids or alcohols in *Frankia* with those previously determined from cyanobacteria.

VI. Conclusions

Our knowledge of *Frankia* physiology and biochemistry has become increasingly sophisticated over the past ten years since cultures first became available. Studies on carbon metabolism are still incomplete although a general outline of carbohydrate metabolism and gluconeogenic pathways has been drawn. Still to be worked out are the contributions to central metabolism of the pentose phosphate pathway during growth on carbohydrates and organic acids. Also to be resolved is the extent to which *Frankia* sp. as a group can grow on carbohydrates. In this regard, it should be noted that one of the problems with such studies is that frankiae often require long induction periods for uptake and utilization of certain substances. The observation that carbohydrates are catabolized internally suggests their potential involvement in N_2 fixation in the vesicle.

Other aspects of carbon metabolism that need clarification include defining the pathway of propionate metabolism in different strains growing under different levels of propionate. In addition, the abundance of long-chain fatty acids or alcohols in vesicles provides an opportunity for the study of their biosynthetic pathways. It also provides an interesting example of convergent evolution in which cyanobacteria and frankiae have solved their O_2 problems in a surprisingly similar manner.

Nitrogen metabolism is central to the actinorhizal symbiosis. It currently appears that frankial nitrogen metabolism, at least in CpI1, will be similar to that found in the rhizobial–legume symbiosis. By analogy, many of the nitrogen regulatory controls in rhizobia may be similar in *Frankia*. This small leap of faith should provide molecular geneticists with reason enough to begin fishing out frankial nitrogen metabolism and regulation genes with heterologous probes from rhizobia or bradyrhizobia. The observation of a second glutamine synthetase in CpI1, and its demonstrated DNA sequence similarity with bradyrhizobial GSII, provides a foundation for speculating on the physiological basis for N_2-fixing root nodule symbioses.

Another area that is ripe for exploration in the realm of physiology and biochemistry of free-living *Frankia* is the extension of what is known in culture to the symbiosis. This has not been an easy transition to make but it is essential that the culture work be followed at some point by work in the plant.

References

Akkermans, A. D. L., Roelofsen, W., Blom, J., Huss-Danell, K., and Harkink, R. (1983). Utilization of carbon and nitrogen compounds by *Frankia* in synthetic media and in root nodules of *Alnus glutinosa, Hippophae rhamnoides*, and *Datisca cannabina*. *Can. J. Bot.* **61,** 2793–2800.

Almassy, R. J., Janson, C. A., Hamlin, R., Xuong, N., and Eisenberg, D. (1986). Novel subunit–subunit interactions in the structure of glutamine synthetase. *Nature (London)* **323,** 304–309.

Benson, D. R., and Eveleigh, D. E. (1979). Ultrastructure of the nitrogen-fixing symbiont of *Myrica pensylvanica* L. (bayberry) root nodules. *Bot. Gaz. (Chicago), Suppl.* **140,** S15–S21.

Benson, D. R., Arp, D. J., and Burris, R. H. (1979). Cell-free nitrogenase and hydrogenase from actinorhizal root nodules. *Science* **205,** 688–689.

Blom, J. (1982). Carbon and nitrogen source requirements of *Frankia* strains. *FEMS Microbiol. Lett.* **13,** 51–55.

Blom, J., and Harkink, R. (1981). Metabolic pathways for gluconeogenesis and energy generation in *Frankia* AvcI1. *FEMS Microbiol. Lett.* **11,** 221–224.

Burggraaf, A. J. P., and Shipton, W. A. (1983). Studies on the growth of *Frankia* isolates in relation to infectivity and nitrogen fixation (acetylene reduction). *Can. J. Bot.* **61,** 2774–2782.

Callaham, D., Del Tredici, P., and Torrey, J. G. (1978). Isolation and cultivation *in vitro* of the actinomycete causing root nodulation in *Comptonia*. *Science* **199,** 899–902.

de Bruijn, F. J., Rossbach, S., Schneider, M., Ratet, P., Messmer, S., Szeto, W. W., Ausubel, F. M., and Schell, J. (1989). *Rhizobium meliloti* 1021 has three differentially regulated loci involved in glutamine biosynthesis, none of which is essential for symbiotic nitrogen fixation. *J. Bacteriol.* **171,** 1673–1682.

Deuel, T. F., and Stadtman, E. R. (1970). Some kinetic properties of *Bacillus subtilis* glutamine synthetase. *J. Biol. Chem.* **245,** 5206–5213.

Edmands, J., Noridge, N. A., and Benson, D. R. (1987). The actinorhizal root nodule symbiont *Frankia* sp. strain CpI1 has two glutamine synthetases. *Proc. Natl. Acad. Sci. U.S.A.* **84,** 6126–6130.

Fontaine, M. F., Lancelle, S. A., and Torrey, J. G. (1984). Initiation and ontogeny of vesicles in cultured *Frankia* sp. strain HFPArI3. *J. Bacteriol.* **160,** 921–927.

Gauthier, D., Diem, H. G., and Dommergues, Y. (1981). *In vitro* nitrogen fixation by two actinomycete strains isolated from *Casuarina* nodules. *Appl. Environ. Microbiol.* **41,** 306–308.

Gottschalk, G. (1986). "Bacterial Metabolism." Springer-Verlag, Berlin and New York.

Haselkorn, R. (1978). Heterocysts. *Annu. Rev. Plant Physiol.* **29,** 319–344.

Hey Ferguson, A., Mitchell, M., and Elbein, A. D. (1973). Trehalose metabolism in germinating spores of *Streptomyces hygroscopicus*. *J. Bacteriol.* **116,** 1084–1085.

Koch, A. L. (1981). Growth measurement. *In* "Manual of Methods for General Bacteriology" (P. Gerhardt, R. G. E. Murray, R. N. Costilow, E. W. Nester, W. A. Wood, N. R. Krieg, and G. B. Phillips, eds.), pp. 179–207. Am. Soc. Microbiol., Washington, D.C.

Lalonde, M., Simon, L., Bousquet, J., and Séguin, A. (1988). Advances in the taxonomy of *Frankia*: Recognition of species *alni* and *elaeagni* and novel subspecies *pommerii* and *vandijkii*. *In* "Nitrogen fixation: Hundred Years After" (H. Bothe, F. J. de Bruijn, and W. E. Newton, eds.), pp. 671–680. Fischer, Stuttgart.

Lambein, F., and Wolk, C. P. (1973). Structural studies on the glycolipids from the envelope of the heterocyst of *Anabaena cylindrica*. *Biochemistry* **12,** 791–798.

Lamont, H. C., Silvester, W. B., and Torrey, J. G. (1988). Nile red fluorescence demonstrates lipid in the envelope of vesicles from N_2-fixing cultures of *Frankia*. *Can. J. Microbiol.* **34,** 656–660.

Lechevalier, M. P. (1984). The taxonomy of the genus *Frankia*. *Plant Soil* **78,** 1–6.

Lechevalier, M. P., Baker, D., and Horrière, F. (1983). Physiology, chemistry, serology, and infectivity of two *Frankia* isolates from *Alnus incana* subsp. *rugosa*. *Can. J. Microbiol.* **61,** 2826–2833.

Lopez, M. F., and Torrey, J. G. (1985a). Enzymes of glucose metabolism in *Frankia* sp. *J. Bacteriol.* **162,** 110–116.

Lopez, M. F., and Torrey, J. G. (1985b). Purification and properties of trehalase in *Frankia* ArI3. *Arch. Microbiol.* **143,** 209–215.

Lopez, M. F., Whaling, C. S., and Torrey, J. G. (1983). The polar lipids and free sugars of *Frankia* in culture. *Can. J. Bot.* **61,** 2834–2842.

Lopez, M. F., Fontaine, M. S., and Torrey, J. G. (1984). Levels of trehalose and glycogen in *Frankia* sp. HFPArI3 (Actinomycetales). *Can. J. Microbiol.* **30,** 746–752.

MacKay, M. A., Norton, R. S., and Borowitzka, L. J. (1984). Organic osmoregulatory solutes in cyanobacteria. *J. Gen. Microbiol.* **130,** 2177–2191.

Martin, G. B., Chapman, K. A., and Chelm, B. K. (1988). Role of the *Bradyrhizobium japonicum* ntrC gene product in differential regulation of the glutamine synthetase II gene (glnII). *J. Bacteriol.* **170,** 5452–5459.

Mazzucco, C., and Benson, D. R. (1984). ^{14}C-Methylammonium transport by *Frankia* sp. strain CpI1. *J. Bacteriol.* **160,** 636–641.

Meesters, T. (1988). "The Function of Vesicles in the Actinomycete *Frankia*." Ph.D., Agric. Univ., Wageningen, The Netherlands.

Meesters, T. M., van Genesen, S. T., and Akkermans, A. D. L. (1985). Growth, acetylene reduction activity and localization of nitrogenase in relation to vesicle formation in *Frankia* strains Cc1.17 and Cp1.2. *Arch. Microbiol.* **143,** 137–142.

Meesters, T. M., Van Vliet, W. M., and Akkermans, A. D. L. (1987). Nitrogenase is restricted to the vesicles in *Frankia* strain $EAN1_{pec}$. *Physiol. Plant* **70,** 267–271.

Murry, M. A., Fontaine, M. S., and Torrey, J. G. (1984). Growth kinetics and nitrogenase induction in *Frankia* sp. HFPArI3 grown in batch culture. *Plant Soil* **78,** 61–78.

Murry, M. A., Zhongze, Z., and Torrey, J. G. (1985). Effect of O_2 on vesicle formation, acetylene reduction, and O_2-uptake kinetics in *Frankia* sp. HFPCcI3 isolated from *Casuarina cunninghamiana*. *Can. J. Microbiol.* **31,** 804–809.

Newcomb, W., and Wood, S. M. (1987). Morphogenesis and fine structure of *Frankia* (Actinomycetales): The microsymbiont of nitrogen-fixing actinorhizal root nodules. *Int. Rev. Cytol.* **109,** 1–88.

Noridge, N. A., and Benson, D. R. (1986). Isolation and nitrogen-fixing activity of *Frankia* sp. strain CpI1 vesicles. *J. Bacteriol.* **166,** 301–305.

Orr, J., and Haselkorn, R. (1982). Regulation of glutamine synthetase activity and synthesis in free-living and symbiotic Anabaena spp. *J. Bacteriol.* **152,** 626–635.

Parsons, R., Silvester, W. B., Harris, S., Gruijters, W. T. M., and Bullivant, S. (1987). *Frankia* vesicles provide inducible and absolute oxygen protection for nitrogenase. *Plant Physiol.* **83,** 728–731.

Quispel, A., Burggraaf, A. J. P., Baerheim Svendsen, A., Schripsema, J., Lugtenburg, J., Erkelens, C., and Baas, W. J. (1988). Identification of an isolation factor for the endophytic phase of *Frankia* in root nodules of *Alnus glutinosa* as dipterocarpol. *In* "Nitrogen Fixation: Hundred Years After" (H. Bothe, F. J. de Bruijn, and W. E. Newton, eds.), p. 697. Fischer, Stuttgart.

Rietzer, L. J., and Magasanik, B. (1987). Ammonia assimilation and the biosynthesis of glutamine, glutamate, aspartate, asparagine, L-alanine, and D-alanine. *In* "*Escherichia coli* and *Salmonella typhimurium* Cellular and Molecular Biology" (F. C. Neidhardt, eds.), pp. 302–320. Am. Soc. Microbiol., Washington, D.C.

Rod, M. L., Alam, K. Y., Cunningham, P. R., and Clark, D. P. (1988). Accumulation of trehalose by *Escherichia coli* K-12 at high osmotic pressure depends on the presence of amber suppressors. *J. Bacteriol.* **170,** 3601–3610.

Sasakawa, H., Hiyoshi, T., and Sugiyama, T. (1988). Immunogold localization of nitrogenase in root nodules of *Elaeagnus pungens* Thunb. *Plant Cell Physiol.* **29,** 1147–1152.

Schultz, N. A. (1989). "Biochemical and Physiological Characteristics of *Frankia* sp. Strain CpI1 Vesicles," Ph.D. thesis. Univ. of Connecticut, Storrs, Connecticut.

Séguin, A., and Lalonde, M. (1989). Detection of pectolytic activity and *pel* homologous sequences in *Frankia*. *Plant and Soil* **118,** 221–229.

Shipton, W. A., and Burggraaf, A. J. P. (1982). A comparison of the requirements for various carbon and nitrogen sources and vitamins in some *Frankia* isolates. *Plant Soil* **69,** 149–161.

Stowers, M. D., Kulkarni, R. K., and Steele, D. B. (1986). Intermediary carbon metabolism in *Frankia*. *Arch. Microbiol.* **143,** 319–324.

Streeter, J. G. (1985). Accumulation of α,α-trehalose by *Rhizobium* bacterium and bacteroids. *J. Bacteriol.* **164,** 78–84.

Streicher, S. L., and Tyler, B. (1981). Regulation of glutamine synthetase by adenylylation in the gram-positive bacterium *Streptomyces cattleya*. *Proc. Natl. Acad. Sci. U.S.A.* **78,** 229–233.

Thevelein, J. M. (1984). Regulation of trehalose mobilization in fungi. *Microbiol. Rev.* **48,** 42–59.

Tisa, L. S., and Ensign, J. C. (1987a). The calcium requirements for functional vesicle development and nitrogen fixation by *Frankia* strains EAN1pec and CpI1. *Arch. Microbiol.* **149,** 24–29.

Tisa, L. S., and Ensign, J. C. (1987b). Isolation and nitrogenase activity of vesicles from *Frankia* sp. strain EAN1pec. *J. Bacteriol.* **169,** 5054–5059.

Tisa, L., McBride, M., and Ensign, J. C. (1983). Studies of growth of *Frankia* isolates in relation to infectivity and nitrogen fixation (acetylene reduction). *Can. J. Bot.* **61,** 2768–2773.

Tjepkema, J. D., Ormerod, W., and Torrey, J. G. (1980a). Vesicle formation and acetylene reduction (nitrogenase activity) in *Frankia* sp. CpI1 in culture. *Can. J. Microbiol.* **27,** 815–823.

Tjepkema, J. D., Ormerod, W., and Torrey, J. G. (1980b). Vesicle formation and acetylene reduction activity in *Frankia* sp. CpI1 cultured in defined nutrient media. *Nature (London)* **287,** 633–635.

Tsai, Y.-L.,and Benson, D. R. (1989). Physiological characteristics of glutamine synthetases I and II of *Frankia* sp. strain CpI1. *Arch. Microbiol.* **152,** 382–386.

Tumer, N. E., Robinson, S. J., and Haselkorn, R. (1983). Different promoters for the *Anabaena* glutamine synthetase gene during growth using molecular or fixed nitrogen. *Nature (London)* **306,** 337–341.

Tunlid, A., Schultz, N. A., Benson, D. R., Steele, D. B., and White, D. C. (1989). Differences in fatty acid composition between vegetative cells and vesicles of *Frankia* sp. Strain CpI1. *Proc. Natl. Acad. Sci. U.S.A.* **86,** 3399–3403.

Wegener, W. S., Reeves, H. C., Rabin, R., and Ajl, S. J. (1968). Alternate pathways of metabolism of short-chain fatty acids. *Bacteriol. Rev.* **32,** 1–26.

Winkenbach, F., Wolk, C. P., and Jost, M. (1972). Lipids of membranes and of the cell envelope in heterocysts of a blue green alga. *Planta* **107,** 69–80.

Xia, T., and Jiao, R. (1986). Studies on glutamine synthetase from *Streptomyces hygroscopicus* var. *jinggangensis*. *Sci. Sin. (Engl. Ed.)* **29B,** 379.

7

The Physiology of Actinorhizal Nodules

Kerstin Huss-Danell
Department of Plant Physiology
University of Umeå
S-901 87 Umeå, Sweden

I. Introduction

An actinorhizal nodule is a highly specialized symbiotic organ made up of both *Frankia* and higher plant cells. Root nodules result from successful infection of a plant root by *Frankia* (Chapter 5). Measured as dry mass, nodules usually comprise some 1–10% of the total biomass of the host plant or tree. Because they live in a nodule, *Frankia* must obtain all nutrients from its host plant. At least in *Alnus*, nodules are not water-absorbing organs (Bond, 1956), and the plant forms the only connection between *Frankia* and soil. There is thus a great need for efficient exchange of metabolites between the host cells and the microsymbiont. In the following sections, some aspects of dormancy and of carbon, nitrogen, and hydrogen metabolism of actinorhizal nodules will be discussed, as well as interactions between host and microsymbiont in nodules. Other recent overviews include those by Dixon and Wheeler (1983, 1986) and by Tjepkema *et al.* (1986).

II. Nodule Dormancy

Actinorhizal nodules are typically perennial structures and, in species from temperate regions, the whole plant, including the nodules, becomes dormant in winter. While in dormancy, the nodules do not grow,

The Biology of *Frankia* and Actinorhizal Plants

nitrogen fixation is not detectable, and symbiotic vesicles are virtually absent although a few hyphae are present and spores may become frequent (Schwintzer *et al.*, 1982; Wheeler *et al.*, 1983). *Frankia* apparently remains viable during winter because isolates have been obtained from nodules collected from frozen soil (Benson, 1982). In spring, nodule growth resumes, and new host cells, produced from the apical meristem, are infected by hyphae. Some hyphae differentiate into symbiotic vesicles. The reserve carbohydrates of entire nodules of *Alnus glutinosa* increase in amount prior to dormancy. Their subsequent decrease during dormancy may support the metabolism of the overwintering nodules (Wheeler and Bowes, 1974; Wheeler *et al.*, 1983). Since *Frankia* remains viable, it can be expected that *Frankia* synthesizes storage compounds for overwintering. Trehalose and glycogen (Benson and Eveleigh, 1979a; Lopez *et al.*, 1984) may possibly serve this role.

The regulation of dormancy of actinorhizal nodules is not fully understood. In *A. glutinosa*, the role of cytokinins and gibberellin-like substances was considered to be unclear although a large increase in cytokinin activity in the spring coincided with renewed cell division in the nodules (Wheeler *et al.*, 1979; Wheeler, 1984). Abscisic acid was considered unlikely as a controlling substance for dormancy (Watts *et al.*, 1987).

The following sections will consider nondormant nodules.

III. Carbon Metabolism

A good supply of carbon is essential for nodules and symbiotic *Frankia*. The carbon compounds are used for maintenance and growth processes, for delivery of energy and reductant in the nitrogen fixation process, and for supply of acceptor molecules in the assimilation of the ammonia produced by nitrogen fixation.

A. Photosynthesis and Supply of Assimilates to the Root Nodules

Root nodules obtain carbon compounds produced in photosynthesis and transported as sucrose in the phloem from the leaves and other green parts of the plant. In the nodule, carbon compounds are transported from the phloem, which forms part of the central vascular strand in the nodule, to noninfected and infected cells. When shoots of *A. glutinosa* seedlings photosynthesized $^{14}CO_2$, radioactivity was mainly localized in the phloem and xylem and in noninfected cells in the nodules 5 hr later. The very small accumulation of radioactivity in infected cells suggests that carbon compounds arriving in these cells are rapidly used (Wheeler and Lawrie, 1976). It is presently not known which specific carbon compounds are transported from the host to symbiotic *Frankia*. Within an infected cell the compounds must traverse a number of barriers, including the plant cell membrane, the capsule, the lipid laminae, the *Frankia* cell wall, and the *Frankia* cell membrane (Fig. 1).

B. Carbon Metabolism of *Frankia* in Root Nodules

Symbiotic *Frankia* needs carbon compounds for growth and for energy to fuel various processes, including nitrogen fixation. Carbon metabolism of symbiotic *Frankia* has been studied for alders (*Alnus glutinosa, A.*

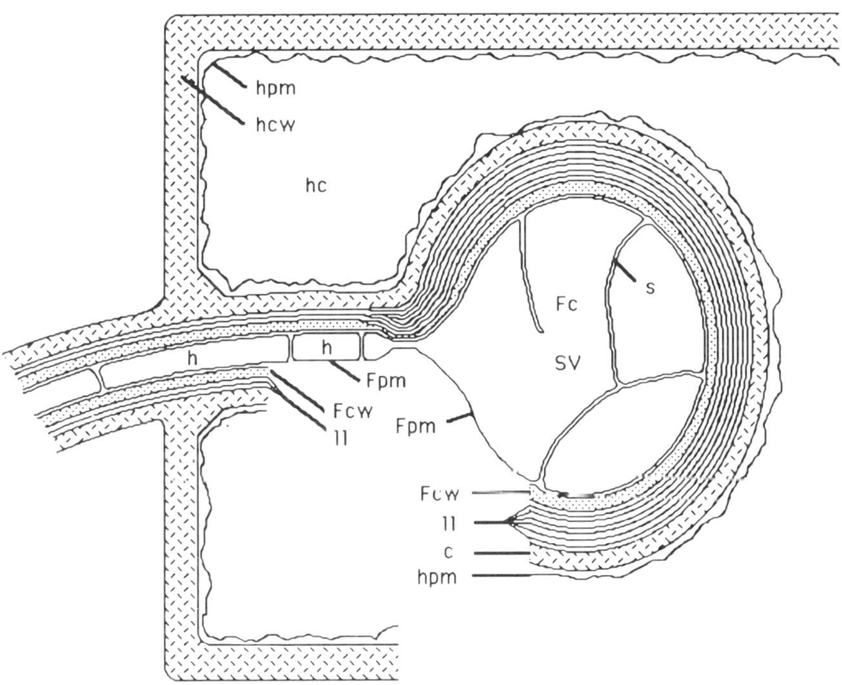

Figure 1 Schematic drawing of an infected *Alnus* nodule cell with a *Frankia* hypha (h) and a symbiotic vesicle (SV). hcw, host cell wall; hpm, host plasma membrane; hc, host cytoplasm; c, capsule (host origin); ll, lipid laminae; Fcw, *Frankia* cell wall; Fpm, *Frankia* plasma membrane; s, septum; Fc, *Frankia* cytoplasm. To simplify the figure, the relative size of the various structures is not drawn to scale. Based on data of Newcomb and Wood (1987) and Abeysekera *et al.* (1990).

incana ssp. *incana*, and *A. rubra*) following inoculation with a few *Frankia* strains or crushed nodule preparations. Also, some of the results from *A. glutinosa* have been tested and verified for *Hippophaë rhamnoides* and *Datisca cannabina* (Huss-Danell *et al.*, 1982a; Akkermans *et al.*, 1983).

1. Methods

Information on carbon metabolism in symbiotic *Frankia* has been obtained from studies of so-called "vesicle clusters." Vesicle clusters are prepared from homogenized root nodules by a filtration procedure (Vikman and Huss-Danell, 1987a) and contain symbiotic vesicles and part of their subtending hyphae (Vikman and Huss-Danell, 1987b).

For studies of metabolism, vesicle clusters should preferably be prepared in a physiologically active state so that their membrane, uptake systems and cellular organization are intact. When nodules are homogenized, either by a wet homogenization or by crushing in liquid nitrogen, *Frankia* is exposed to a variety of plant compounds, including tannins and proteases that are destructive to proteins and enzyme activities, and protecting agents are therefore needed in the preparations. In spite of these difficulties, vesicle clusters have been used for studies of carbon metabolism by measurements of respiration and enzyme activities, and also in studies of nitrogen (Section IV) and hydrogen (Section V) metabolism.

High purity in vesicle cluster preparations is necessary since contaminating mitochondria and other plant material may contribute to recorded enzyme activities. In vesicle clusters from *D. cannabina* nodules, contaminating mitochondria were seen but appeared to be largely destroyed (Akkermans *et al.*, 1983). Recent work on *A. incana* ssp. *incana* showed that vesicle clusters can be obtained with high purity such that *Frankia* constitutes at least 98% of the particle volume in the preparations (Vikman and Huss-Danell, 1987b). In such vesicle clusters the volume ratio for vesicles to hyphae was estimated to be 3 : 1. The enzyme 6-phosphogluconate dehydrogenase was shown to be NAD dependent in vesicle clusters but NADP dependent in roots of *A. incana* ssp. *incana* (Vikman and Huss-Danell, 1987a). This difference in cofactor dependency was used in respiration studies to show that contamination from plant cytosol was nondetectable in vesicle clusters.

2. Aerobic Metabolism

Symbiotic *Frankia* from *Alnus* spp. appears to have an aerobic metabolism, as concluded from presence of cytochromes *a*, *b*, *c*, and *o* (Ching *et al.*, 1983) and cyanide-sensitive oxygen consumption in vesicle clusters (Vikman and Huss-Danell, 1987a). Cyanide-inhibited O_2-consumption in vesicle clusters was not affected by salicylhydroxamic acid (Vikman and Huss-Danell, 1987a), an inhibitor of the alternative respiration pathway in plant mitochondria.

3. Sugars as Carbon Source

Since photosynthates enter the nodules as sucrose and extracts of *A. glutinosa* nodules contain sucrose, glucose, and fructose (Wheeler, 1971), it is logical to investigate the possible role of sugars as carbon sources for symbiotic *Frankia*. Respiration of vesicle clusters was shown to be stimulated by sucrose, trehalose, maltose, glucose, and fructose (Lopez *et al.*, 1986). The phosphorylated sugars glucose 6-phosphate and 6-phosphogluconate also stimulated respiration in the presence of NAD (Vikman and Huss-Danell, 1987a). Several of the necessary enzymes (hexokinase, glucose-6-phosphate dehydrogenase, 6-phosphogluconate dehydrogenase, and pyruvate kinase) were shown to be active in vesicle clusters (Lopez and Torrey, 1985; Vikman and Huss-Danell, 1987a).

The ability to respire hexoses does not prove that hexoses are the carbon compounds which *Frankia* receives from the host plant. The role of hexose metabolism might also be to degrade carbohydrates synthesized and stored inside *Frankia*. Examples of such compounds are glycogen and trehalose (Lopez *et al.*, 1984).

It is possible that symbiotic *Frankia* obtains sugars via degradation of pectic substances or polysaccharides (Chapter 4), which are constituents of the capsule (Fig. 1) surrounding *Frankia* in *Alnus* (Lalonde and Knowles, 1975; Abeysekera *et al.*, 1990).

4. Carboxylic Acids as Carbon Source

Of the enzymes involved in the tricarboxylic acid cycle, the enzymes isocitrate dehydrogenase, succinate dehydrogenase, fumarase, and malate dehydrogenase are active in vesicle clusters (Akkermans *et al.*, 1981). Respiration in vesicle clusters is stimulated by succinate (Akkermans *et al.*, 1981), a combination of malate, glutamate, and NAD (Huss-Danell *et al.*, 1982a), and by NADH (Akkermans and Roelofsen, 1980; Vikman and Huss-Danell, 1987a). The need for a combination of malate, glutamate, and NAD, rather than malate or glutamate alone, to sustain a continued stimulation of respiration suggested the operation of a malate–aspartate shuttle between host and microsymbiont (Akkermans *et al.*, 1981). The presence of the necessary enzymes malate dehydrogenase and glutamate-oxaloacetate aminotransferase in both *Frankia* cells and host plant cells has been confirmed (Akkermans *et al.*, 1981). However, the presence of the necessary carriers in such a shuttle remains to be shown.

5. Lipids as Carbon Source

In early studies on isolation and growth of pure-cultured *Frankia*, lipids are often cited as necessary carbon compounds (Chapter 3). When *Frankia* AvcI1 is grown in a medium containing the lipid compound Tween 80, the glyoxylate cycle enzymes isocitrate lyase and malate synthase are active (Blom and Harkink, 1981). However, when the same *Frankia* strain was inoculated onto *A. glutinosa* and vesicle clusters were prepared from these nodules, no activity of the two enzymes could be detected (Huss-Danell *et al.*, 1982a). This suggests that lipids are not a likely carbon source for *Frankia* AvcI1 in symbiosis.

6. Further Studies

The results on lipids point to differences in physiology between pure-cultured and symbiotic stages of *Frankia*. A further difference between symbiotic and pure-cultured *Frankia* was reported for strain ArI3 (Lopez *et al.*, 1986). Although both forms respired sugars, only the pure-cultured form respired propionate, succinate, pyruvate, and malate. It should not be presumed, therefore, that data from pure-cultured *Frankia* are directly applicable to the symbiotic stage. The need for such caution has also been shown by studies on other symbiotic systems, for example, pure-cultured and symbiotic nitrogen-fixing cyanobacteria (Stewart *et al.*, 1980).

The failure of a number of substrates to stimulate respiration in vesicle clusters (Vikman and Huss-Danell, 1987a) does not necessarily show that these substrates cannot be metabolized but may well be due to impaired uptake systems for these compounds. Further studies are clearly needed to reveal which compounds *Frankia* obtains from its host plant, and these studies should include a wider range of hosts and microsymbiont strains. It would also be of interest to determine whether the various enzyme activities are localized in symbiotic vesicles or in hyphae, since vesicle clusters comprise a mixture of these *Frankia* cell types. Such information is important for studies on possible links between carbon metabolism and the energy-demanding nitrogen fixation process in symbiotic *Frankia*.

C. CO_2-Fixation in Root Nodules

Carbon dioxide fixation is common in plant roots. It is therefore not surprising that CO_2-fixation occurs in root nodules from all actinorhizal plants studied to date, that is, *A. glutinosa* (Schubert *et al.*, 1981; McClure *et al.*, 1983) and *A. hirsuta, A. incana* ssp. *incana, A. rubra, A. sinuata, Casuarina montana, Colletia cruciata,* and *Datisca cannabina* (unpublished observations). The net incorporation of CO_2 into nodules of *A. glutinosa* is catalyzed by phosphoenolpyruvate (PEP) carboxylase and carbamyl phosphate synthetase (CPS). The PEP carboxylase reaction produces oxaloacetate from PEP and HCO_3^-. Oxaloacetate is then easily converted to malate and to amino acids like aspartate and glutamate. The CPS reaction produces carbamyl phosphate from an amino donor, CO_2, and ATP. Carbamyl phosphate is then used for the synthesis of citrulline (McClure *et al.*, 1983). In *A. glutinosa* nodules, PEP carboxylase was localized in the cytosol of infected cells by an immunocytochemical reaction with antibodies against the enzyme from *Sorghum* leaves (Perrot-Rechenmann *et al.*, 1981).

Pulse-labeling experiments on CO_2-fixation suggest the presence of at least two pools of malate in *A. glutinosa* nodules. A major pool is thought to be inactive and of unknown function, while a smaller pool is thought to be rapidly used in amino acid synthesis (McClure *et al.*, 1983). It remains to be determined to what extent CO_2-fixation in actinorhizal nodules is linked to the ability of *Frankia* vesicle clusters to respire organic acids.

IV. Nitrogen Metabolism

A. Nitrogen Fixation

Atmospheric nitrogen is reduced to ammonium by *Frankia*, and the fixed nitrogen is used by the host plant. Actinorhizal plants can, therefore, grow in soils devoid of ammonium and nitrate. Various aspects of nitrogen fixation research were recently summarized (Bothe *et al.*, 1988).

1. Nitrogenase

Knowledge about the biological nitrogen fixation process is based on studies from a variety of bacteria, mainly nonsymbiotic forms such as *Azotobacter, Clostridium,* and *Klebsiella*. The reduction of N_2 is catalyzed by the nitrogenase enzyme complex and requires large amounts of energy

(ATP) and reductant (Fig. 2). The nitrogenase enzyme complex consists of two proteins. The so-called dinitrogenase reductase is commonly referred to as the Fe-protein because of its content of nonheme iron. It is a dimeric protein with 63 kDa molecular mass. The other protein, the so-called dinitrogenase, is referred to as the MoFe-protein because of its content of molybdenum and iron. It consists of two subunits of 55 kDa and two subunits of 60 kDa molecular mass (Burris *et al.*, 1980).

The Fe-protein accepts an electron from an electron donor and complexes with two Mg-ATP molecules. The reduced Fe-protein molecule then associates with a MoFe-protein molecule to form an active enzyme complex. An electron is passed on to the MoFe-protein and Mg-ADP and phosphate is released. The two proteins then dissociate, and the Fe-protein is ready to recommence the electron transport to a MoFe-protein molecule. When a MoFe-protein molecule has become sufficiently reduced the substrate N_2 will be reduced to NH_3. The process of reducing a mole of N_2 is comparatively slow because of the numerous associations

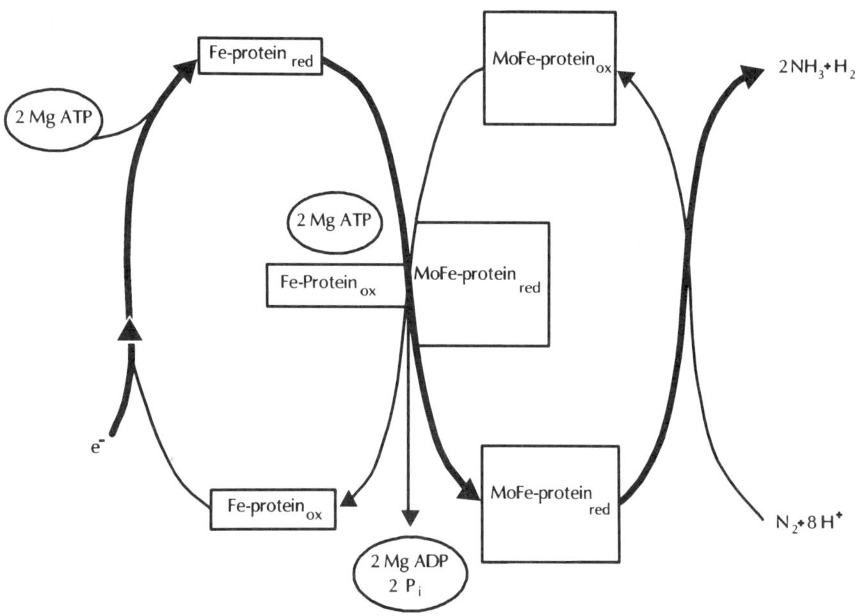

$$N_2 + 8e^- + 16\ MgATP + 16\ H^+ \rightarrow 2NH_3 + H_2 + 16\ Mg\ ADP + 16\ P_i$$

Figure 2 Simplified scheme of nitrogenase reaction. MoFe-protein contains two independent sites for Fe-protein, but for simplicity only one is indicted here. As shown in the equation, reduction of one mole N_2 requires MoFe-protein to obtain $8e^-$ from Fe-protein, that is, the electron transfer cycle must occur eight times.

and dissociations. To maintain a reasonable nitrogen fixation rate, the slow process is compensated for by having a large amount of nitrogenase, up to 5–30% of total cell protein (Haaker and Klugkist, 1987).

The reduction of protons to H_2 always occurs concomitantly with the reduction of N_2, with a minimum of one H_2 formed per N_2 reduced (Simpson and Burris, 1984). The relative flow of electrons through nitrogenase that results in N_2 reduction is described by a value called relative efficiency (RE) (Schubert and Evans, 1976), defined as RE = 1 − (rate of H_2 evolution in air/rate of H_2 evolution in 80% Ar + 20% O_2 + CO_2). In this calculation the rate of C_2H_2 reduction can be substituted for the rate of H_2 evolution in Ar+O_2+CO_2. The determination of RE is frequently complicated by the presence of an uptake hydrogenase that oxidizes all or part of the H_2 formed by nitrogenase (Section V). N_2 reduction, but not proton reduction, is inhibited by CO and by H_2. Nitrogenase can also reduce a number of artificial substrates of which acetylene (ethyne) is the best known. It is widely used for measurements of nitrogenase activity by the acetylene reduction assay (Chapter 13).

2. Nitrogenase in Symbiotic Frankia

From the limited information available so far (Benson *et al.*, 1979), there are no differences in the nitrogenase reaction in symbiotic *Frankia* as compared with other nitrogen-fixing organisms (see Chapters 6 and 10). The sources of ATP and reductant for nitrogenase activity are not yet identified in actinorhizal nodules *in vivo*. However, it has been possible to measure nitrogenase activity in root nodule homogenates prepared from *Alnus* ssp., *H. rhamnoides, Shepherdia canadensis,* and *Myrica* ssp. under anaerobic conditions when supplied with exogenous Mg-ATP and with $Na_2S_2O_4$ as reductant (Akkermans *et al.*, 1977; van Straten *et al.*, 1977; Benson and Eveleigh, 1979b; Ching *et al.*, 1983; Huss Danell and Ahlqvist, 1984). Substantial endogenous nitrogenase activity has not yet been obtained in root nodule homogenates and the nitrogenase reaction has not yet been coupled to natural reductants, such as NADH or reduced ferredoxin.

3. Localization of Nitrogenase within Nodules

In pure-cultured *Frankia*, immunogold labeling of electron microscope preparations localized nitrogenase in *Frankia* vesicles but not in hyphae (Meesters, 1987). In actinorhizal nodules, there are several lines of indirect evidence that nitrogenase is localized in the symbiotic vesicles. (i) *Alnus glutinosa* nodules showed strong tetrazolium reduction activity in the symbiotic vesicles (Akkermans, 1971). (ii) The frequency of symbiotic vesicles coincided approximately with nitrogenase activity in apical, middle, and basal parts of *A. glutinosa* nodules (Becking, 1977). (iii) In *M. faya* (Mian *et al.*, 1976), Elaeagnaceae (Baker *et al.*, 1980), *Comptonia peregrina* (VandenBosch and Torrey, 1983), and *A. glutinosa* (Hahn *et al.*, 1988; Kurdali *et al.*, 1989), ineffective (no nitrogenase activity) root nodules lacked symbiotic vesicles while effective root nodules were rich in symbiotic vesicles. (iv) In young, newly infected plants of *A. glutinosa*, the onset of nitrogenase activity was associated with the formation of symbiotic vesicles and was not recorded in the younger phases of nodule development when *Frankia* was seen only as hyphae (Mian and Bond, 1978). (v) In seasonal studies of *M. gale* (Schwintzer *et al.*, 1982) and *A. glutinosa* (Wheeler *et al.*, 1983), nitrogenase activity was first seen when symbiotic vesicles had developed, and ceased when vesicles disappeared in autumn. However, lack of correlation between presence of symbiotic vesicles and nitrogenase activity has been reported for *Gymnostoma papuanum* (Racette and Torrey, 1988).

Recently, more direct evidence for the localization of nitrogenase in symbiotic vesicles was provided for *Elaeagnus pungens*, but the immunogold labeling study could not exclude presence of nitrogenase in hyphae of *Frankia* in the nodules (Sasakawa *et al.*, 1988). The precise localization of nitrogenase within actinorhizal nodules thus merits further study, and especially in *Casuarina* and *Allocasuarina* nodules, where typical symbiotic vesicles are not seen (Newcomb and Wood, 1987). The role of symbiotic vesicles in the exclusion of oxygen from nitrogenase is discussed in Chapter 8.

4. Energy Requirements and Relation to Photosynthesis

It is generally agreed that nitrogenase activity is dependent on newly formed photosynthates transported from the green shoot to the root nodules (Wheeler, 1971). When young *A. incana* ssp. *incana* plants are grown on N_2 as their only source of nitrogen there is a parallel increase in nitrogenase activity and in shoot net CO_2-assimilation (Fig. 3). Similar relationships were also reported for greenhouse-grown *A. glutinosa* (Gordon and Wheeler, 1978; Dawson and Gordon, 1979). Various treatments that are likely to reduce carbon supply to the nodules, like removal of shoot or leaves, darkening, and stem-ringing, all result in decreased nitrogenase activity (Wheeler, 1971; Wheeler and Bowes, 1974; Huss-Danell and Sellstedt, 1985; Sundström and Huss-Danell, 1987). Attempts have been made to correlate nitrogenase activity with sugar and starch content of nodules (Wheeler and Lawrie, 1976; Wheeler *et al.*, 1983). However, nitrogenase activity does not necessarily need to

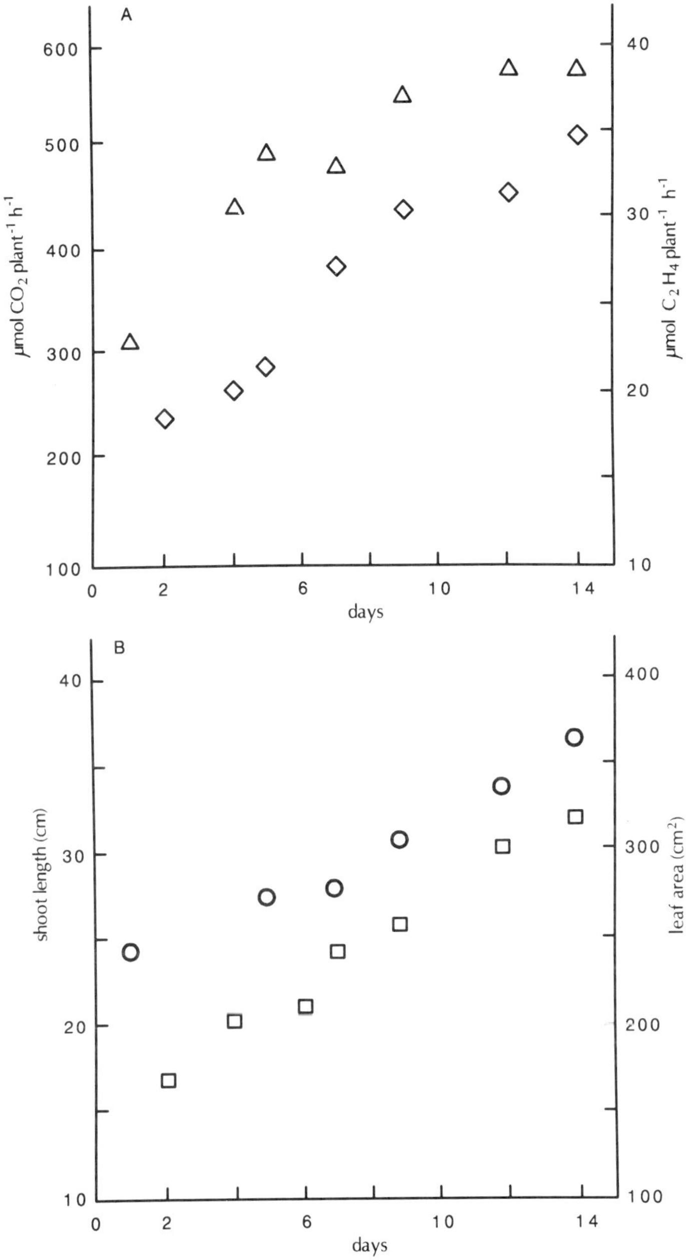

Figure 3 Photosynthesis and nitrogenase activity (A), and shoot length and leaf area (B) of a young *Alnus incana* ssp. *incana* plant under growth chamber conditions. ◇, nitrogenase activity measured as acetylene reduction in the intact potted plant; △, net CO_2 assimilation in shoot measured by infrared gas analysis; ○, shoot length; □, leaf area. Unpublished data of A. Sellstedt and K. Huss-Danell.

be correlated with changes in the pool size of compounds in the nodules. The flux through the pool rather than the pool size may determine nitrogenase activity (Vessey *et al.*, 1988). There may also be more than one pool of a compound, with a small pool that is rapidly turned over and another, large pool that is mainly used as storage.

The precise energy requirement for nitrogenase activity in a root nodule is difficult to estimate, as nitrogenase activity is only one of several energy-consuming host and microsymbiont activities. Nevertheless, ratios between net CO_2 evolution (respiration) and nitrogenase activity may be taken as a measure of energy cost. Such ratios for a number of actinorhizal nodules are approximately the same (3–9 mol CO_2/mol C_2H_4) as for a number of legume–*Rhizobium* symbioses (Tjepkema and Winship, 1980). For *A. rubra* nodules the energy cost is stable in the temperature range 16 to 26°C but increases markedly below 16°C as nitrogenase activity decreases more than respiration. At a stable temperature (20°C) energy cost is hardly affected by O_2 tension between 5 and 20 kPa O_2 (Winship and Tjepkema, 1985). These data apply to short-term changes and it is possible that during longer times adaptation to temperature and oxygen conditions may occur. No seasonal variation in energy cost was found in field-grown nodules of *M. gale*, except for the very beginning and the very end of the season, when nitrogenase activity is very low (Schwintzer and Tjepkema, 1983).

5. Nitrogen Fixation Rates

It is almost impossible to characterize actinorhizal nodules as having a certain nitrogen fixation activity. Nitrogen fixation rates vary considerably depending on plant species–*Frankia* combination, plant and nodule age, growth conditions in the field or in the laboratory, method of measurement, time of the day and season of the year for short-term measurements, and how the rates are expressed. Nevertheless, the range of nitrogen fixation rates reported for *Frankia* symbioses coincide with those for *Rhizobium*–legume symbioses (Torrey, 1978). Expression of nitrogenase activity on a nodule mass basis may introduce unwanted variation into the measures since the nodules can be several years old and thus have varying amounts of non-nitrogen-fixing tissue. To avoid this phenomenon, nitrogenase activity has been expressed per nodule lobe, per plant, or in relation to plant size measures such as leaf area or shoot length.

6. Effects of Environmental Factors

a. Long-Term Effects. Because of the intimate relationship between a root nodule and the remaining parts of a plant, any factor that influences the physiology of the plant is also likely to affect the nodules. Li temperature, water, nutrients, soil pH, salinity, and the like all affect the growth of plants in various ways. Effects of such abiotic factors are reviewed at length by Dixon and Wheeler (1983).

Among biotic factors, presence of mycorrhizae was shown to improve phosphate uptake by *A viridis* ssp. *viridis* (Mejstrik and Benecke, 1969). Nitrogen-fixing plants are believed to have a relatively high demand for phosphorous, and mycorrhizae may, therefore, be of great importance in the field.

Of the essential mineral nutrients, molybdenum, cobalt, and nitrogen are of special interest for nitrogen fixation. Molybdenum is part of nitrogenase (Section IV,A,1), and accordingly root nodules of *A. glutinosa* plants were found to have a molybdenum concentration in the nodules about six times as high as that in the roots (Becking, 1970). Cobalt is the only mineral element known to be essential only for nodulated nitrogen-fixing plants, where it is needed for vitamin B_{12} in the microsymbiont (Dixon and Wheeler, 1983).

Combined nitrogen not only inhibits nodule formation (Chapter 4) but also inhibits the nitrogenase activity of nodules already formed. In field as well as in most laboratory experiments, these two effects of combined nitrogen occur concurrently (Stewart and Bond, 1961; Rodriguez-Barrueco *et al.*, 1970; Bond and Mackintosh, 1975; Ingestad, 1980). However, a continuous supply of combined nitrogen in small doses, which were continuously increased to match the consumption of the plants, stimulated both growth and nitrogen fixation in *A. incana* ssp. *incana* seedlings (Ingestad, 1980). Short-term studies of nitrogen effects, specifically on nitrogenase activity, are described in the next section.

b. Short-Term Effects. When young *A. incana* ssp. *incana* plants are grown under stable environmental conditions, nitrogenase activity is correlated with plant size, leaf area, and photosynthesis (Fig. 3). However, the close correlations observed are disturbed when the environment of the plants is changed. Nitrogenase activity decreased in response to several factors, for example, addition of combined nitrogen (Huss-Danell *et al.*, 1982b; Huss-Danell and Hahlin, 1988), drought stress (Sundström and Huss-Danell, 1987), darkening and leaf removal (Huss-Danell and Sellstedt, 1985), and sudden changes in temperature or oxygen (Huss-Danell *et al.*, 1987; Rosendahl and Huss-Danell, 1988).

The mechanisms for decreased activity are not fully understood. One possibility is that *Frankia* receives a decreased supply of carbon compounds, and this could result in reduced availability of energy for nitrogenase, and thereby reduced nitrogenase activity. Another possibility is that nitrogenase activity is limited by a reduced amount of active nitrogenase. To distinguish between these possibilities, measurements were made of nitrogenase activity *in vivo* (intact nodulated plants) and *in vitro* (root nodule homogenates) in *A. incana* ssp. *incana* plants subjected to drought stress or prolonged darkening or additions of ammonium chloride. The *in vitro* incubation of the nodule homogenates included reductant and ATP (Section IV,A,2). This assay should distinguish between a possible decreased energy supply or lack of active nitrogenase as a cause of lower activity. Compared to control plants, nitrogenase activity in the treated plants decreased in both *in vivo* and *in vitro* assays (Huss-Danell and Sellstedt, 1985; Sundström and Huss-Danell, 1987; Huss-Danell and Hahlin, 1988). The results obtained thus indicate reduced amounts of active nitrogenase as a major cause of lower activity in stressed plants. It remains to be clarified whether the nitrogenase proteins are still present in their normal amounts but are partially inactivated, or whether the amount of nitrogenase proteins actually decreases.

In the case of ammonium additions to *A. incana* ssp. *incana*, a strongly reduced content of photosynthetic ^{14}C was found in vesicle clusters prepared from the nodules, but the reduced amount of ^{14}C was not significantly correlated with the decrease in nitrogenase activity (Huss-Danell and Hahlin, 1988). The experiments described accumulation rather than flux of ^{14}C. It is possible that addition of combined nitrogen to the root system alters the carbon translocation pattern, and that a reduced carbon supply to nodules and *Frankia* therein initiates several reactions resulting in inactivation and/or reduced synthesis of nitrogenase. After ammonium was supplied to *A. incana* ssp. *incana* plants for four days, degeneration of symbiotic vesicles was observed, thus indicating a strong effect on *Frankia* metabolism in the nodules (Huss-Danell *et al.*, 1982b).

c. Diurnal Variation. Diurnal variation in nitrogenase activity is reported from greenhouse or growth chamber studies on *Alnus* spp. and *M. gale* (Wheeler, 1969, 1971; Wheeler and Lawrie, 1976; Sellstedt and Huss-Danell, 1986) as well as from field studies of *Alnus* spp. (Johnsrud, 1978; Tripp *et al.*, 1979). A natural diurnal variation contains a light–dark cycle but also a lower temperature during darkness. To distinguish light–dark effects from temperature effects, both shoot and root temperature should be controlled separately from the light regime. In spite of a roughly constant temperature, presence of diurnal variation in *A. glutinosa* and *M. gale* (Wheeler, 1969) as well as absence of diurnal variation in *M. gale* (Schwintzer, 1979; Tjepkema, 1985) was reported. It is possible that other factors, like plant size and preceding growth conditions, can influence the supply of carbon from stored reserves and thereby the dependency of nitrogenase activity on newly formed assimilates.

d. Seasonal Variation. Seasonal variation in nitrogen fixation is even more complex. Actinorhizal plants from temperate regions are mainly deciduous trees or shrubs. In a number of studies, nitrogenase activity was first detected at the time of bud burst in spring, showed a maximum in the middle of the summer, and decreased to zero at leaf-fall. In *M. gale*, seasonal variation in nitrogenase activity was shown to be correlated with the presence of symbiotic vesicles in the nodules (Schwintzer *et al.*, 1982). During summer, when leaves are developed, there are variations in activity that can be due to variation in weather conditions. A correlation between nitrogenase activity and number of hours of sunshine was noted for young *A. incana* ssp. *incana* under field conditions (Huss-Danell *et al.*, 1989). In 20- to 30-year-old trees of *A. glutinosa*, summer nitrogenase activity was low when development of fruits was rapid but increased when maturation of seeds was complete (Pizelle, 1984). Such a variation in nitrogenase activity can be explained by competition between nodules and developing fruits for assimilates.

B. Assimilation and Transport of Fixed Nitrogen

Nitrogen fixation results in formation of ammonium. Because of its toxicity at high concentrations, ammonium is not a transport form of nitrogen in plants, rather ammonium is assimilated into amino acids and amides for transport.

1. Ammonia Assimilation

The most common route for assimilation of ammonia into amino acids in plants is via the combined action of glutamine synthetase (GS) and glutamate synthase (GOGAT; glutamine oxo-glutarate aminotransferase). GS catalyzes the addition of one mole ammonium to one mole of glutamate, which results in the formation of one mole of the amide glutamine. GOGAT catalyzes the formation of two moles glutamate from one mole of glutamine and one mole of 2-oxo-glutarate. The glutamate formed can again be used in the GS-catalyzed incorporation of ammonium into glutamine or be used in further transformations.

There is also another possible route for ammonia assimilation in plants. The enzyme glutamate dehydrogenase (GDH) catalyzes the reversible reaction in which ammonia is combined with 2-oxo-glutarate to form glutamate. Compared to GS, GDH has a low affinity for ammonium. Also, GDH has a lower affinity for ammonium than for glutamate. The GDH-catalyzed reaction therefore operates mostly in the deaminating direction but, at high concentrations of ammonium, in the direction of ammonium assimilation.

GS has clearly been shown to be present in pure cultures of *Frankia* (Chapter 6). In nodule homogenates of *A. glutinosa,* GS and GDH activity was associated with plant fractions but not with vesicle clusters (Blom *et al.*, 1981). Failure to detect GOGAT activity in *A. glutinosa* nodules has been ascribed to possible inactivation of GOGAT by some compound released during nodule homogenization (Blom, 1982).

Presence of GS in *A glutinosa* nodules was confirmed by both activity measurements and immunological studies (Hirel *et al.*, 1982). In addition to a major isoform of GS, with the same chromatographic behavior as root GS, a minor GS isoform was found and represented $\leq 3\%$ of total nodule GS activity. GS was localized in the cytosol of the larger inner cortical cells (Hirel *et al.*, 1982). However, the antibodies used were raised against spinach chloroplast GS and it remains to be shown whether symbiotic *Frankia* also contains GS, but with antigenic sites differing from those of plant GS. It is also not clear whether ammonium from nitrogen fixation is assimilated in the GS–GOGAT or GDH pathway or both. If symbiotic *Frankia* really lacks the ability to assimilate ammonia formed in its own nitrogen fixation process, *Frankia* becomes dependent on its host plant for both carbon and nitrogen compounds.

When detached nodules of *A. glutinosa* were supplied with exogenous $^{13}NH_4^+$, the distribution of radioactivity in various products indicated that GDH rather than GS was involved in the assimilation of exogenous ammonium (Schubert *et al.*, 1981).

2. Transport of Fixed Nitrogen

After ammonia assimilation, fixed nitrogen is exported from the nodules as amino acids or amides. Analyses of nodule extracts or xylem fluid show that the dominating nitrogenous compound in actinorhizal nodules varies among host genera. Citrulline dominated in most alder species (*A. glutinosa, A. viridis* ssp. *crispa, A. incana* ssp. *incana, A. nitida, A. rubra*) studied (Miettinen and Virtanen, 1952; Leaf *et al.*, 1958; Gardner and Leaf, 1960; Wheeler and Bond, 1970; Blom *et al.*, 1981; Schubert *et al.*, 1981; Hafeez *et al.*, 1984; Schubert, 1986), while glutamine, asparagine, and glutamate were all slightly more common than citrulline in nodules of *A. inokumai* (Wheeler and Bond, 1970). Asparagine dominated in xylem sap of *Hippophaë rhamnoides* (Wheeler and Bond, 1970), *Myrica* spp., *Elaeagnus* spp., and *Ceanothus* spp. (Wheeler and Bond, 1970; Schubert, 1986), and *Comptonia* spp. (Schubert, 1986). In *Coriaria* spp. and *Datisca* spp., glutamic acid, glutamine, arginine, and asparagine were reported to be the major amino compounds (Wheeler and Bond, 1970; Hafeez *et al.*, 1984; Schubert, 1986). In nodules of *Casuarina cunninghamiana*, asparagine and glutamine dominated and no citrulline was detected (Wheeler and Bond, 1970), while in *Casuarina equisetifolia*, citrulline was found to be the major nitrogenous compound in xylem sap, in both nodulated and nonnodulated plants (Walsh *et al.*, 1984).

The presence of considerable amounts of ornithine carbamyl transferase (OCT) in *A. glutinosa* nodules (Martin *et al.*, 1983) is in good accordance with the high concentrations of citrulline in alder nodules. OCT catalyzes the synthesis of citrulline from ornithine and carbamyl phosphate. Two isoforms of OCT were found, but their precise location in *A. glutinosa* nodules is not known (Martin *et al.*, 1982). Based on a histochemical technique, OCT was ascribed to host cell mitochondria (Scott *et al.*, 1981) and OCT activity measurements in nodule extracts also ascribed OCT to plant fractions and not to vesicle clusters of *A. glutinosa* nodules (Blom *et al.*, 1981). It is likely, therefore, that formation of citrulline is a plant characteristic. Even if *Frankia* can form citrulline, the presence of *Frankia* is not necessary for citrulline formation in nodules, since citrulline is also found in nonnodulated plants of *Alnus* (Miettinen and Virtanen, 1952) and *Casuarina* (Walsh *et al.*, 1984).

C. Nitrogen Fixation Compared to Ammonium and Nitrate Uptake

Of the three nitrogen sources available for plants, ammonium is regarded as the least expensive in terms of energy cost. Nitrate has to be reduced to ammonium in an energy-demanding reaction catalyzed by nitrate reductase. Nitrogen fixation is energy demanding because of the need for ATP and reductant in the nitrogenase reaction and because of the cost for nodule growth and maintenance. A high energy cost for nitrogen uptake and utilization can, in theory, result in reduced growth and biomass production of the plant.

When comparing different nitrogen sources, each type of nitrogen should preferably be given to plants at the same rate rather than comparing increasing nitrogen fixation with a more or less constant concentration of combined nitrogen. Also, frequent pH adjustments of nutrient solutions are important (Troelstra *et al.*, 1985). Comparisons of nitrogen fixation with increasing additions of ammonium (Sellstedt and Huss-Danell, 1986) and nitrate (Sellstedt, 1986) were made for *A. incana* ssp. *incana*. Daily nitrogen fixation rates were calculated from repeated measurements of acetylene reduction and hydrogen evolution in one group of intact plants. Other groups of nonnodulated plants of the same clone were given either ammonium or nitrate at the same daily rate as the nodulated plants fixed their nitrogen. Compared to ammonium-fed plants, the nitrogen-fixing plants developed more biomass and they also contained more nitrogen. This was ascribed to the fact that all ammonium given was not always taken up and that ammonium-grown plants excreted some 10% of the nitrogen taken up (Sellstedt and Huss-Danell, 1986). In the comparison of nitrogen-fixing plants with nitrate-grown plants, there was no significant difference in biomass production. Nitrate-grown plants took up virtually all nitrate added and excreted almost no nitrogen (Sellstedt, 1986). Nitrogen-fixing alders did not excrete detectable amounts of nitrogen. Apparently, the energy cost for nitrogen fixation is not high enough that substantial reductions in biomass production result.

V. Hydrogen Metabolism

In actinorhizal nodules, as in other nitrogen-fixing systems, H_2 is formed by nitrogenase concurrent with the reduction of N_2 to ammonia (Fig. 2; Benson *et al.*, 1979; Roelofsen and Akkermans, 1979). However, usually little or no H_2 is detected when actinorhizal nodules are placed in a cuvette with air (Schubert and Evans, 1976; Mirza *et al.*, 1987; Winship *et al.*, 1987; Sellstedt, 1989). This is due to the action of an uptake hydrogenase that oxidizes H_2 (Roelofsen and Akkermans, 1979; Benson *et al.*, 1980).

A. Hydrogenase

In the uptake hydrogenase reaction, electrons from H_2 are passed down an electron transport chain, resulting in ATP production and the final consumption of O_2 (Dixon, 1968). In vesicle clusters of actinorhizal nodules, hydrogenase activity can be driven by a range of artificial electron acceptors of which phenazine methosulfate was the most efficient among those tested (Benson *et al.*, 1980). Various aspects of hydrogenase in rhizobia were recently reviewed (Evans *et al.*, 1987).

1. Beneficial Effects of Hydrogenase

There are three potential advantages of having an uptake hydrogenase in a nitrogen-fixing organism (Dixon, 1972). First, O_2 is consumed. This is important as O_2 inactivates nitrogenase and can repress nitrogenase synthesis. In pure-cultured *Frankia*, H_2 stimulated nitrogenase activity at supraoptimal oxygen concentrations, which suggests that hydrogenase activity provides O_2 protection of nitrogenase (Murry and Lopez, 1989). However, the role of hydrogenase in oxygen exclusion from nitrogenase in nodules is somewhat uncertain. Only 6 to 7% of total O_2 consumption was calculated to be due to hydrogenase in *Rhizobium* bacteroids (Dixon, 1972). Hydrogenase activity in actinorhizal nodule homogenates, measured *in vitro* with artificial electron acceptors, was inhibited by 2 μM O_2 (Benson *et al.*, 1980).

Second, H_2 is consumed and this can help to prevent buildup of high H_2 concentrations near nitrogenase. It is important to keep the concentration of H_2 low, as H_2 is an inhibitor of N_2 reduction. Whether diffusion of H_2 out of the nodule is rapid enough to keep H_2 concentration in root nodules sufficiently low is still an open question (Dixon and Wheeler, 1986).

Third, some energy in the form of ATP is gained without using carbon compounds, but the gain is smaller than the energy cost for producing H_2 in the nitrogen fixation process. The value of energy gain may be greatest under carbon-limited conditions.

2. Occurrence and Characteristics of Hydrogenase

So far, actinorhizae that have been examined have, with one exception, shown hydrogenase activity. This is in sharp contrast to legume nodules, where many natural *Rhizobium* strains without hydrogenase activity have been reported. The exceptional actinorhiza lacking hydrogenase activity is a symbiosis between *A. incana* ssp. *incana* and a crushed nodule inoculum ("local source") of *Frankia* from Sweden (Sellstedt and Huss-Danell, 1984; Sellstedt *et al.*, 1986). The phenotypic character has been stable in numerous inoculation cycles over more than six years. A comparison was made between this symbiosis lacking hydrogenase activity and symbioses between the same clone of *A. incana* ssp. *incana* inoculated with *Frankia* strains AvcI1 and CpI1 (Sellstedt *et al.*, 1986). Plants with the latter two symbioses showed higher biomass and higher nitrogen content and nitrogen percentage than plants inoculated with the "local source" of *Frankia,* thus indicating that hydrogenase has some benefit. However, the symbiosis without hydrogenase had spore-containing nodules while the two other symbioses did not. The presence of spores may decrease the efficiency of nitrogen fixation and biomass production (Chapter 9). The potential benefit of an uptake hydrogenase in symbiotic *Frankia* must therefore await final evaluation until *Frankia* strains differing only with respect to hydrogenase become available.

Hydrogenase activity in actinorhizal nodules is inactivated to a varying extent by a conventional acetylene reduction assay. Nodules of *A. rubra* and *A. incana* ssp. *incana* are more sensitive than nodules of *Elaeagnus angustifolia* and *A. incana* ssp. *rugosa*, and nodules of *Ceanothus americanus* and *Datisca glomerata* are not affected (Winship *et al.*, 1987). The inhibition is due to acetylene rather than the ethylene formed in an acetylene reduction assay (Sellstedt and Winship, 1988). Inhibition of hydrogenase by acetylene was also reported for pure-cultured *Frankia* (Murry and Lopez, 1989), *Azotobacter* (Smith *et al.*, 1976), and cyanobacteria (Bothe *et al.*, 1977) but not for symbiotic *Rhizobium* (Ruiz-Argüeso *et al.*, 1979).

Hydrogenase activity may also be influenced by the season since lower activity was found in field-grown *A. glutinosa* nodules harvested in late autumn compared to summer-harvested nodules (Roelofsen and Akkermans, 1979).

VI. Host–Microsymbiont Interactions

The host and *Frankia* affect each other's physiology. A major plant response to the presence of *Frankia* is of course all the steps that result in a root nodule (Chapter 4). In the following sections some examples of interactions within already formed nodules are presented.

A. Proteins and Enzyme Activities

1. Hemoglobin and Nodulins

In *Rhizobium*–legume symbioses both partners collaborate in the formation of leghemoglobin with *Rhizobium* making the heme group and the host producing the protein. Actinorhizal nodules contain various amounts of hemoglobin, but the role of *Frankia* and host in hemoglobin formation is not known. In legumes, leghemoglobin is one of some forty so-called nodulins, that is, proteins encoded by host genes that are expressed during development of the symbiosis and are only found in root nodules, not in uninfected roots or other parts of the plant (van Kammen, 1984). It has recently been reported that during development of root nodules in *A. incana* ssp. *incana*, the plant synthesized new polypeptides in the nodules and that these polypeptides were not found in free-living *Frankia* or in uninfected roots (Séguin and Lalonde, 1988).

2. Activity of Nitrogenase, Hydrogenase, and other *Frankia* Enzymes

Nitrogenase activity in root nodules is determined by the plant genotype as well as the *Frankia* genotype (Gordon and Wheeler, 1978; Huss-Danell, 1980; Dillon and Baker, 1982; Carpenter *et al.*, 1984; Reddell and Bowen, 1985; Simon *et al.*, 1985; Sellstedt and Winship, 1987; St-Laurent and Lalonde, 1987; Hahn *et al.*, 1988; Sellstedt, 1988). Ineffective nodules, that is, those with no expression of nitrogenase activity, are reported from several actinorhizal genera (Mian *et al.*, 1976; Baker *et al.*, 1980; VandenBosch and Torrey, 1983; Reddell and Bowen, 1985; Hahn *et al.*, 1988; Kurdali *et al.*, 1989). Different specific nitrogenase activity in *A. glutinosa* and *A. rubra* inoculated with the same *Frankia* could not be explained by differences in percentage of nodule cells containing symbiotic vesicles (Wheeler *et al.*, 1981). A spore-containing crushed nodule inoculum from *A. incana* ssp. *incana* resulted in effective nodules on *A. incana* ssp. *incana* but ineffective nodules on *A. glutinosa*. Grafts between the two alder species indicated that the compatibility between *Alnus* and *Frankia* was restricted to the root system (Kurdali *et al.*, 1989).

Hydrogenase activity in *Casuarina glauca* and *C. obesa* nodules appeared to be influenced by both host and *Frankia* genotype (Sellstedt and Winship, 1987).

Enzymes involved in lipid metabolism are active in pure-cultured *Frankia* AvcI1 but not in symbiosis with *A. glutinosa* (Section III,B,5), and respiration of some organic acids occurs in pure-cultured but not symbiotic *Frankia* ArI3 (Section III,B,6).

B. Optimization of Symbiotic Combinations

There is great potential for selecting more efficient nitrogen-fixing symbioses for various environments. Possibilities for isolating and propagating *Frankia* (Chapter 3) are steadily increasing. Many host plants, like *Alnus* (Huss-Danell *et al.*, 1980; Huss-Danell, 1981), *Casuarina* (Lundquist and Torrey, 1984), and *Elaeagnus*, *Hippophaë*, and *Myrica* (unpublished observations) can be propagated with green or woody cuttings. Micropropagation techniques are also available (Chapter 11). *Alnus* genotypes can be hybridized (Hall and Maynard, 1979), and further studies on population genetics will hopefully provide a better basis for host genotype selection (Chapter 12).

To evaluate host–*Frankia* combinations it seems important to cover as long a time period as possible, since differences among young plants may be altered later on (Simon *et al.*, 1985; Sellstedt, 1988). It is also important to know that mineral nitrogen and soil factors can influence the result considerably, and that *Frankia* can adapt to certain soil factors (MacKay *et al.*, 1987; Sheppard *et al.*, 1988). This has implications for the practical use of actinorhizal plants. In nature, soil is rarely completely free of combined nitrogen. Often, there is an indigenous *Frankia* population in soil even at sites lacking the host plant (Huss-Danell and Frej, 1986). Existence of nodules containing more than one *Frankia* strain

(Benson and Hanna, 1983) indicates simultaneous presence of several *Frankia* strains in the soil. Interactions of host–*Frankia* combinations in the field thus become more complex than those experienced to date in studies with single, pure strains under controlled conditions with small amounts of combined nitrogen available in sterilized substrates.

VII. Concluding Remarks

Actinorhizal nodules are complex structures, and their physiological activities are the net result of interactions between *Frankia* and its host plant. Both partners can influence the physiology of the other, and it is obvious that future research is greatly needed to provide detailed information on these events. The regulation of various activities, in both short- and long-term perspectives, needs to be precise to allow a nitrogen-fixing root nodule to operate optimally in a constantly changing environment. The very diverse nature of host plant species also contributes to the broad range of physiological properties recorded for actinorhizal nodules. This is a complication in the understanding of the physiology of *Frankia* symbioses but also offers an opportunity for foresters, horticulturists, plant breeders, and others to select actinorhizae having the most appropriate characteristics.

Acknowledgments

Financial support from the Swedish Council for Forestry and Agricultural Research and the Swedish Natural Science Research Council is gratefully acknowledged.

References

Abeysekera, R. M., Newcomb, W., Silvester, W. B., and Torrey, J. G. (1990). A freeze–fracture electron microscopic study of *Frankia* in *Alnus incana* nodules grown at three oxygen tensions. *Can. J. Microbiol.*, in press.

Akkermans, A. D. L. (1971). "Nitrogen Fixation and Nodulation of *Alnus* and *Hippophaë* under Natural Conditions," Ph.D. thesis. State Univ., Leiden, The Netherlands.

Akkermans, A. D. L., and Roelofsen, W. (1980). Symbiotic nitrogen fixation by actinomycetes in *Alnus*-type root nodules. *In* "Nitrogen Fixation" (W. D. P. Stewart and J. R. Gallon, eds.), pp. 279–299. Academic Press, New York.

Akkermans, A. D. L., van Straten, J., and Roelofsen, W. (1977). Nitrogenase activity of nodule homogenates of *Alnus glutinosa*: A comparison with the *Rhizobium*–pea system. *In* "Recent Developments in Nitrogen Fixation" (W. Newton, J. R. Postgate, and C. Rodriguez-Barrueco, eds.), Proc. 2nd Int. Symp. Nitrogen Fixation, pp. 591–603. Academic Press, New York.

Akkermans, A. D. L., Huss-Danell, K., and Roelofsen, W. (1981). Enzymes of the tricarboxylic acid cycle and the malate–aspartate shuttle in the N_2-fixing endophyte of *Alnus glutinosa*. *Physiol. Plant.* **53,** 289–294.

Akkermans, A. D. L., Roelofsen, W., Blom, J., Huss-Danell, K., and Harkink, R. (1983). Utilization of carbon and nitrogen compounds by *Frankia* in synthetic media and in root nodules of *Alnus glutinosa, Hippophaë rhamnoides,* and *Datisca cannabina*. *Can. J. Bot.* **61,** 2793–2800.

Baker, D., Newcomb, W., and Torrey, J. G. (1980). Characterization of an ineffective actinorhizal microsymbiont, *Frankia* sp. EuI1 (Actinomycetales). *Can. J. Microbiol.* **26,** 1072–1089.

Becking, J. H. (1970). Plant–endophyte symbiosis in non-leguminous plants. *Plant Soil* **32,** 611–654.

Becking, J. H. (1977). Endophyte and association establishment in non-leguminous nitrogen-fixing plants. *In* "Recent Developments in Nitrogen Fixation" (W. Newton, J. R. Postgate, and C. Rodriguez-Barrueco, eds.), Proc. 2nd Int. Symp. Nitrogen Fixation, pp. 551–567. Academic Press, New York.

Benson, D. R. (1982). Isolation of *Frankia* strains from alder actinorhizal root nodules. *Appl. Environ. Microbiol.* **44,** 461–465.

Benson, D. R., and Eveleigh, D. E. (1979a). Ultrastructure of the nitrogen-fixing symbiont of Myrica pensylvanica L. (bayberry) root nodules. *Bot. Gaz. (Chicago), Suppl.* **140,** S15–S21.

Benson, D. R., and Eveleigh, D. E. (1979b). Nitrogen-fixing homogenates of *Myrica pensylvanica* (bayberry) non-legume root nodules. *Soil Biol. Biochem.* **11,** 331–334.

Benson, D. R., and Hanna, D. (1983). *Frankia* diversity in an alder stand as estimated by sodium dodecyl sulfate–polyacrylamide gel electrophoresis of whole-cell proteins. *Can. J. Bot.* **61,** 2919–2923.

Benson, D. R., Arp, D. J., and Burris, R. H. (1979). Cell-free nitrogenase and hydrogenase from actinorhizal root nodules. *Science* **205,** 688–689.

Benson, D. R., Arp, D. J., and Burris, R. H. (1980). Hydrogenase in actinorhizal root nodules and root nodule homogenates. *J. Bacteriol.* **142,** 138–144.

Blom, J. (1982). "Carbon and Nitrogen Metabolism of Free-Living *Frankia* and of *Frankia–Alnus* Symbioses," Ph.D. thesis. Agric. University, Wageningen, The Netherlands.

Blom, J., and Harkink, R. (1981). Metabolic pathways for gluconeogenesis and energy generation in *Frankia* AvcI1. *FEMS Microbiol. Lett.* **11,** 221–224.

Blom, J., Roelofsen, W., and Akkermans, A. D. L. (1981). Assimilation of nitrogen in root nodules of alder (*Alnus glutinosa*). *New Phytol.* **89,** 321–326.

Bond, G. (1956). Some aspects of translocation in root nodule plants. *J. Exp. Bot.* **7,** 387–394.

Bond, G., and Mackintosh, A. H. (1975). Effect of nitrate-nitrogen on the nodule symbioses of *Coriaria* and *Hippophaë*. *Proc. R. Soc. London, Ser. B* **190,** 199–209.

Bothe, H., Tennigkeit, J., Eisbrenner, G., and Yates, M. G. (1977). The hydrogenase–nitrogenase relationship in the blue-green alga *Anabaena cylindrica*. *Planta* **133,** 237–242.

Bothe, H., de Bruijn, F. J., and Newton, W. E. (eds.) (1988). "Nitrogen Fixation: Hundred Years After," 878 pp. Fischer, Stuttgart.

Burris, R. H., Arp, D. J., Benson, D. R., Emerich, D. W., Hageman, R. V., Ljones, T., Ludden, P. W., and Sweet, W. J. (1980). The biochemistry of nitrogenase. *In* "Nitrogen Fixation" (W. D. P. Stewart and J. R. Gallon, eds.), pp. 37–54. Academic Press, New York.

Carpenter, C. V., Robertson, L. R., Gordon, J. C., and Perry, D. A. (1984). The effect of four new *Frankia* isolates on growth and nitrogenase activity in clones of *Alnus rubra* and *Alnus sinuata*. *Can. J. For. Res.* **14,** 701–706.

Ching, T. M., Monaco, P. A., and Ching, K. K. (1983). Energy status and cytochromes in isolated endophytic vesicle clusters of red alder root nodules. *Can. J. For. Res.* **13,** 921–928.

Dawson, J. O., and Gordon, J. C. (1979). Nitrogen fixation in relation to photosynthesis in *Alnus glutinosa. Bot. Gaz. (Chicago), Suppl.* **140,** S70–S75.

Dillon, J. T., and Baker, D. (1982). Variations in nitrogenase activity among pure-cultured *Frankia* strains tested on actinorhizal plants as an indication of symbiotic compatibility. *New Phytol.* **92,** 215–219.

Dixon, R. O. D. (1968). Hydrogenase in pea root nodule bacteroids. *Arch. Mikrobiol.* **62,** 272–283.

Dixon, R. O. D. (1972). Hydrogenase in legume root nodule bacteroids: Occurrence and properties. *Arch. Mikrobiol.* **85,** 193–201.

Dixon, R. O. D., and Wheeler, C. T. (1983). Biochemical, physiological and environmental aspects of symbiotic nitrogen fixation. *In* "Biological Nitrogen Fixation in Forest Ecosystems: Foundations and Applications" (J. C. Gordon and C. T. Wheeler, eds.), pp. 107–171. Nijhoff/Dr. W. Junk, The Hague.

Dixon, R. O. D., and Wheeler, C. T. W. (1986). "Nitrogen Fixation in Plants." Blackie, Glasgow and London.

Evans, H. J., Harker, A. R., Papen, H., Russell, S. A., Hanus, F. J., and Zuber, M. (1987). Physiology, biochemistry, and genetics of the uptake hydrogenase in rhizobia. *Annu. Rev. Microbiol.* **41,** 335–361.

Gardner, I. C., and Leaf, G. (1960). Translocation of citrulline in *Alnus glutinosa. Plant Physiol.* **35,** 948–950.

Gordon, J. C., and Wheeler, C. T. (1978). Whole plant studies on photosynthesis and acetylene reduction in *Alnus glutinosa. New Phytol.* **80,** 179–186.

Haaker, H., and Klugkist, J. (1987). The bioenergetics of electron transport to nitrogenase. *FEMS Microbiol. Rev.* **46,** 57–71.

Hafeez, F., Chaudhary, A. H., and Akkermans, A. D. L. (1984). Physiological studies on N_2-fixing root nodules of *Datisca cannabina* L. and *Alnus nitida* Endl. from Himalaya region in Pakistan. *Plant Soil* **78,** 129–146.

Hahn, D., Starrenburg, M. J. C., and Akkermans, A. D. L. (1988). Variable compatibility of cloned *Alnus glutinosa* ecotypes against ineffective *Frankia* strains. *Plant Soil* **107,** 233–243.

Hall, R. B., and Maynard, C. A. (1979). Considerations in the genetic improvement of alder. *In* "Symbiotic Nitrogen Fixation in the Management of Temperate Forests" (J. C. Gordon, C. T. Wheeler, and D. A. Perry, eds.), pp. 322–344. For. Res. Lab., Oregon State Univ., Corvallis, Oregon.

Hirel, B., Perrot-Rechenmann, C., Maudinas, B., and Gadal, P. (1982). Glutamine synthetase in alder (*Alnus glutinosa*) root nodules. Purification, properties and cytoimmunochemical localization. *Physiol. Plant.* **55,** 197–203.

Huss-Danell, K. (1980). Nitrogen fixation and biomass production in clones of *Alnus incana. New Phytol.* **85,** 503–511.

Huss-Danell, K. (1981). Clonal differences in rooting of *Alnus incana* leafy cuttings. *Plant Soil* **59,** 193–199.

Huss-Danell, K., and Ahlqvist, A.-S. (1984). Nitrogenase activity in root nodule homogenates of *Alnus incana. Plant Soil* **78,** 159–170.

Huss-Danell, K., and Frej, A.-K. (1986). Distribution of *Frankia* in soils from forest and afforestation sites in northern Sweden. *Plant Soil* **90,** 407–418.

Huss-Danell, K., and Hahlin, A.-S. (1988). Nitrogenase activity decay and energy supply in *Frankia* after addition of ammonium to the host plant *Alnus incana. Physiol. Plant.* **74,** 745–751.

Huss-Danell, K., and Sellstedt, A. (1985). Nitrogenase activity in response to darkening and defoliation of *Alnus incana*. *J. Exp. Bot.* **36,** 1352–1358.

Huss-Danell, K., Eliasson, L., and Öhberg, I. (1980). Conditions for rooting of leafy cuttings of *Alnus incana*. *Physiol. Plant.* **49,** 113–116.

Huss-Danell, K., Roelofsen, W., Akkermans, A. D. L., and Meijer, P. (1982a). Carbon metabolism of *Frankia* spp. in root nodules of *Alnus glutinosa* and *Hippophaë rhamnoides*. *Physiol. Plant.* **54,** 461–466.

Huss-Danell, K., Sellstedt, A., Flower-Ellis, A., and Sjöström, M. (1982b). Ammonium effects on function and structure of nitrogen-fixing root nodules of *Alnus incana* (L.) Moench. *Planta* **156,** 332–340.

Huss-Danell, K., Winship, L. J., and Hahlin, A.-S. (1987). Loss and recovery of nitrogenase in *Alnus incana* nodules exposed to low oxygen and low temperature. *Physiol. Plant.* **70,** 355–360.

Huss-Danell, K., Lundquist, P.-O., and Ekblad, A. (1989). Growth and acetylene reduction activity by intact plants of *Alnus incana* under field conditions. *Plant Soil* **118,** 61–73.

Ingestad, T. (1980). Growth, nutrition, and nitrogen fixation in grey alder at varied rate of nitrogen addition. *Physiol. Plant.* **50,** 353–364.

Johnsrud, S. C. (1978). Nitrogen fixation by root nodules of *Alnus incana* in a Norwegian forest ecosystem. *Oikos* **30,** 475–479.

Kurdali, F., Capellano, A., Moiroud, A., and Domenach, A. M. (1989). Study of the contribution of the shoot and/or root of *Alnus* sp. in the compatibility between the host and a Sp+ *Frankia* strain using a grafting technique. *Plant Soil* **113,** 101–109.

Lalonde, M., and Knowles, R. (1975). Ultrastructure, composition and biogenesis of the encapsulation material surrounding the endophyte in *Alnus crispa* var. *mollis* root nodules. *Can. J. Bot.* **53,** 1951–1971.

Leaf, G., Gardner, I. C., and Bond, G. (1958). Observations on the composition and metabolism of the nitrogen-fixing root nodules of *Alnus*. *J. Exp. Bot.* **9,** 320–331.

Lopez, M. F., and Torrey, J. G. (1985). Enzymes of glucose metabolism in *Frankia* sp. *J. Bacteriol.* **162,** 110–116.

Lopez, M. F., Fontaine, M. S., and Torrey, J. G. (1984). Levels of trehalose and glycogen in *Frankia* sp. HFP ArI3 (Actinomycetales). *Can. J. Microbiol.* **30,** 746–752.

Lopez, M. F., Young, P., and Torrey, J. G. (1986). A comparison of carbon source utilization for growth and nitrogenase activity in two *Frankia* isolates. *Can. J. Microbiol.* **32,** 353–358.

Lundquist, R., and Torrey, J. G. (1984). The propagation of *Casuarina* species from rooted stem cuttings. *Bot. Gaz. (Chicago)* **145,** 378–384.

MacKay, J., Simon, L., and Lalonde, M. (1987). Effect of substrate nitrogen on the performance of *in vitro* propagated *Alnus glutinosa* clones inoculated with Sp+ and Sp− *Frankia* strains. *Plant Soil* **103,** 21–31.

Martin, F., Hirel, B., and Gadal, P. (1982). Sur l'activité enzymatique ornithine carbamyl transférase des actinorhizes d'*Alnus glutinosa* (L.) Gaertn. *C. R. Hebd. Seances Acad. Sci. Série III* **295,** 557–559.

Martin, F., Hirel, B., and Gadal, P. (1983). Purification and properties of ornithine carbamoyl transferase 1 from *Alnus glutinosa* root nodules. *Z. Pflanzenphysiol.* **111,** 413–422.

McClure, P. R., Coker, G. T., and Schubert, K. R. (1983). Carbon dioxide fixation in roots and nodules of *Alnus glutinosa*. I. Role of phosphoenolpyruvate carboxylase and carbamyl phosphate synthetase in dark CO_2 fixation, citrulline synthesis, and N_2 fixation. *Plant Physiol.* **71,** 652–657.

Meesters, T. M. (1987). Localization of nitrogenase in vesicles of *Frankia* sp. Cc1.17 by immunogoldlabelling on ultrathin cryosections. *Arch. Microbiol.* **146,** 327–331.

Mejstrik, V., and Benecke, U. (1969). The ectotrophic mycorrhizas of *Alnus viridis* (Chaix)

D.C. and their significance in respect to phosphorus uptake. *New Phytol.* **68,** 141–149.

Mian, S., and Bond, G. (1978). The onset of nitrogen fixation in young alder plants and its relation to differentiation in the nodular endophyte. *New Phytol.* **80,** 187–192.

Mian, S. Bond, G., and Rodriguez-Barrueco, C. (1976). Effective and ineffective root nodules in *Myrica faya*. *Proc. R. Soc. London, Ser. B* **194,** 285–293.

Miettinen, J. K., and Virtanen, A. I. (1952). The free amino acids in the leaves, roots, and root nodules of the alder (*Alnus*). *Physiol. Plant.* **5,** 540–557.

Mirza, M. S., Chaudhary, A. H., and Akkermans, A. D. L. (1987). Hydrogen uptake by the nitrogen fixing root nodules and nodule homogenates of *Coriaria nepalensis*. *Pak. J. Bot.* **19,** 237–241.

Murry, M. A., and Lopez, M. F. (1989). Interaction between hydrogenase, nitrogenase and respiratory activities in a *Frankia* isolate from *Alnus rubra*. *Can. J. Microbiol.*, **35,** 636–641.

Newcomb, W., and Wood, S. M. (1987). Morphogenesis and fine structure of *Frankia* (Actinomycetales): The microsymbiont of nitrogen-fixing actinorhizal root nodules. *Int. Rev. Cytol.* **109,** 1–88.

Perrot-Rechenmann, C., Vidal, J., Maudinas, B., and Gadal, P. (1981). Immunocytochemical study of phosphoenolpyruvate carboxylase in nodulated *Alnus glutinosa*. *Planta* **153,** 14–17.

Pizelle, G. (1984). Seasonal variations of the sexual reproductive growth and nitrogenase activity (C_2H_2) in mature *Alnus glutinosa*. *Plant Soil* **78,** 181–184.

Racette, S. E., and Torrey, J. G. (1988). *Proc. Int. Conf. Frankia Actinorhizal Plants, 7th,* Univ. of Connecticut, Storrs, Connecticut, August 7–10, 1988. Abstract only.

Reddell, P., and Bowen, G. D. (1985). *Frankia* source affects growth, nodulation and nitrogen fixation in *Casuarina* species. *New Phytol.* **100,** 115–122.

Rodriguez-Barrueco, C., Mackintosh, A. H., and Bond, G. (1970). Some effects of combined nitrogen on the nodule symbioses of *Casuarina* and *Ceanothus*. *Plant Soil* **33,** 129–139.

Roelofsen, W., and Akkermans, A. D. L. (1979). Uptake and evolution of H_2 and reduction of C_2H_2 by root nodules and nodule homogenates of *Alnus glutinosa*. *Plant Soil* **52,** 571–578.

Rosendahl, L., and Huss-Danell, K. (1988). Effects of elevated oxygen tensions on acetylene reduction in *Alnus incana–Frankia* symbioses. *Physiol. Plant.* **74,** 89–94.

Ruiz-Argüeso, T., Emerich, D. W., and Evans, H. J. (1979). Characteristics of the H_2 oxidizing system in soybean nodule bacteroids. *Arch. Microbiol.* **121,** 199–206.

Sasakawa, H., Hiyoshi, T., and Sugiyama, T. (1988). Immunogold localization of nitrogenase in root nodules of *Elaeagnus pungens* Thunb. *Plant Cell Physiol.* **29,** 1147–1152.

Schubert, K. R. (1986). Products of biological nitrogen fixation in higher plants: Synthesis, transport, and metabolism. *Annu. Rev. Plant Physiol.* **37,** 539–574.

Schubert, K. R., and Evans, H. J. (1976). Hydrogen evolution: A major factor affecting the efficiency of nitrogen fixation in nodulated symbionts. *Proc. Natl. Acad. Sci. U.S.A.* **73,** 1207–1211.

Schubert, K. R., Coker, G. T., III, and Firestone, R. B. (1981). Ammonia assimilation in *Alnus glutinosa* and *Glycine max*. Short-term studies using [^{13}N]ammonium. *Plant Physiol.* **67,** 662–665.

Schwintzer, C. R. (1979). Nitrogen fixation by *Myrica gale* root nodules in a Massachusetts wetland. *Oecologia* **43,** 283–294.

Schwintzer, C. R., and Tjepkema, J. D. (1983). Seasonal pattern of energy use, respiration, and nitrogenase activity in root nodules of *Myrica gale*. *Can. J. Bot.* **61,** 2937–2942.

Schwintzer, C. R., Berry, A. M., and Disney, L. D. (1982). Seasonal patterns of root nodule growth, endophyte morphology, nitrogenase activity and shoot development in *Myrica gale*. *Can. J. Bot.* **60,** 746–757.

Scott, A., Gardner, I. C., and McNally, S. F. (1981). Localisation of citrulline synthesis in the alder root nodule and its implication in nitrogen fixation. *Plant Cell Rep.* **1,** 21–22.

Séguin, A., and Lalonde, M. (1988). *Int. Congr. Plant Mol. Biol., 2nd,* Jerusalem, November 13–18, 1988. Abstract only.

Sellstedt, A. (1986). Nitrogen and carbon utilization in *Alnus incana* fixing N_2 or supplied with NO_3^- at the same rate. *J. Exp. Bot.* **37,** 786–797.

Sellstedt, A. (1988). Nitrogenase activity, hydrogen evolution and biomass production in different *Casuarina* symbioses. *Plant Soil* **105,** 33–40.

Sellstedt, A. (1989). Occurrence and activity of hydrogenase in symbiotic *Frankia* from field-collected *Alnus incana. Physiol. Plant.* **75,** 304–308.

Sellstedt, A., and Huss-Danell, K. (1984). Growth, nitrogen fixation and relative efficiency of nitrogenase in *Alnus incana* grown in different cultivation systems. *Plant Soil* **78,** 147–158.

Sellstedt, K., and Huss-Danell, K. (1986). Biomass production and nitrogen utilization by *Alnus incana* when grown on N_2 or NH_4^+ made available at the same rate. *Planta* **167,** 387–394.

Sellstedt, A., and Winship, L. J. (1987). Hydrogen metabolism of *Casuarina* root nodules: A comparison of two inoculum sources. *Physiol. Plant.* **70,** 367–372.

Sellstedt, A., and Winship, L. J. (1988). Effect of acetylene reduction assays on hydrogenase in symbiotic *Frankia. In* "Nitrogen Fixation: Hundred Years After" (H. Bothe, F. J. de Bruijn, and W. E. Newton, eds.), p. 69. Fischer, Stuttgart.

Sellstedt, A., Huss-Danell, K., and Ahlqvist, A.-S. (1986). Nitrogen fixation and biomass production in symbioses between *Alnus incana* and *Frankia* strains with different hydrogen metabolism. *Physiol. Plant.* **66,** 99–107.

Sheppard, L. J., Hooker, J. E., Wheeler, C. T., and Smith, R. I. (1988). Glasshouse evaluation of the growth of *Alnus rubra* and *Alnus glutinosa* on peat and acid brown earth soils when inoculated with four sources of *Frankia. Plant Soil* **110,** 187–198.

Simon, L., Stein, A., Côté, S., and Lalonde, M. (1985). Performance of *in vitro* propagated *Alnus glutinosa* (L.) Gaertn. clones inoculated with Frankiae. *Plant Soil* **87,** 125–133.

Simpson, F. B., and Burris, R. H. (1984). A nitrogen pressure of 50 atmospheres does not prevent evolution of hydrogen by nitrogenase. *Science* **224,** 1095–1097.

Smith, L. A., Hill, S., and Yates, M. G. (1976). Inhibition by acetylene of conventional hydrogenase in nitrogen-fixing bacteria. *Nature (London)* **262,** 209–210.

Stewart, W. D. P., and Bond, G. (1961). The effect of ammonium nitrogen on fixation of elemental nitrogen in *Alnus* and *Myrica. Plant Soil* **14,** 347–359.

Stewart, W. D. P., Rowell, P., and Rai, A. N. (1980). Symbiotic nitrogen-fixing cyanobacteria. *In* "Nitrogen Fixation" (W. D. P. Stewart and J. R. Gallon, eds.), pp. 239–277. Academic Press, New York.

St-Laurent, L., and Lalonde, M. (1987). Isolation and characterization of *Frankia* strains isolated from *Myrica gale. Can. J. Bot.* **65,** 1356–1363.

Sundström, K.-R., and Huss-Danell, K. (1987). Effects of water stress on nitrogenase activity in *Alnus incana. Physiol. Plant.* **70,** 342–348.

Tjepkema, J. D. (1985). Utilization of photosynthate for nitrogen fixation in seedlings of *Myrica gale* and *Alnus rubra. In* "Nitrogen Fixation and CO_2 Metabolism" (P. W. Ludden and J. E. Burris, eds.), pp. 183–192. Elsevier, Amsterdam.

Tjepkema, J. D., and Winship, L. J. (1980). Energy requirement for nitrogen fixation in actinorhizal and legume root nodules. *Science* **209,** 279–281.

Tjepkema, J. D., Schwintzer, C. R., and Benson, D. R. (1986). Physiology of actinorhizal nodules. *Annu. Rev. Plant Physiol.* **37,** 209–232.

Torrey, J. G. (1978). Nitrogen fixation by actinomycete-nodulated angiosperms. *BioScience* **28,** 586–592.

Tripp, L. N., Bezdicek, D. F., and Heilman, P. E. (1979). Seasonal and diurnal patterns and rates of nitrogen fixation by young red alder. *For. Sci.* **25,** 371–380.

Troelstra, S. R., van Dijk, K., and Blacquière, T. (1985). Effects of N source on proton excretion, ionic balance and growth of *Alnus glutinosa* (L.) Gaertner: Comparison of N_2 fixation with single and mixed sources of NO_3 and NH_4. *Plant Soil* **84,** 361–385.

van Kammen, A. (1984). Plant genes involved in nodulation and symbiosis. *In* "Advances in Nitrogen Fixation Research" (C. Veeger and W. E. Newton, eds.), pp. 587–588. Nijhoff/Dr. W. Junk, The Hague.

van Straten, J., Akkermans, A. D. L., and Roelofsen, W. (1977). Nitrogenase activity of endophyte suspension derived from root nodules of *Alnus, Hippophaë, Shepherdia* and *Myrica* spp. *Nature (London)* **266,** 257–258.

VandenBosch, K. A., and Torrey, J. G. (1983). Host–endophyte interactions in effective and ineffective nodules induced by the endophyte of *Myrica gale*. *Can. J. Bot.* **61,** 2898–2909.

Vessey, J. K., Walsh, K. B., and Layzell, D. B. (1988). Can a limitation in phloem supply to nodules account for the inhibitory effect of nitrate on nitrogenase activity in soybean? *Physiol. Plant.* **74,** 137–146.

Vikman, P.-Å., and Huss-Danell, K. (1987a). Capacity for hexose respiration in symbiotic *Frankia* from *Alnus incana*. *Physiol. Plant.* **70,** 349–354.

Vikman, P.-Å., and Huss-Danell, K. (1987b). Purity of *Frankia* preparations from root nodules of *Alnus incana*. *Physiol. Plant.* **71,** 489–494.

Walsh, K. B., Ng, B. H., and Chandler, G. E. (1984). Effects of nitrogen nutrition on xylem sap composition of Casuarinaceae. *Plant Soil* **81,** 291–293.

Watts, S. H., Wheeler, C. T., and Hillman, J. R. (1987). Abscisic acid and dormancy in root-nodulated *Alnus glutinosa* (L.) Gaertn. *New Phytol.* **105,** 459–468.

Wheeler, C. T. (1969). The diurnal fluctuation in nitrogen fixation in the nodules of *Alnus glutinosa* and *Myrica gale*. *New Phytol.* **68,** 675–682.

Wheeler, C. T. (1971). The causation of the diurnal changes in nitrogen fixation in the nodules of *Alnus glutinosa*. *New Phytol.* **70,** 487–495.

Wheeler, C. T. (1984). *Frankia* and its symbiosis in non-legume (actinorhizal) root nodules. *In* "Current Developments in Biological Nitrogen Fixation" (N. S. Subba Rao, eds.), pp. 173–195. Oxford Univ. Press IBH Publ., New Delhi, India.

Wheeler, C. T., and Bond, G. (1970). The amino acids of non-legume root nodules. *Phytochemistry* **9,** 705–708.

Wheeler, C. T., and Bowes, B. G. (1974). Effects of light and darkness on nitrogen fixation by root nodules of *Alnus glutinosa* in relation to their cytology. *Z. Pflanzenphysiol.* **71,** 71–75.

Wheeler, C. T., and Lawrie, A. C. (1976). Nitrogen fixation in root nodules of alder and pea in relation to the supply of photosynthetic assimilates. *In* "Symbiotic Nitrogen Fixation in Plants" (P. S. Nutman, ed.), pp. 497–509. Cambridge Univ. Press, Cambridge, England.

Wheeler, C. T., Henson, I. E., and McLaughlin, M. E. (1979). Hormones in plants bearing actinomycete nodules. *Bot. Gaz. (Chicago), Suppl.* **140,** S52–S57.

Wheeler, C. T., McLaughlin, M. E., and Steele, P. (1981). A comparison of symbiotic nitrogen fixation in Scotland in *Alnus glutinosa* and *Alnus rubra*. *Plant Soil* **61,** 169–188.

Wheeler, C. T., Watts, S. H., and Hillman, J. R. (1983). Changes in carbohydrates and nitrogenase compounds in the root nodules of *Alnus glutinosa* in relation to dormancy. *New Phytol.* **95,** 209–218.

Winship, L. J., and Tjepkema, J. D. (1985). Nitrogen fixation and respiration by root nodules of *Alnus rubra* Bong.: Effects of temperature and oxygen concentration. *Plant Soil* **87,** 91–107.

Winship, L. J., Martin, K. J., and Sellstedt, A. (1987). The acetylene reduction assay inactivates root nodule uptake hydrogenase in some actinorhizal plants. *Physiol. Plant.* **70,** 361–366.

8

Oxygen Regulation and Hemoglobin

Warwick B. Silvester and Sharon L. Harris
University of Waikato
Private Bag
Hamilton, New Zealand

John D. Tjepkema
University of Maine
Orono, Maine, U.S.A.

I. Introduction

The nitrogenase complex, whether in free-living or symbiotic systems, is uniquely sensitive to free oxygen. This oxygen sensitivity is central to the ecology and physiology of all diazotrophs and many of the fascinating, varied lifestyles (including symbiosis) of these organisms can be explained in terms of oxygen protection and oxygen avoidance strategies.

Gallon (1981) identified five theoretical generic "strategies" that a nitrogen-fixing organism could adopt to protect its nitrogenase from oxygen and showed that all possible strategies are observable in nature. At one end of the spectrum the anaerobes totally avoid the problem by expressing nitrogenase only in the absence of oxygen, while at the other end some aerobes simply balance nitrogenase synthesis with inactivation. The majority of strategies, however, are intermediate and involve an external barrier to oxygen diffusion, which, when coupled with the high oxygen utilization required for nitrogenase, maintains a steep oxygen gradient.

Organisms that rely on external barriers to gas diffusion may be further subdivided into two groups. The first group relies on the external

The Biology of *Frankia* and Actinorhizal Plants

environment, either ecological or symbiotic, to provide the diffusion resistance layer. Free-living microaerophiles that exist in water, soil, and decaying organic matter are among the most widespread, but *Rhizobium* in the legume nodule has been most studied. In this case the nodule cortex provides a highly resistive barrier to gas diffusion, thus producing a very low internal oxygen environment (Tjepkema and Yocum, 1974). The second group includes those that can express nitrogenase in air and therefore have an intrinsic diffusion resistance mechanism. This group is very much the minority and includes only heterocystous cyanobacteria and *Frankia.* Both of these possess specialized thick-walled cells that apparently provide the required diffusion resistance to maintain active nitrogenase when growing at atmospheric pO_2.

In contrast to legume nodules, actinorhizal nodules have *Frankia* as their diazotrophic symbiont and the sites of oxygen diffusion resistance have not yet been unequivocally identified. This chapter summarizes the range of oxygen responses that have so far been described in actinorhizal nodules and attempts to clarify the oxygen dilemma. Before looking at nodule physiology, however, it will be important to describe the oxygen response of *Frankia* in culture.

II. Response of *Frankia* to Oxygen

One of the really surprising results of the reisolation of *Frankia* in 1978 (Callaham *et al.*, 1978) was the discovery that nitrogen fixation is maximum at or about atmospheric oxygen levels (Tjepkema *et al.*, 1980; Gauthier *et al.*, 1981). This is in strong contrast to *Rhizobium,* which, while able to grow in air on nitrogen-containing media, is able to express nitrogenase only when external oxygen levels are extremely low (Tjepkema and Evans, 1975). The implication of this is that *Frankia* has an intrinsic oxygen protection mechanism while *Rhizobium,* whether in culture or in symbiosis, relies on an external limitation of oxygen diffusion. The site of nitrogenase in aerobic cultures of *Frankia* has been clearly identified as the vesicle (Chapter 7), which possesses a unique multilaminated outer envelope that may be the site of oxygen protection (Torrey and Callaham, 1982).

The hypothesis that nitrogenase in cultures of *Frankia* is a diffusion-limited system, in which oxygen supply is limited by some properties of the vesicle, is supported by work on the kinetics of oxygen and acetylene uptake by *Frankia* cultures. The K_m for C_2H_2 uptake by most N_2-fixing organisms is of the order of 0.6 kPa, which is similar to that of isolated nitrogenase (Hardy *et al.*, 1973). However, the apparent K_m for

Frankia (Murry *et al.*, 1984) is approximately 2 kPa. Murry *et al.* (1984) also showed that the acetylene K_m does not vary when V_{max} is altered through changing the pO_2 of the assay. These results are consistent with a diffusion-limited system in which the apparent K_m is the result of combined diffusion and enzyme kinetics. In addition, it has been shown (Murry *et al.*, 1984) that the oxygen uptake kinetics of repressed and derepressed *Frankia* are significantly different. In NH_4^+-grown (repressed) cultures, which lack vesicles, oxygen uptake shows simple kinetics, saturating at 18 μM O_2 (K_m = 1μM O_2). In derepressed cultures oxygen uptake shows a biphasic curve with two apparent K_m values, one at 1μM O_2 and the other at 170 μM O_2 (V_{max} = 350 μM O_2). The conclusions from these experiments are that the high acetylene K_m and the second high O_2 K_m relate to the activities of the vesicle and that the vesicle wall provides a passive barrier to oxygen diffusion in *Frankia*.

Air-grown cultures of *Frankia* show a nitrogenase optimum at or about 20 kPa O_2 and this has now been shown to be an adaptive feature of *Frankia* physiology. When *Frankia* is grown at a variety of pO_2 levels, ranging from 0.3 to 70 kPa O_2 in stirred or gas-sparged cultures, it is able to adapt nitrogenase activity to the ambient pO_2 (Fig. 1; Murry *et al.*, 1985).

Parsons *et al.* (1987) hypothesized that if the adaptation to external pO_2 was related to diffusion resistance of the vesicle wall (envelope) then the enormous range of pO_2 over which the organism can adapt must result in vast changes in wall properties. They used dark-field microscopy to observe vesicles, and their results together with the observations of Murry *et al.* (1985) show that the visual properties of the vesicle wall change in proportion to the ambient oxygen. At very low pO_2 (0.3 kPa) vesicles are not formed, at 2 kPa O_2 vesicles have very thin walls, while at 70 kPa O_2 the vesicle stands out as a very bright thick-walled cell (Fig. 2).

The multilaminate outer envelope of the vesicle is the most obvious contender for the observed dark-field results and preliminary observations on freeze-fractured vesicles (Parsons *et al.*, 1987) indicate that there are more layers at high pO_2 than at low pO_2. Use of Nile red fluorescent stain, specific for lipid, confirms that the bright dark-field images of vesicles are due to lipid (Lamont *et al.*, 1988). Chemical analysis shows that vesicles contain more neutral lipids and glycolipids than vegetative hyphae and that the neutral lipids are enriched with long-chain fatty acids or alcohols (Chapter 6). Thus the vesicle envelope is apparently composed of lipid arranged in monolayers that provide an adaptive oxygen barrier.

Early work on induction of nitrogenase and vesicles in *Frankia* strain

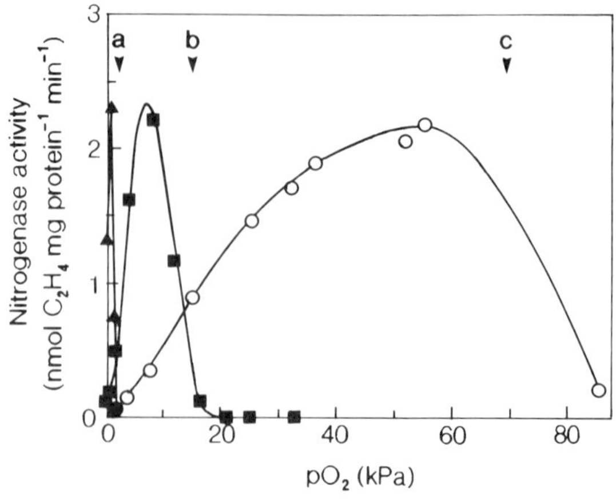

Figure 1 Effect of pO_2 on nitrogenase activity in *Frankia* grown at three O_2 levels. *Frankia* was grown at 2 (▲), 16 (■), and 70 (○) kPa O_2 and assayed for nitrogenase activity at various pO_2 levels from 0.4 to 82 kPa O_2. The arrows marked a, b, and c indicate the pO_2 level at which the culture was grown. Reproduced with permission from Parsons *et al.* (1987).

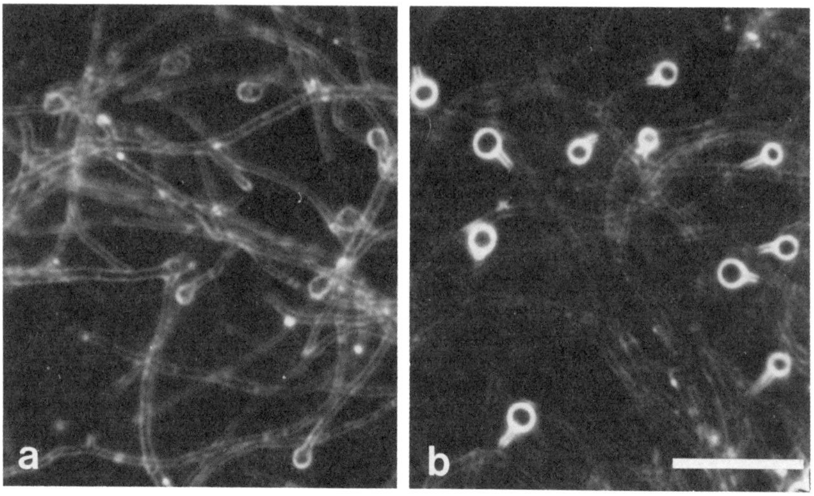

Figure 2 Dark-field microscope photos of *Frankia* (CcI3) grown in stirred culture at 3 kPa O_2 (a) and at 80 kPa O_2 (b). Bar is 10 μm.

CpI1 (Tjepkema *et al.*, 1981) indicated an absolute requirement for one of the organic acids: malate, fumarate, or succinate. No other carbohydrate source was able to support vesicle formation. More recently, however, Lopez *et al.* (1986) showed that the utilization of carbon substrates by ArI_3 and CcI_3 was related to the pO_2 at which they were assayed. Cells utilizing an organic acid (propionate in this case) had maximum nitrogenase activity at 15–20 kPa O_2. However, cells able to respire glucose or trehalose supported lower nitrogenase activity with an oxygen optimum of less than 10 kPa O_2. They also showed that derepressed growth on trehalose was much faster at lower oxygen than at atmospheric. One conclusion that may be drawn from this work is that the higher rate of respiration, associated with the utilization of the organic acid, provides a lower internal pO_2 and thus better control of oxygen at the higher ambient pO_2.

The evidence summarized here indicates that free-living *Frankia* has a highly adaptive intrinsic oxygen control system located in the vesicle. In nodules, the host tissue may also play a role in oxygen control. This is the topic of the remainder of the chapter.

III. Plant Growth and Nodule Development in Response to Varying Oxygen

A. Plant Growth

Several studies have investigated the effect of varying the pO_2 in the root environment on growth of actinorhizal plants. MacConnell (1959) worked with *Alnus glutinosa* and demonstrated that nodulated plants showed a 50% growth reduction at 5 kPa O_2 while unnodulated plants were only 15% reduced. Both nodulated and unnodulated plants showed severely reduced growth at 2 kPa O_2. In contrast to this we have grown nodulated *Myrica gale, Alnus incana* ssp. *rugosa,* and *Coriaria arborea* in aeroponics systems with root environments at 5, 21, and 40 kPa O_2 with no significant effects on growth (Silvester *et al.*, 1988a,b; Silvester and Harris, 1989). The difference between the earlier work by MacConnell and this later work is likely to be that the nodules of MacConnell's plants were immersed in water while the later work employed aeroponics. The general conclusion that nodulated plants appear able to adapt all their root functions, including nitrogenase, to a wide variety of oxygen levels above and below atmospheric accords well with earlier work on legumes (Criswell *et al.*, 1976).

B. Nodule Structure and Development

Nodules of legumes and actinorhizal plants, while showing some external similarities, differ enormously in many details of structure and function. Legume nodule structure is remarkably uniform over the many species studied so far (Fig. 3). All have a central zone of infected tissue containing hemoglobin and bounded by a tightly packed, diffusion-limiting inner cortex (Bergersen, 1982). Vascular tissue lies outside this limiting layer and the nodule may or may not be ventilated by lenticels. The universal presence of the diffusion-limiting layer in the inner cortex surrounding a hemoglobin-containing infected zone sets the legume and *Parasponia* nodule apart as a unique organ adapted to the sensitive control of a low-oxygen internal environment.

In contrast, the actinorhizal nodule is much more diverse in structure, both externally and internally, reflecting the diverse origins of actinorhizal plants. Although the general arrangement of tissues is that of a lateral root (Newcomb and Wood, 1987), there are many variations on the theme and in no case has a continuous, tightly packed, gas diffusion barrier been identified within the cortex of actinorhizal nodules.

The range of nodule structure is illustrated in Fig. 3 and shows that actinorhizal nodules are usually bounded by a periderm that may be up to eight cell layers thick in *Coriaria* and *Datisca* (Silvester and Harris, 1989). This periderm may be penetrated by lenticels distributed over the nodule surface as in *Alnus* or as one single elongate lenticel in *Coriaria*. The genera *Casuarina, Datisca, Myrica, Comptonia,* and *Gymnostoma* lack lenticels and have roots growing upward from nodule lobes. These roots have large air spaces and are implicated as oxygen diffusion pathways for the nodules (Bond, 1952). The nodule roots of *Myrica* have been studied in some detail, especially in relation to its habit of growing partially submerged in bogs. Tjepkema (1978) showed that when *Myrica* nodules are submerged, the nodule roots provide the major pathway for oxygen diffusion to the nodule. Despite the ability of nodule roots to transport oxygen, spring flooding of *Myrica* plants strongly restricts their growth mainly because of nitrogen limitation (Schwintzer, 1985).

C. Oxygen Effects on Nodule Anatomy

The ability of actinorhizal plants to sustain growth and nitrogenase activity over a wide range of applied external oxygen concentrations suggests considerable adaptation, either in nodule structure or in *Frankia* structure or physiology, to maintain an optimum oxygen concentration and flux rate. Several species of actinorhizal plants have been studied

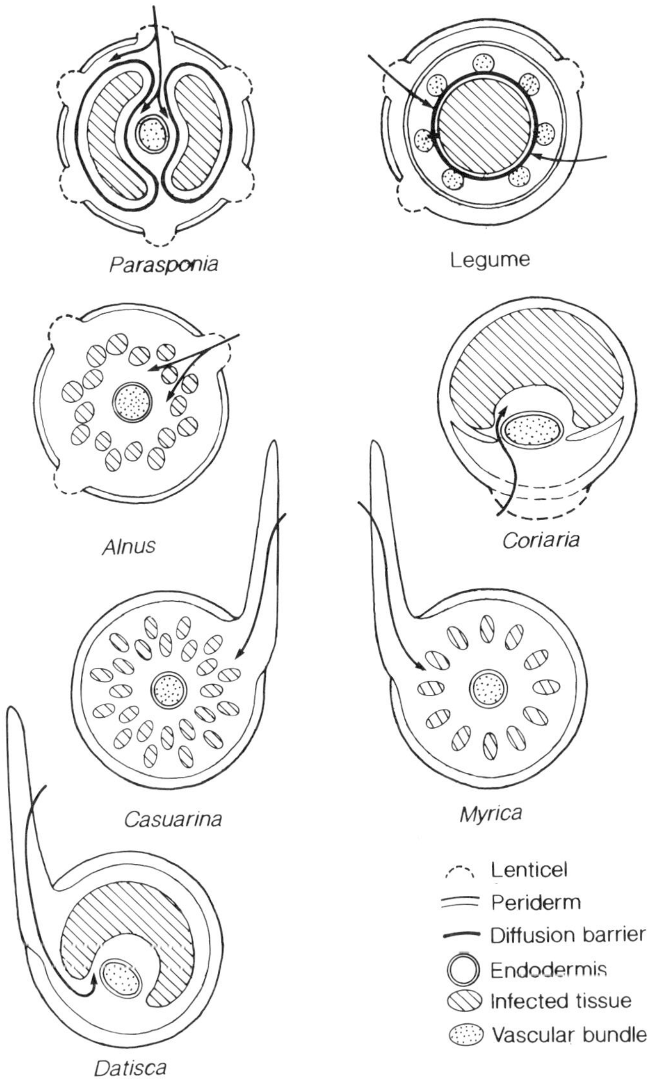

Figure 3 Diagrammatic representation of nodule structure demonstrating main nodule types. *Parasponia* and legume are illustrative of typical legume physiologies dominated by an external diffusion resistance. Arrows represent putative major oxygen diffusion pathways. Note that nodule roots attach at the nodule lobe apex.

using the hypothesis that, if anatomical features of the nodules are responsible for maintaining these oxygen conditions, these should adapt in response to applied oxygen changes and be observable.

Myrica gale is perhaps the most extreme case in which nodules are able to function beneath water when their negatively geotropic nodule roots are intact. In natural wetlands these nodule roots vary in length from 32 mm long in stagnant standing water to 7 mm long in freely drained soils and adapt well to the ambient environment (Sprent and Scott, 1979). When grown aeroponically at various pO_2 levels (Silvester *et al.*, 1988a), root length showed an equally dramatic adaptation from 62 mm at 2 kPa O_2 to 6 mm at 40 kPa O_2. The conclusion is that the nodule roots act as variable oxygen antennae and the nodule itself, which has a thin periderm lacking lenticels, may be relatively impermeable to oxygen. Other effects of oxygen on nodule anatomy include an increase in air spaces and an increase in the size of infected cells at lower oxygen concentrations.

The effects of varying pO_2 on nodule structure in *Alnus* are not as dramatic as in *Myrica* (Silvester *et al.*, 1988b). Lenticels were much increased in size at low pO_2 as were intercellular spaces. However, there were no apparent changes in nodule anatomy proportional to the pO_2 levels applied. The major change noted in *Alnus* was in the vesicle structure. Vesicles from nodules at high pO_2 had apparently much thicker walls than at low pO_2. These changes parallel responses in vesicles that have already been noted for *Frankia* in culture (Parsons *et al.*, 1987) and provide a plausible explanation for the lack of apparent large effects on nodule structure. However, a word of caution needs to be sounded as attempts to correlate the dark-field images of vesicles in nodules with the number of lipid layers show there is a nonsignificant increase in layers at high pO_2 (Abeysekera *et al.*, 1990).

Finally, *Coriaria arborea* shows even less modification of nodule anatomy to applied pO_2. The only obvious responses were decreased lenticel size and increased periderm thickness at high pO_2 (Silvester and Harris, 1989).

IV. Internal Oxygen Environment in Nodules

The internal oxygen environment of actinorhizal and legume nodules has been determined using oxygen microelectrode probes and information on the distribution of gas spaces within nodules obtained from vacuum infiltration of nodules with India ink.

Early use of these techniques on legume nodules, for example, soybean, confirmed the presence of the cortical diffusion barrier since ink does not penetrate this layer of tightly packed cortical cells. Microelectrode measurements showed that the pO_2 in the outer cortex is approximately atmospheric, while pO_2 in the central bacteroid zone is less than 1 kPa O_2 (Tjepkema and Yocum, 1974).

India ink infiltration of actinorhizal nodules indicates that these are much more aerated than legume nodules and when used in combination with microscopic studies this shows that there are two basic conditions. Nodules such as *Alnus, Coriaria,* and *Datisca* (Tjepkema, 1979; Tjepkema *et al.*, 1988; Silvester and Harris, 1989) appear to be well aerated with a continuity of air spaces leading from lenticels to the surfaces of infected cells or, in the case of *Alnus,* groups of infected cells. In strong contrast to this, the nodules of *Casuarina* possess a discontinuity in the diffusion pathway (Zeng *et al.*, 1989) and it is suggested that the infected cells are in a low-oxygen environment. *Myrica* appears somewhat intermediate between these two situations. While India ink diffuses throughout *Myrica* nodules it does not readily penetrate between infected cells, and microelectrode studies indicate that the pockets of infected cells in *Myrica* are sites of low pO_2 (Tjepkema, 1983a). Berg and McDowell (1988) have shown that the infected cells of *Casuarina* nodules are surrounded by a lignified wall produced by the adjacent uninfected cell. They postulate that this wall layer may be a site of diffusion resistance and point to reports of wall thickening in *Myrica* and *Comptonia* as being indicative of a more widespread phenomenon. The site of diffusion resistance in *Casuarina* has been further studied by observing the temperature response of nodules of *C. cunninghamiana* (Tjepkema and Murry, 1989). They conclude that *Frankia* does not contribute to diffusion resistance in this species and believe that there is a low Q_{10} boundary outside the infected cell.

Although a complete and systematic study has not been made, and despite the wide range of nodule anatomies among the actinorhizal plants, there appears to be only minor resistance to gas diffusion outside the infected cell zone. The general patterns of gas diffusion are indicated in Fig. 3. This hypothesis is further enhanced by the position of vascular tissue in various nodules. Phloem loading and unloading are strongly dependent on respiratory metabolism requiring oxygen. In keeping with this, the vascular bundles in legume nodules are outside the infected tissue in the aerated cortex. In actinorhizal nodules the stele is interior to the infected zone except in *Coriaria* and *Datisca*, where it is closely associated with the infected tissue.

V. Oxygen Effects on Nodule Function

A. Introduction

The temptation to invoke one simple model to explain oxygen responses in actinorhizal plants may not be valid in view of the wide range of nodule morphologies observed. Considerable variation occurs between families in some key morphological characteristics (Fig. 3) and, perhaps most significantly, in the possession of vesicles. At one end of the spectrum the majority of families in which *Frankia* in symbiosis has characteristic spherical vesicles, usually disposed around the periphery of the host cell, for example, *Alnus*. At the other end of the spectrum is the complete absence of vesicles in nodules of the *Casuarina* group. The extreme variability found in actinorhizal plants emphasizes the fact that nodule structure is a host-determined phenomenon with *Frankia* operating with apparently equal efficiency throughout this wide range of structures.

With such a wide diversity of nodule anatomy and physiology, it is impossible to generalize about these phenomena. The following account deals with the four genera that have been studied (*Alnus, Casuarina, Myrica,* and *Coriaria*), covering the broad range of anatomy.

B. Nitrogenase Activity in Response to pO_2 in Intact Plants

The use of intact plants and continuous-flow cuvette systems has enabled the dynamics of nitrogenase activity and accompanying respiration to be followed in a number of species. In general these results have confirmed that actinorhizal nodules are oxygen limited up to approximately ambient pO_2 and that both respiration and nitrogenase increase rapidly and in parallel up to approximately 20–25 kPa O_2. Results for *Alnus rubra* (Fig. 4; Winship and Tjepkema, 1985) show that at above ambient pO_2 levels the respiration rate continues to rise but nitrogenase activity drops off rapidly. Similar results for nitrogenase have been observed in other species (Rosendahl and Huss-Danell, 1988; Silvester *et al.*, 1988b), in *Myrica gale* (Silvester *et al.*, 1988a), in *Coriaria arborea* (Silvester and Harris, 1989), and in *Casuarina cunninghamiana* (Tjepkema and Murry, 1989). However, the details of each case lead us to believe that the pathways of gas diffusion and locations of diffusion resistance are significantly different across the range of species.

C. Transient Effects

Two important transient effects occur in root nodules that have important implications for an understanding of oxygen effects. The first is the so-called acetylene-induced decline, which was first observed in legume nodules (Minchin *et al.*, 1983) and is also apparent in some actinorhizal nodules (Tjepkema *et al.*, 1988; Silvester and Harris, 1989). The second is the dramatic drop and recovery in nitrogenase that occurs in response to step shifts in oxygen.

The acetylene-induced decline (Minchin *et al.*, 1983) is a drop in nitrogenase activity following exposure to acetylene. In legume nodules the response occurs within several minutes of exposure to acetylene and activity equilibrates after 10–20 min to a new level considerably below the preacetylene rate. Actinorhizal plants also show a decline in nitrogenase following exposure to acetylene but in many cases the decline is spontaneously recovered (Chapter 13; Tjepkema *et al.*, 1988; Silvester and Winship, 1990) to levels of 80% or more of the predecline rate. Typical results for *Myrica gale* are shown in Fig. 5. However, it is significant that two species, *Datisca glomerata* (Tjepkema *et al.*, 1988) and *Coriaria arborea* (Silvester and Harris, 1989), show an acetylene decline with no recovery. These two genera, while totally unrelated (*Datisca* is in Cucurbitaceae and *Coriaria* in Coriariaceae), show a similar but quite distinct nodule structure. The phenomenon of acetylene-induced decline and recovery has not been extensively studied in actinorhizal nodules but as it is generally spontaneously reversed it has been termed an acetylene-induced transient (Silvester and Winship, 1990).

The oxygen transient is a dramatic response in actinorhizal nodules and is manifest as a sharp drop followed by a recovery in nitrogenase activity when nodules are exposed to a higher pO_2. *Myrica gale* shows quite dramatic transient responses (Fig. 6). In this case an intact plant was assayed in a flow-through cuvette and subjected to pO_2 levels from 6 to 33 kPa O_2. Except at low pO_2 each step rise in pO_2 is accompanied by a transient drop in nitrogenase activity. The phenomenon has also been observed in *Alnus* (Rosendahl and Huss-Danell, 1988; Silvester *et al.*, 1988b), *Casuarina* (Silvester and Winship, 1990), and *Coriaria* (Silvester and Harris, 1989). Attempts to locate the site of action of the transient effect have demonstrated that *Frankia* in culture also shows a transient response to oxygen. It is now considered that the transient response to increased pO_2 is an expression of conformational protection of nitrogenase in *Frankia* (Silvester and Winship, 1990) and the following working hypothesis is proposed. The wave of oxygen that passes into the nodule following a step increase in external pO_2 results in an instantaneous switch-off of nitrogenase (conformational protection). Increased respiration mops up the extra oxygen and, as the internal pO_2 drops back to normal, nitrogenase activity returns.

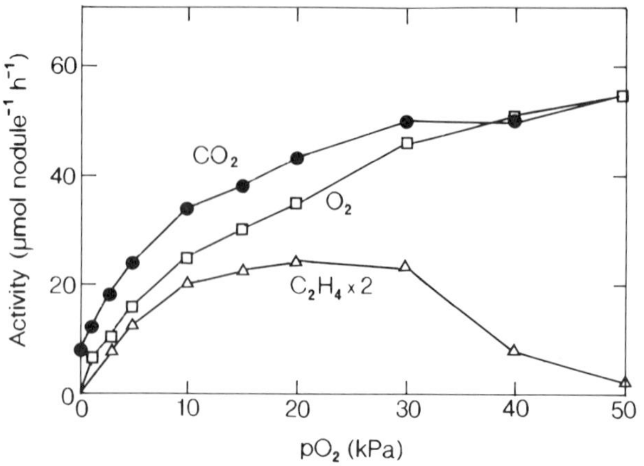

Figure 4 Effect of changes in oxygen concentration on nitrogenase activity, oxygen uptake, and CO_2 evolution in an attached *Alnus rubra* Bong. nodule. Reproduced with permission from Winship and Tjepkema (1985).

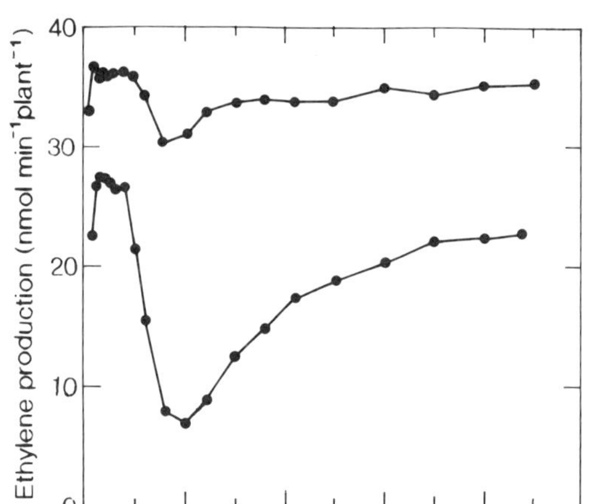

Figure 5 Acetylene-induced decline in *Myrica gale*. Nitrogenase activity of intact plants was assayed using a flow-through system. Results are for two plants selected from five replicates to show the greatest and least fluctuation in rate. Reproduced with permission from Tjepkema *et al.* (1988).

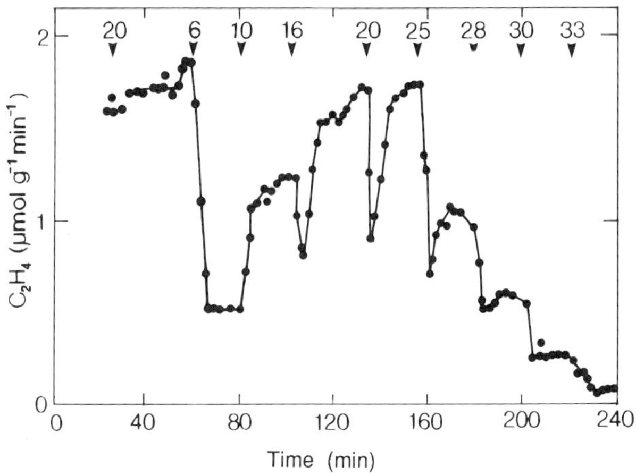

Figure 6 Nitrogenase response curve of *Myrica gale* plant grown with root system at 21 kPa O_2. The plant was equilibrated at 20 kPa O_2 and then submitted to step changes in pO_2 as indicated by arrows. Reproduced with permission from Silvester *et al.* (1988a).

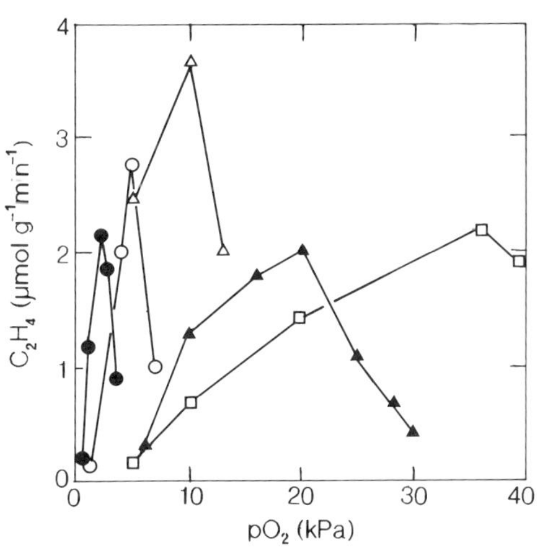

Figure 7 Nitrogenase activity of *Myrica gale* plants at various pO_2 levels, following growth with root systems at defined pO_2. Plants were grown at 2 (●), 5 (○), 10 (△), 21 (▲), and 40 (□) kPa O_2. Reproduced with permission from Silvester *et al.* (1988a).

D. Physiological Adaptation to Varying pO_2

Several species of actinorhizal plants have been grown at, above, and below ambient pO_2 to detect changes in both nodule anatomy and nitrogenase activity. In *Myrica gale* the result is a clear-cut, long-term adaptation that is obvious in nodule anatomy and physiological response. Nitrogenase activity shows marked optima at or near the pO_2 at which the nodules were grown (Fig. 7) and the changes in pO_2 optimum were accompanied by distinct changes in nodule-root length and thickness and in nodule aeration (Silvester *et al.*, 1988a).

The ability of actinorhizal nodules to rapidly adapt to above and below ambient pO_2 has been studied since the advent of flow-through assays of nitrogenase. Traditional batch assays of activity at various pO_2 levels give a result with a more or less marked optimum at near ambient pO_2 (Wheeler *et al.*, 1979). However, when nodules are taken to lower or higher ambient pO_2 in a flow-through cuvette, nitrogenase shows an initial transient and may also rise slowly to near maximum rates over the next 20–40 min. This is true for *Alnus incana* ssp. *rugosa* (Silvester *et al.*, 1988b) and is most marked in *Coriaria arborea* (Silvester and Harris, 1989). In the latter species it is most difficult to obtain a realistic pO_2 optimum for nitrogenase as nitrogenase activity continuously adapts to achieve a maximum activity. This short-term adaptation has been linked to a change in diffusion resistance of the nodule, which modifies gas exchange properties of nodules.

VI. Hemoglobins in Actinorhizal Nodules

A. Occurrence of Hemoglobins

There has long been evidence for the presence of hemoglobin and elevated concentrations of total heme in actinorhizal nodules (see Davenport, 1960). However, the extent of hemoglobin occurrence in these nodules is not yet clear. We have estimated hemoglobin concentration in actinorhizal nodules by measuring total and CO-reactive heme (Tjepkema and Asa, 1987). Thus far, concentrations of heme approaching those of legume nodules have been found only in *Casuarina cunninghamiana* and *Myrica gale* (Table I). For these two species it is clear that the major hemoprotein is in fact a hemoglobin. The hemoglobin from *Myrica gale* has been purified and forms a stable complex with O_2, as is characteristic of all hemoglobins (Pathirana, 1989). The hemoglobin from *Casuarina glauca* has been extensively characterized and has an amino acid sequence similar to that of hemoglobins from other plant sources (Fleming *et al.*, 1987; Appleby *et al.*, 1988).

Total and CO-reactive heme in other genera of actinorhizal nodules is much lower than in *Casuarina* and *Myrica*. Nonetheless, part of this heme may be due to hemoglobin. Davenport (1960) concluded that there was a hemoglobin in *Alnus*, and Tjepkema (1983b) found spectral evidence for hemoglobins in slices of various actinorhizal nodules other than *Casuarina* and *Myrica*.

We have also detected measurable quantities of CO-reactive heme in a wide variety of roots as well as in nodules (Tjepkema and Asa, 1987). In fact, the concentration of total and CO-reactive heme in some actinorhizal nodules is not very much higher than the concentration in roots (Table I). Whether any of the CO-reactive heme in roots is due to hemoglobins is unknown, since our assay is not specific for hemoglobin and also could detect other CO-reactive hemoproteins that have absorption maxima at 416 to 420 nm (Tjepkema and Asa, 1987). However, Appleby *et al.* (1988) have presented conclusive proof for the presence of hemoglobin in nonnodulated roots from *Parasponia* and *Trema*. *Parasponia* is a member of the Ulmaceae that forms root nodules in symbiosis with *Rhizobium*, while *Trema* is a nonnodulating relative of *Parasponia*.

B. Function of Hemoglobin

As discussed in Section IV, there is a barrier to oxygen diffusion exterior to the cytoplasm of the infected cells in nodules of *Casuarina* and *Myrica*. Coupled with a high rate of respiration by *Frankia* in the infected cells, this diffusion barrier results in a low concentration of oxygen in the infected cells. Under these conditions, hemoglobin could facilitate the transport of O_2 from the outer surface of the cytoplasm to the respiratory sinks (*Frankia* cells and host mitochondria) distributed throughout the cytoplasm. Such facilitated transport of O_2 is the generally accepted function of hemoglobin in legume nodules (Appleby, 1984).

The function of hemoglobin in other actinorhizal nodules is less clear. For hemoglobins to be significant in the facilitation of O_2 transport, the concentration of hemoglobin must be high relative to the concentration of O_2. Appleby *et al.* (1988) have pointed out that this condition is not met for the hemoglobin that they observe in roots. Instead they suggest that the hemoglobin in roots may serve as an oxygen sensor, being involved in the shift between oxidative and fermentative plant metabolism. Likewise, even if all of the CO-reactive heme in actinorhizal nodules is a hemoglobin, the concentration in many actinorhizal nodules is

Table I Total and CO-Reactive Heme Contents of Nodules and Roots (±SE)

Species	Total heme (nmol g^{-1}, fresh wt)	CO-Reactive heme (nmol g^{-1}, fresh wt)	C_2H_2 reduction rate (μmol h^{-1} g^{-1} fresh wt)	Reference[a]
Lupinus albus (legume)	197 ± 11	174 ± 1	—	1
Myrica gale	127 ± 11	103 ± 7	43 ± 4	1
Casuarina cunninghamiana	96 ± 23	80 ± 4	67 ± 2	1
	110 to 120	83	—	2
Casuarina glauca	—	32	—	3
Comptonia peregrina	32 ± 2	17 ± 1	16 ± 1	1
Alnus rubra	24 ± 1	15 ± 1	43 ± 2	1
Alnus glutinosa	49	—	—	2
Ceanothus americanus	16 ± 3	12 ± 1	49 ± 3	1
Hippophae rhamnoides	13 ± 1	8 ± 0.6	25 ± 3	4
	42	—	—	2
Elaeagnus angustifolia	17 ± 1	6 ± 0.5	20 ± 2	4
Coriaria myrtifolia	11 ± 2	4 ± 0.4	22 ± 1	4
Purshia tridentata	8 ± 1	2 ± 0.4	23 ± 1	4
Datisca glomerata	13 ± 1	—	48 ± 2	1
C. americanus, roots only	4.3 ± 0.4	1.4 ± 0.2	—	1
Zea mays, roots	4.1 ± 0.3	1.7 ± 0.1	—	1
Actinorhizal root cortex	5 to 7	—	—	2

[a]References: (1) Tjepkema and Asa (1987); (2) Davenport (1960); (3) Fleming *et al.* (1987); (4) J. D. Tjepkema (unpublished observations).

too low to have a function in oxygen transport, especially in the case of *Coriaria, Purshia,* and *Datisca* (Table I). The situation with respect to *Alnus* is less clear. The concentration of CO-reactive heme in *Alnus* is 10 times greater than that observed in roots. If most of this is a hemoglobin and is concentrated in a relatively small fraction of the total nodule volume, it could play a role in oxygen transport under conditions of low O_2 concentration.

From the preceding it is clear that many or most actinorhizal nodules differ greatly from legume nodules with respect to hemoglobin. While *Casuarina* and *Myrica* have high concentrations of heme that approach those of legumes, the nodules of other actinorhizal genera have much lower concentrations. In spite of low heme concentrations, these nodules still have high rates of nitrogenase activity (Table I). The differences between legume and actinorhizal nodules are probably due to differences in the microsymbionts. Rhizobia require low concentrations of O_2 in the host cytoplasm in order to fix nitrogen, while frankiae can adapt to variable concentrations of O_2. Hemoglobin may only be needed in actinorhizal nodules when the concentration of O_2 in the host cytoplasm is low because of the presence of an external diffusion barrier.

VII. Conclusions

The root nodule, whether legume or actinorhizal, is an organ that is modified to optimize carbohydrate and oxygen supply to the bacterial symbiont. While other oxygen-consuming mechanisms such as superoxide dismutase, catalase, peroxidase, and hydrogenase may all play a role in maintaining the internal oxygen balance of root nodules, there is little doubt that respiration coupled with a diffusion barrier represents the main element in the equation of oxygen protection.

In the absence of a well-defined cortical barrier in actinorhizal nodules and in view of their apparent ventilation via lenticels or "breathing roots," we postulate that the major limitation to oxygen diffusion probably resides in host cell walls, in host cytoplasm, or in *Frankia* vesicles. Modeling of nodule properties by network simulation analysis (Winship and Silvester, 1989) has confirmed that resistance at these sites can account for all the known physiological responses, including oxygen transients.

There is little doubt that there is no single physiological mechanism of oxygen control in actinorhizal nodules, but a variety consistent with the variety of host plants that control the internal anatomy. Present evidence, although not completely verified, leads us to postulate that the *Casuarina* group represents one end of a continuum of response. In this case *Frankia* lacks vesicles in symbiosis, hemoglobin is present in high concentrations, and it is possible that the host cell wall thickenings or cell packing may play a major role in diffusion resistance. *Alnus*, and related genera in which the vesicle is prominent, may represent the other end of the continuum, where the vesicle envelope provides the major resistance.

It is also apparent from this review that there is no one mechanism of protecting nitrogenase from oxygen that is more efficient than any other. If in the final analysis it is proven that the vesicle is the major site of oxygen protection in *Alnus* nodules this will have major implications for any possible development of new symbioses. It is evident from studies on the legume symbiosis that a large component of the relationship between host and *Rhizobium* is concerned with oxygen protection, for example, hemoglobin development and cortical tissue resistance. If neither of these is needed the chance of matching or modifying *Frankia* to form new symbioses with other plant groups in either a leaf or root symbiosis is greatly enhanced.

References

Abeysekera, R. M., Newcomb, W., Silvester, W. B., and Torrey, J. G. (1990). A freeze–fracture electron microscopic study of *Frankia* in *Alnus incana* nodules grown at three oxygen tensions. *Can. J. Microbiol.*, in press.

Appleby, C. A. (1984). Leghemoglobin and *Rhizobium* respiration. *Annu. Rev. Plant Physiol.* **35,** 443–478.

Appleby, C. A., Bogusz, D., Dennis, E. S., and Peacock, W. J. (1988). A role for hemoglobin in all plant roots? *Plant Cell* **11,** 359–367.

Berg, R. H., and McDowell, L. (1988). Cytochemistry of the wall of infected cells in *Casuarina* actinorhizae. *Can. J. Bot.* **66,** 2038–2047.

Bergersen, F. J. (1982). "Root Nodules of Legumes: Structure and Function." Res. Stud. Press, Chichester, England.

Bond, G. (1952). Some features of root growth in nodulated plants of *Myrica gale* L. *Ann. Bot. (London)* **16,** 467–475.

Callaham, D., Del Tredici, P., and Torrey, J. G. (1978). Isolation and cultivation *in vitro* of the actinomycete causing root nodulation in *Comptonia*. *Science* **199,** 899–902.

Criswell, J. G., Havelka, U. D., Quebedeaux, B., and Hardy, R. W. F. (1976). Adaptation of nitrogen fixation by intact soybean nodules to altered pO_2. *Plant Physiol.* **58,** 622–625.

Davenport, H. E. (1960). Haemoglobin in the root nodules of *Casuarina cunninghamiana*. *Nature (London)* **186,** 653–654.

Fleming, A. I., Wittenberg, J. B., Wittenberg, B. A., Dudman, W. F., and Appleby, C. A. (1987). The purification, characterization and ligand-binding kinetics of hemoglobins from root nodules of the non-leguminous *Casuarina glauca–Frankia* symbiosis. *Biochim. Biophys. Acta.* **911,** 209–220.

Gallon, J. R. (1981). The oxygen sensitivity of nitrogenase: A problem for biochemists and microorganisms. *Trends Biochem. Sci. (Pers. Ed.)* **6,** 19–23.

Gauthier, D. L., Diem, H. G., and Dommergues, Y. (1981). *In vitro* nitrogen fixation by two actinomycete strains isolated from *Casuarina* nodules. *Appl. Environ. Microbiol.* **41,** 306–308.

Hardy, R. W. F., Burns, R. C., and Holsten, R. D. (1973). Application of the acetylene–ethylene assay for measurement of nitrogen fixation. *Soil Biol. Biochem.* **5,** 47–81.

Lamont, H. C., Silvester, W. B., and Torrey, J. G. (1988). Nile red fluorescence demonstrates lipid in the envelope of vesicles from N_2-fixing cultures of *Frankia*. *Can. J. Microbiol.* **34,** 656–660.

Lopez, M. F., Murry, M. A., and Torrey, J. G. (1986). Effect of oxygen on substrate utilization for nitrogen fixation and growth in *Frankia* spp. *Arch. Microbiol.* **145,** 209–214.

MacConnell, J. T. (1959). The oxygen factor in the development and function of the root nodules of alder. *Ann. Bot. (London)* **23,** 261–268.

Minchin, F. R., Witty, J. F., Sheehy, J. E., and Muller, M. (1983). A major error in the acetylene reduction assay: Decreases in nodular nitrogenase activity under assay conditions. *J. Exp. Bot.* **34,** 641–649.

Murry, M. A., Fontaine, M. S., and Tjepkema, J. D. (1984). Oxygen protection of nitrogenase in *Frankia* sp. HFPArI3. *Arch. Microbiol.* **139,** 162–166.

Murry, M. A., Zhang, Z., and Torrey, J. G. (1985). Effect of oxygen on vesicle formation, acetylene reduction and oxygen-uptake kinetics in *Frankia* sp. HFPCcI3 isolated from *Casuarina cunninghamiana*. *Can. J. Microbiol.* **31,** 804–809.

Newcomb, W. R., and Wood, S. (1987). Morphogenesis and fine structure of *Frankia* (Actinomycetales): The microsymbiont of nitrogen-fixing actinorhizal root nodules. *Int. Rev. Cytol.* **109,** 1–88.

Parsons, R., Silvester, W. B., Harris, S., Gruijters, W. T. M., and Bullivant, S. (1987). *Frankia* vesicles provide inducible and absolute protection for nitrogenase. *Plant Physiol.* **83,** 728–731.

Pathirana, M. S. (1989). "Occurrence of Hemoglobin in the Nitrogen-Fixing Root Nodules of *Myrica gale*," M.S. thesis. Univ. of Maine, Orono, Maine.

Rosendahl, L., and Huss-Danell, K. (1988). Effects of elevated oxygen tensions on acetylene reduction in *Alnus incana–Frankia* symbioses. *Physiol. Plant.* **74,** 89–94.

Schwintzer, C. R. (1985). Effect of spring flooding on endophyte differentiation, nitrogenase activity, root growth and shoot growth in *Myrica gale*. *Plant Soil* **87,** 109–124.

Silvester, W. B., and Harris, S. L. (1989). Nodule structure and nitrogenase activity of *Coriaria arborea* in response to varying pO_2. *Plant Soil* **118,** 97–109.

Silvester, W. B., and Winship, L. J. (1990). Transient responses of nitrogenase to acetylene and oxygen by actinorhizal nodules and cultured *Frankia*. *Plant Physiol.*, in press.

Silvester, W. B., Whitbeck, J., Silvester, J. K., and Torrey, J. G. (1988a). Growth, nodule morphology and nitrogenase activity of *Myrica gale* grown with roots at various oxygen levels. *Can. J. Bot.* **66,** 1762–1771.

Silvester, W. B., Silvester, J. K., and Torrey, J. G. (1988b). Adaptation of nitrogenase to varying oxygen tension and the role of the vesicle in root nodules of *Alnus incana* subsp. *rugosa*. *Can. J. Bot.* **66,** 1772–1779.

Sprent, J. I., and Scott, R. (1979). The nitrogen economy of *Myrica gale* and its possible significance for the afforestation of peat soils. *In* "Symbiotic Nitrogen Fixation in the Management of Temperate Forests" (J. C. Gordon, C. T. Wheeler, and D. A. Perry, eds.), pp. 234–242. For. Res. Lab., Oregon State Univ., Corvallis, Oregon.

Tjepkema, J. D. (1978). The role of diffusion from the shoots and nodule roots in nitrogen fixation by root nodules of *Myrica gale*. *Can. J. Bot.* **56,** 1365–1371.

Tjepkema, J. D. (1979). Oxygen relations in leguminous and actinorhizal nodules. *In* "Symbiotic Nitrogen Fixation in the Management of Temperate Forests" (J. C. Gordon, C. T. Wheeler, and D. A. Perry, eds.), pp. 175–186. For. Res. Lab., Oregon State Univ., Corvallis, Oregon.

Tjepkema, J. D. (1983a). Oxygen concentration within the nitrogen-fixing root nodules of *Myrica gale* L. *Am. J. Bot.* **70,** 59–63.

Tjepkema, J. D. (1983b). Hemoglobins in the nitrogen-fixing root nodules of actinorhizal plants. *Can. J. Bot.* **61,** 2924–2929.

Tjepkema, J. D., and Asa, D. J. (1987). Total and CO-reactive heme content of actinorhizal nodules and the roots of some non-nodulated plants. *Plant Soil* **100,** 225–236.

Tjepkema, J. D., and Evans, H. J (1975). Nitrogen fixation by free-living *Rhizobium* in a defined liquid medium. *Biochim. Biophys. Acta* **65,** 625–628.

Tjepkema, J. D. and Murry, M. A. (1989). Respiration and nitrogenase activity in nodules of *Casuarina cunninghamiana* and cultures of *Frankia* sp HFP020203: Effects of temperature and partial pressure of O_2. *Plant Soil* **118,** 111–118.

Tjepkema, J. D., and Yocum, C. S. (1974). Measurement of oxygen partial pressure within soybean nodules by oxygen microelectrodes. *Planta* **119,** 351–360.

Tjepkema, J. D., Omerod, W., and Torrey, J. G. (1980). Vesicle formation and acetylene reduction activity in *Frankia* sp CpI1 cultured in defined nitrogen media. *Nature (London)* **287,** 633–635.

Tjepkema, J. D., Omerod, W., and Torrey, J. G. (1981). Factors affecting vesicle formation and acetylene reduction (nitrogenase activity) in *Frankia* sp. CpI1. *Can. J. Microbiol.* **27,** 815–823.

Tjepkema, J. D., Schwintzer, C. R., and Monz, C. A. (1988). Time course of acetylene reduction in nodules of five actinorhizal genera. *Plant Physiol.* **86,** 581–583.

Torrey, J. G., and Callaham, D. (1982). Structural features of the vesicle of *Frankia* sp. CpI1 in culture. *Can. J. Microbiol.* **28,** 749–757.

Wheeler, C. T., Gordon, J. C., and Ching, T. M. (1979). Oxygen relations of the root nodules of *Alnus rubra* Bong. *New Phytol.* **82,** 449–457.

Winship, L. J., and Silvester, W. B. (1989). Modeling nitrogenase-associated gas exchange in actinorhizal root nodules. *In* "Applications of Constant and Steady-State Methods to Root Biology" (J. G. Torrey and L. J. Winship, eds.), pp. 121–146. Kluwer Academic, Dordrecht, The Netherlands.

Winship, L. J., and Tjepkema, J. D. (1985). Nitrogen fixation and respiration by root nodules of *Alnus rubra* Bong.: Effects of temperature and oxygen concentration. *Plant Soil* **87,** 91–107.

Zeng, S., Tjepkema, J. D., and Berg, R. H. (1989). Gas diffusion pathway in nodules of *Casuarina cunninghamiana*. *Plant Soil* **118,** 119–123.

9

Spore-Positive and Spore-Negative Nodules

Christa R. Schwintzer
Department of Botany and Plant Pathology
University of Maine
Orono, Maine, U.S.A.

Two types of nitrogen-fixing nodules have been recognized in actinorhizal plants based on the extent of spore formation by *Frankia* within nodules (van Dijk and Merkus, 1976). One contains numerous spores and is designated sp^+, for spore-positive. The other contains no or only a few spores and is designated sp^- (van Dijk and Merkus, 1976; van Dijk, 1978). Recently some workers have divided nodules into three types: sp^+, containing numerous spores; sp^-, containing no spores; and "intermediate," containing only a very few sporangia (Smolander and Sundman, 1987). Most, if not all, "intermediate" nodules are probably sp^- nodules in the broader sense.

Sp^- nodules are substantially more common that sp^+ nodules. At present sp^+ nodules have been described in only 9 of the 25 known genera of actinorhizal plants (Table I). Moreover, within the taxa in which they do occur, sp^+ nodules are generally less common than sp^- nodules.

Sp^+ nodules have been reported extensively only in the genus *Alnus* and in *Myrica gale* (Table II). Evidence for the existence of sp^+ nodules in other taxa is based on very limited observations. As more nodules are examined for presence of sporangia and spores, especially field-collected materials, additional genera and species with sp^+ nodules will probably be found.

The Biology of *Frankia* and Actinorhizal Plants

Table I Actinorhizal Plants in Which Sp$^+$ Nodules Have Been Observed[a]

Family	Genus and species	Reference[b]
Betulaceae	*Alnus glutinosa* (L.) Gaertn.	Wheeler *et al.* (1981), Weber (1986)
	Alnus rubra Bong.	Wheeler *et al.* (1981)
	Alnus viridis ssp. *crispa* (Ait.) Turr.	Holman and Schwintzer (1987)
	Alnus incana ssp. *incana* (L.) Moench.	Weber (1986), Daniere *et al.* (1986)
	Alnus incana ssp. *rugosa* (Duroi.) Claus.	Holman and Schwintzer (1987)
	Alnus cordata (Lois.) Desf.	
Myricaceae	*Comptonia peregrina* (L.) Coult.	
	Myrica cerifera L.	
	Myrica gale L.	Kashanski and Schwintzer (1987)
Rhamnaceae	*Ceanothus velutinus* Dougl.	
Elaeagnaceae	*Elaeagnus umbellata* Thunb.	
	Hippophae rhamnoides L.	
Casuarinaceae	*Casuarina cunninghamiana* Miq.	P. A. Rosbrook (personal communication)
	Casuarina equisetifolia Forst. & Forst.	
	Casuarina glauca Sieb. ex. Spreng	P. A. Rosbrook (personal communication)
	Casuarina obesa Miq.	P. A. Rosbrook (personal communication)
Rosaceae	*Purshia tridentata* (Pursh.) D.C.	
	Dryas drummondii Rich.	

[a]Modified from Torrey (1987).
[b]References as in the article by Torrey (1987), except for additional references shown here.

I. Factors Determining Nodule Spore Type

Considerable evidence indicates that the presence or absence of sporulation within nodules is largely controlled by the *Frankia* strain that infected the host (Tjepkema *et al.*, 1986; Lalonde *et al.*, 1988). However, it has recently been proposed that the environment within the nodule determines whether or not spores are formed (Torrey, 1987). Interestingly, sp$^-$ strains sporulate freely on most media when grown in pure culture (Burggraaf *et al.*, 1981; Baker, 1982; Normand and Lalonde, 1982). Thus both sp$^+$ and sp$^-$ strains have the genetic capacity to sporulate but in sp$^-$ strains sporulation is suppressed within nodules.

Experimental evidence that sporulation within nodules is a genetic

Table II Proportion of Sp$^-$ Nodules in the Field

Host species	Location	% sp$^-$	Reference[a]
Alnus glutinosa	The Netherlands, 150 ha	66	1
	The Netherlands, 650 ha	91	1
	Finland, 12 sites	91	2
Alnus incana ssp.	Finland, 17 sites	18	2
incana	France, 3 sites	0	3
Alnus incana ssp.	Quebec, 11 sites	68	4
rugosa	Maine, 29 sites	76	5
Alnus viridis ssp. *crispa*	Quebec, 14 sites	93	4
Comptonia peregrina	Maine, 31 sites	100	6
Myrica gale	Maine, 34 sites	61	7
Myrica pensylvanica	Maine, 8 sites	100	6

[a] References: (1) van Dijk (1984); (2) Weber (1986); (3) Daniere *et al.* (1986); (4) Normand and Lalonde (1982); (5) Holman and Schwintzer (1987); (6) Schwintzer (1989); (7) Kashanski and Schwintzer (1987).

trait of the microsymbiont includes the following: in *Alnus glutinosa, A. cordata,* and *A. incana* ssp. *incana,* the nodule type is consistently determined by the spore type of the inoculum, that is, inoculum prepared from sp$^+$ nodules gives rise to sp$^+$ nodules and inoculum prepared from sp$^-$ nodules gives rise to sp$^-$ nodules. This cannot be changed by host seed source or by manipulation of the nitrogen content or the pH of the nutrient solution in which the plants are growing (van Dijk, 1978). In *Comptonia peregrina, Myrica gale,* and *Myrica cerifera,* cross-inoculation experiments under uniform environmental conditions show that the spore type of the nodules is determined by the inoculum and that this is true within and between these related species. For example, crushed nodule inoculum prepared from sp$^+$ *Comptonia peregrina* nodules gives rise to sp$^+$ nodules on *Myrica gale* as well as on *Comptonia peregrina* (VandenBosch and Torrey, 1985). Moreover, *Myrica gale* seedlings grown under uniform conditions and inoculated with five sp$^-$ and five sp$^+$ strains of *Frankia* consistently produced nodules of the same spore type as the inoculum used to induce the nodules (Monz and Schwintzer, 1989).

Although the spore type of nodules is apparently determined largely by the genotype of the *Frankia* strain, the number of sporangia formed within sp$^+$ nodules can be affected by the environment in which the host is growing. For example, in field-grown *Myrica gale* nodules, mature sporangia are more numerous in September than in June in the lobes produced during the current growing season (VandenBosch and Torrey, 1985); and in *Alnus glutinosa* seedlings growing in water culture, sporangia are more numerous at moderate pH than at low pH (van Dijk, 1978). Moreover, *Frankia* strain AiSp$^+$H from Finland showed only relatively weak development of sporangia in nodules on *Alnus incana* ssp. *incana* in water culture, whereas strain AiSp$^+$ showed strong development of sporangia (van Dijk *et al.*, 1988). It is possible that different sp$^+$ *Frankia* strains respond to environmental influences on the host to differing degrees.

Sp$^-$ strains are relatively easy to isolate and grow in pure culture (Chapter 3) but to date there are no strains in culture that consistently induce sp$^+$ nodules (Torrey, 1987) even though numerous attempts have been made to isolate frankiae from sp$^+$ nodules. However, some frankiae have been isolated from sp$^+$ nodules but these consistently induce sp$^-$ nodules (VandenBosch and Torrey, 1984; Hooker and Wheeler, 1987). These isolations have been made with difficulty (Burggraaf *et al.*, 1981; Daniere *et al.*, 1986; Hooker and Wheeler, 1987) and the resulting strains are less vigorous than isolates from sp$^-$ nodules (Normand and Lalonde, 1982). A variety of ineffective *Frankia* strains have also been isolated from sp$^+$ nodules of *Alnus glutinosa* (Hahn *et al.*, 1988). More than one strain of *Frankia* can be found within a single nodule (Chapters 3 and 5) and the strains that have been isolated from sp$^+$ nodules are probably not the dominant strains found in those nodules. This is indicated by the low success rate in these isolations and failure of most of the resulting isolates to induce sp$^+$ nodules. However, sp$^+$ nodules clearly contain abundant infective *Frankia* particles because crushed nodule inoculum prepared from sp$^+$ nodules is highly infective (Houwers and Akkermans, 1981; van Dijk, 1984; Sheppard *et al.*, 1988).

In studies of sp$^+$ and sp$^-$ nodules, sp$^-$ nodules are usually induced with *Frankia* strains in pure culture although they may also be induced with inoculum prepared from crushed nodules. In contrast, sp$^+$ nodules can only be induced with crushed nodule inoculum because no pure-cultured sp$^+$ *Frankia* strains are available.

II. Location of Spores within Nodules and Timing of Spore Development

Spores develop within sporangia that form either intrahyphally or terminally in both pure culture and within root nodule cells (Torrey, 1987). Each sporangium contains numerous, thick-walled, spherical or angular spores about 1.0 μm in diameter. Upon maturation, the sporangial wall disintegrates and the spores are released (Torrey, 1987).

The overall tissue organization of sp$^+$ nodules is similar to that of sp$^-$

nodules (van Dijk and Merkus, 1976; VandenBosch and Torrey, 1985). Within each nodule lobe there is an age gradient. The microsymbiont and the host cells near the apex of the nodule lobe are younger than those near the base. In sp^{+} nodules of *Alnus glutinosa, Comptonia peregrina,* and *Myrica gale,* sporangia begin to develop a short distance behind the nodule lobe apex in the region where vesicles have just matured (van Dijk and Merkus, 1976; VandenBosch and Torrey, 1985). The host cell senesces following the onset of sporangial differentiation and is dead when the sporangia are mature. Mature sporangia are found in the region of vesicle-containing cells and become increasingly common toward the base of the nodule lobe. Sporangia persist within the cells in the lower regions of the nodule lobe where the vesicles and hyphae have senesced and all infected cells have died (van Dijk and Merkus, 1976; VandenBosch and Torrey, 1985). Development of sporangia within vesicle-containing cells reduces vesicle longevity and hastens the death of the host cell. Sporangia do not form in all vesicle-containing cells and the degree of sporulation, and hence vesicle suppression, is not the same in all sp^{+} nodules (van Dijk, 1978; Wheeler *et al.*, 1981; VandenBosch and Torrey, 1985; Kashanski and Schwintzer, 1987).

Sporangia begin to form relatively early in the development of sp^{+} nodules on seedlings. In *Comptonia peregrina,* vesicles differentiate by 3 weeks after inoculation, young sporangia are first discernible by 4 weeks, and mature sporangia are present at 6 weeks (VandenBosch and Torrey, 1985). *Myrica gale* nodules show a similar pattern (VandenBosch and Torrey, 1985) with sporangia clearly visible at 6, 8, and 10 weeks after inoculation (Monz and Schwintzer, 1989). Moreover, in *Alnus glutinosa, A. cordata,* and *A. incana* ssp. *incana* seedlings, sporangia can be detected 6 weeks after inoculation (van Dijk, 1978).

In overwintering sp^{+} nodules in the field, sporangia begin to form in the current season's tissue in late spring and become increasingly numerous as the summer progresses until they are the predominant form of the microsymbiont in autumn. Although the nodules of actinorhizal plants are perennial, the vesicle-containing cells live for only one season and are newly produced each year (Schwintzer *et al.*, 1982). In *Myrica gale,* vesicles first differentiate in May and continue to differentiate through August. As in seedlings, sporangial differentiation follows the formation of vesicles. Sporangia are first seen in the current season's tissue in early June and become more prevalent in July and August when they are most numerous in the oldest tissues at the base of the nodule lobe. In September and October, sporangia become still more common, extending into all the formerly vesicle-containing tissue

(Schwintzer *et al.*, 1982; VandenBosch and Torrey, 1985). The sporangia persist in the dead, formerly vesicle-containing cells through the winter and are consistently present in the previous year's tissue throughout the following growing season (Schwintzer *et al.*, 1982; VandenBosch and Torrey, 1985).

III. Methods of Determining Nodule Spore Type

Nodules may be examined microscopically for presence of sporangia and spores in either hand sections of nodules or in microtome sections. To ensure that sporangia will be found if present, sections should be taken from the base of peripheral nodule lobes in the region containing the oldest infected cells. In the case of overwintering nodules collected early in the season, sections should be taken from the previous year's tissue. Free-hand sections are easily cut with a razor blade and are then mounted fresh in dilute Fabil reagent (Noel, 1964; van Dijk, 1978) or lactophenol-cotton blue (Torrey, 1987) and examined at 400× magnification. Nodules may be stored in plastic bags under refrigeration for up to three months prior to examination (Holman and Schwintzer, 1987).

Free-hand sections from sp$^+$ nodules often contain many sporangia and numerous free spores that are liberated during sectioning and are easily seen as a cloud around the edges of the tissue. Nodules from which such sections were taken can be quickly identified as sp$^+$, although care must be taken not to confuse the debris that is often found around sections of field nodules with spores. When sporangia are less numerous or absent, enough sections should be examined to survey a minimum of 50 clearly visible infected cells. We consider nodules to be sp$^+$ if more than one sporangium is found, otherwise they are sp$^-$.

Nodules can also be examined for spore type in 12-μm-thick sections cut from fixed material with a microtome (Smolander and Sundman, 1987). Twenty or more sections from each lobe are examined to determine nodule spore type.

IV. Physiology of Sp$^+$ and Sp$^-$ Nodules

Sp$^+$ and sp$^-$ nodules differ physiologically. In general, sp$^-$ nodules seem to be superior symbioses and are more effective than sp$^-$ nodules in supporting growth of the host plant.

A. Nitrogenase Activity

Measurements of nitrogenase activity in actinorhizal nodules have been made almost exclusively with the acetylene reduction method (see Chapter 13). It has recently been shown that acetylene reduction activity is not constant following addition of acetylene but shows a brief stable maximum followed by a decline to a minimum that is followed by a partial recovery (Silvester *et al.*, 1988a,b; Tjepkema *et al.*, 1988). All measurements made prior to 1988 were made without regard for the acetylene-induced decline and thus include an unknown quantitative error. However, when comparisons of sp^+ and sp^- nodules are made under controlled conditions and using nodules of similar age, the relative values are probably reliable.

Sp^+ and sp^- nodules probably do not differ in the extent of the acetylene induced decline. In *Myrica gale* nodules the time course of acetylene reduction following addition of acetylene is similar for both sp^+ and sp^- nodules. Moreover, concurrent nodule respiration also shows essentially the same pattern in both sp^+ and sp^- nodules (Monz and Schwintzer, 1989).

Specific nitrogenase activity (nitrogenase activity per unit weight of nodule) is generally greater in sp^- nodules than in sp^+ nodules, although there is considerable variation among strains belonging to the same spore type and some sp^- nodules have very low specific activities. Specific activity of sp^- nodules in *Alnus rubra* and *Alnus incana* ssp. *incana* is substantially greater than that of sp^+ nodules (Sellstedt *et al.*, 1986; Wheeler *et al.*, 1986). In *Alnus glutinosa*, mean nitrogenase activities are greater in sp^- than in sp^+ nodules but the difference is not statistically significant (van Dijk, 1978). In *Comptonia peregrina*, specific activity of nodules induced with a sp^- strain is much greater than in nodules containing a sp^+ strain (VandenBosch and Torrey, 1984). Specific nitrogenase activity has also been compared in *Myrica gale* nodules induced by four sp^- and five sp^+ strains obtained from *Myrica gale* (Monz and Schwintzer, 1989). Here the specific nitrogenase activity of the sp^- nodules as a group is 32% greater than that of the sp^+ nodules (Monz and Schwintzer, 1989).

B. Energy Cost of Nitrogen Fixation

1. CO_2/C_2H_4 Evolved Ratios and Biomass Invested in Nodules

The bulk of the energy cost of nitrogen fixation is due to nodule respiration, which can be approximated by the ratio of carbon dioxide evolved to ethylene evolved (CO_2/C_2H_4; one ethylene is evolved per acetylene reduced) (Tjepkema and Winship, 1980). Although data are available only for *Myrica gale* and *Comptonia peregrina*, sp^+ nodules apparently have higher rates of respiration per unit of nitrogenase activity than sp^- nodules. In *Myrica gale* the CO_2/C_2H_4 ratios of the sp^+ nodules induced by five sp^+ strains are 19% greater, as a group, than those induced by five sp^- strains (Monz and Schwintzer, 1989). Results for single sp^+ and sp^- strains in *Comptonia perigrina* are similar except that the difference between sp^+ and sp^- nodules is substantially larger than in *Myrica gale* (VandenBosch and Torrey, 1984).

Plants with sp^+ nodules incur additional energy costs of nitrogen fixation by investing more plant biomass (carbon) in nodules than plants with sp^- nodules. This increased biomass compensates in part for the lower specific nitrogenase activities characteristic of sp^+ nodules. Thus nitrogenase activity per plant in plants with sp^+ nodules is more comparable to that of plants with sp^- nodules than would be expected based on specific nitrogenase activity alone. In *Myrica gale* plants with nodules induced by five sp^+ and five sp^- strains of *Frankia*, the percentage of biomass invested in nodules is on average 45% higher in sp^+ than in sp^- plants (Monz and Schwintzer, 1989). Greater biomass in sp^+ nodules has also been observed in *Alnus rubra* in a comparison of single sp^+ and sp^- strains (Wheeler *et al.*, 1986).

2. Hydrogen Evolution

Hydrogen evolution is a possible means of energy loss from nodules (Chapter 7). In general, evolution of more than trace amounts of hydrogen is rare among actinorhizal nodules (Tjepkema *et al.*, 1986; Chapter 7) but the spore type of most of the nodules tested is not known. Among nodules where the spore type is known, both sp^+ and sp^- nodules evolve little or no hydrogen in *Comptonia peregrina* nodules induced by single *Frankia* strains (VandenBosch and Torrey, 1984) and *Myrica gale* nodules induced by five *Frankia* strains each (Monz and Schwintzer, 1989). However, in *Alnus incana* ssp. *incana*, nodules induced by a single sp^+ *Frankia* strain from Sweden evolve hydrogen freely (Sellstedt *et al.*, 1986). Attempts to find additional *Frankia* strains that evolve substantial amounts of hydrogen have been unsuccessful. Thus almost all actinorhizal nodules regardless of spore type lose only insignificant amounts of energy via hydrogen evolution.

C. Host Biomass Production and Effect of Plant Age

The favorable physiological characteristics of sp^- nodules described in the preceding, namely, greater specific nitrogenase activity and lower energy cost of nitrogen fixation, might be expected to be accompanied by greater host plant growth (biomass production) under nitrogen-limiting conditions. In fact this occurs. In *Myrica gale* plants with nodules induced by five sp$^+$ and five sp$^-$ strains of *Frankia*, the sp$^-$ nodules as a group supported 16% more host biomass than the sp$^+$ nodules (Monz and Schwintzer, 1989). In addition, sp$^-$ nodules also support greater host biomass production in *Alnus glutinosa* (Hall *et al.*, 1979), *Alnus incana* ssp. *incana* (Sellstedt *et al.*, 1986), *Alnus rubra* (Malcolm *et al.*, 1985), and *Comptonia peregrina* (VandenBosch and Torrey, 1984).

Plants nodulated with various strains of *Frankia* may show greater differences early in their development than later (Simon *et al.*, 1985; Sellstedt, 1988; Monz and Schwintzer, 1989). For example, in *Myrica gale*, differences between seedlings bearing a sp$^-$ strain and a sp$^+$ strain are substantially greater at 6 weeks after inoculation than at 10 weeks. At 6 weeks, plants nodulated with the sp$^-$ strain have 45% more total biomass than plants nodulated with the sp$^+$ strain, but at 10 weeks the sp$^-$ plants are only 19% larger. Comparable changes also occur in the relative growth rate, specific nitrogenase activity, and total nitrogen fixed per week (Monz and Schwintzer, 1989). These changes may be caused by limitation of growth by some factor other than nitrogen fixation, such as self-shading as the plants become larger. The possibility of changes in the relative performance of plants bearing different *Frankia* strains in the course of plant development should be given careful consideration when comparing strain differences based on single harvest dates.

On average, sp$^-$ *Frankia* strains are more desirable for practical applications than sp$^+$ strains because they exhibit greater specific nitrogenase activity, have a lower energy cost of nitrogenase activity, support the production of more host biomass, and are readily grown in culture. These characteristics make them good choices when selecting strains to be used in inoculating plants for various applications, including forestry, land reclamation, and amenity planting.

V. Ecological Distribution of Sp$^+$ and Sp$^-$ Nodules

In the field, sp$^-$ nodules are much more common than sp$^+$ nodules (Table II). At present, sp$^+$ nodules have been reported with substantial frequency only in the genus *Alnus* and in *Myrica gale*. Even in these taxa sp$^-$ nodules usually predominate except in *Alnus incana* ssp. *incana*. Individual sites having only sp$^-$ nodules are common, as are mixed sites with both spore types. Individual sites having only sp$^+$ nodules are relatively uncommon except in *Alnus incana* ssp. *incana* (Daniere *et al.*, 1986; Weber, 1986; Holman and Schwintzer, 1987; Kashanski and

Schwintzer, 1987). When both spore types are found at the same site they intermingle closely (Kashanski and Schwintzer, 1987) and can occur on the same plant (van Dijk, 1978).

A. Distribution of Sp^+ and Sp^- Nodules in the Field

This discussion is organized by host species. In *Alnus glutinosa,* both spore types occur in the Netherlands (van Dijk, 1978, 1984) and Finland (Weber, 1986) and sp^+ nodules are common in Scotland (Wheeler *et al.*, 1986). In the Netherlands the distribution of spore type at a 150-ha site on the island of Voorne varies with the age of the substrate. Sp^+ nodules are common in the old dune region, where alders are found in hedges and in abandoned fields. In contrast, sp^- nodules dominate the young dune region, which was stabilized only after 1910 (van Dijk, 1984). There is no relationship between the distribution pattern and pH between pH 5 and 8 (van Dijk, 1984). At a 0.1-ha site in a peat bog both types are closely intermingled and are frequently found close together on the same root (van Dijk, 1978). In Finland the nodules on *Alnus glutinosa* are predominantly sp^- with 75% of the sites examined being exclusively sp^- (Weber, 1986).

In *Alnus incana* ssp. *incana,* sp^+ nodules predominate in France (Daniere *et al.*, 1986; Kurdali *et al.*, 1989) and Finland (Weber, 1986). In Finland both spore types are found but sp^+ nodules predominate with many sites being exclusively sp^+. Spore type of the nodules at an experimental field varies with the presence or absence of forest soil. All nodules are sp^- on *Alnus incana* ssp. *incana* that have spontaneously invaded the field. But sp^+ nodules are dominant in a nearby spruce–*Alnus incana* ssp. *incana* forest and on plots in the field where forest soil has been added (Weber, 1986).

In *Alnus incana* ssp. *rugosa* both spore types occur in Quebec and Maine with sp^- nodules predominating (Normand and Lalonde, 1982; Holman and Schwintzer, 1987). In Maine the distribution of spore type varies with pH. Sp^+ nodules become more common at pH < 4.0. The distribution of spore type also varies with the length of time that a given site has been colonized by alders. At disturbed areas, including abandoned logging roads, roadside banks, sand and gravel pits, and other sites with exposed mineral soil only recently available for colonization by alder, the nodules are almost exclusively sp^-. In contrast, sp^+ nodules are common at streamsides and lakeshores that have long been available for colonization by alder (Holman and Schwintzer, 1987).

In *Myrica gale,* both spore types occur in Maine with sp^- nodules predominating (Kashanski and Schwintzer, 1987). Overall, sp^+ sites and mixed sites with both spore types dominate in the southern interior and coastal regions, areas with lower elevations and longer growing seasons. Sp$^-$ sites are found primarily in northern and western Maine, which have higher elevations and shorter growing seasons (Fig. 1). In all habitats occurrence of sp$^+$ nodules is positively correlated with the average number of frost-free days per year ($r = 0.57$ for 13 peatlands, $r = 0.71$ for 21 lakeshores and streamsides). In addition, at the lakeshore and

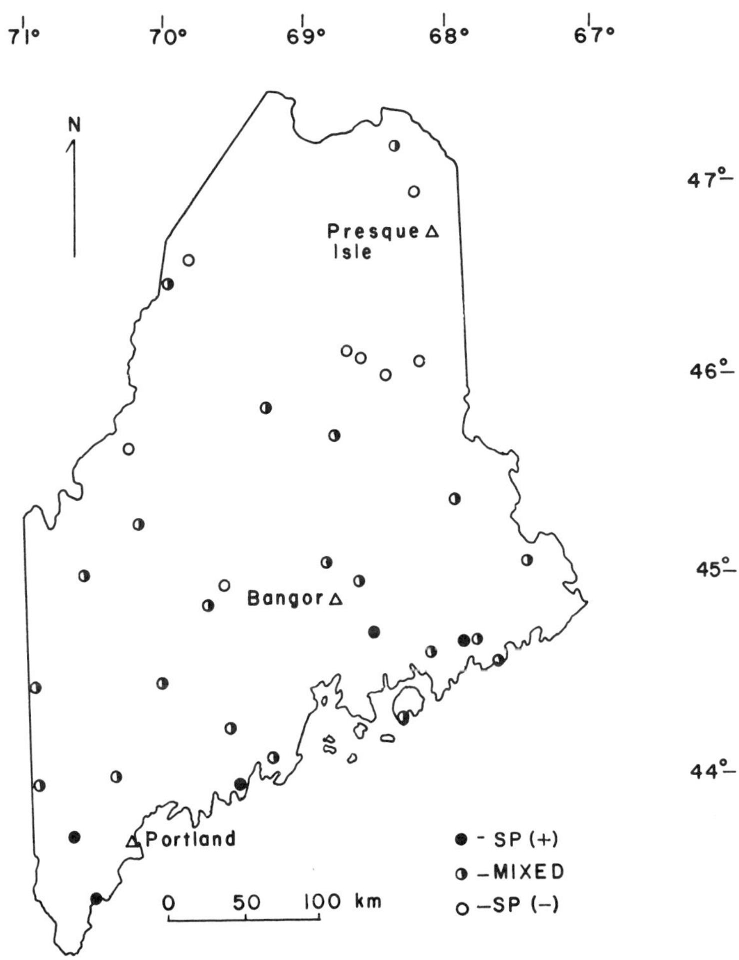

Figure 1 Map of Maine showing distribution of sp$^+$ and sp$^-$ nodules in *Myrica gale*. Reproduced with permission from Kashanski (1986).

streamside sites, occurrence of sp$^+$ nodules is also positively correlated with the percentage organic matter in the soil ($r = 0.56$) and negatively correlated with pH ($r = 0.60$). As in *Alnus incana* ssp. *rugosa*, sp$^+$ nodules become increasingly common at low pH (Fig. 2).

In *Comptonia peregrina* and *Myrica pensylvanica* the nodules are almost exclusively sp$^-$ (Schwintzer, 1989). In *Comptonia peregrina*, nodules at 31 sites in Maine, 7 in Massachusetts, and 2 in Michigan are exclusively sp$^-$. But at a site on Martha's Vineyard, an offshore island in Massachusetts, 30% of the nodules are sp$^+$ (VandenBosch and Torrey, 1985). In Maine, *Comptonia peregrina* is found primarily on recently disturbed sites that are relatively open and have coarse-textured soils with pH ranging from 3.8 to 6.1 (Schwintzer, 1989). Most of these populations probably became established only recently from buried seed following exposure of mineral soil (Schwintzer, 1989).

In soils devoid of host plants, *Frankia* strains that induce sp$^-$ nodules predominate. In fact most such soils produce only sp$^-$ nodules on test

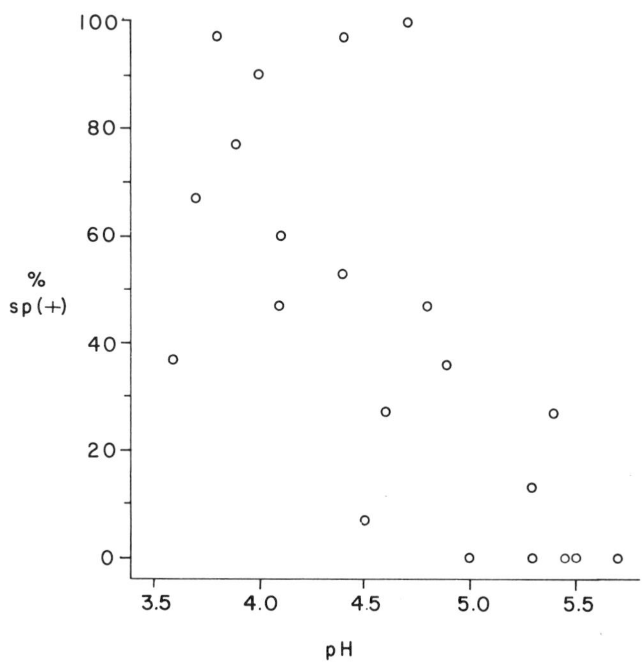

Figure 2 Relationship between percentage sp$^+$ nodules and pH at 21 lakeshore and streamside *Myrica gale* sites. Reproduced with permission from Kashanski and Schwintzer (1987).

plants. In the Netherlands, all nodules induced by soils from a birch stand are sp$^-$ (van Dijk, 1984). In Finland, nodules induced by soils from six sites, including spruce, pine, and birch forests, are all sp$^-$ in the broader sense. However, soils from a pine stand containing only a very small number of infective particles induce sp$^+$ nodules (Smolander and Sundman, 1987).

B. Possible Mechanisms Responsible for the Observed Patterns

Several factors probably interact to determine the distribution of sp$^+$ and sp$^-$ nodules in the field. Factors that may be involved include: host selection of *Frankia* strains (Weber, 1986), length of time that the host has been present on the site (Holman and Schwintzer, 1987), soil pH (Holman and Schwintzer, 1987; Kashanski and Schwintzer, 1987), and other, as yet unidentified factors.

Host selection may be a major factor in determining nodule spore type distribution in some instances. The host has been said to "select" certain *Frankia* strains (Weber, 1986) because some *Frankia* strains are able to form effective (nitrogen-fixing) nodules in all host species tested within a given genus or host-specificity group whereas others are not (VandenBosch and Torrey, 1983; Baker, 1987; Weber *et al.*, 1987; van Dijk, 1988). Some sp$^+$ strains have only a limited host range. For example, a sp$^+$ strain type, found in sp$^+$ nodules and soils in Finland and designated AiSp$^+$, forms effective nodules in *Alnus incana* ssp. *incana*, ineffective nodules in *Alnus glutinosa*, and no nodules at all in *Alnus nitida* (van Dijk *et al.*, 1988). *Alnus nitida* is not inherently incompatible with all sp$^+$ strains because strains type AgSp$^+$Holland forms effective nodules in this species (van Dijk *et al.*, 1988). Interestingly there is also a sp$^+$ strain type in France that induces effective nodules in *Alnus incana* ssp. *incana* but not in *Alnus glutinosa* (Domenach *et al.*, 1988). Similarly a sp$^+$ strain from *Myrica gale* forms effective nodules in *Myrica gale* but only ineffective nodules in *Comptonia peregrina* (VandenBosch and Torrey, 1983). Thus some hosts such as *Alnus glutinosa* may be incompatible with widely distributed sp$^+$ strains and consequently bear primarily sp$^-$ nodules. In contrast, other hosts, such as *Alnus incana* ssp. *incana*, are compatible with these strains and thus bear primarily sp$^+$ nodules. Host selection may be an important factor in the predominance of sp$^-$ nodules in *Alnus glutinosa* and sp$^+$ nodules in *Alnus incana* ssp. *incana* in Finland (Weber, 1986; Weber *et al.*, 1987). It may also be important in the great preponderance of sp$^-$ nodules in *Comptonia peregrina* (Schwintzer, 1989).

The length of time that the host has been present on the site may also be an important factor in determining nodule spore type distribution (Holman and Schwintzer, 1987). There is considerable evidence that some frankiae can grow saprophytically in soil (Chapter 1) and that these strains are predominantly sp^- (Weber, 1986; Holman and Schwintzer, 1987; Smolander and Sundman, 1987). Thus sp^- frankiae are presumably already present in soils at many sites that are being newly colonized by host plants and consequently sp^- nodules would predominate at such sites. This mechanism may explain the presence of predominantly sp^- nodules on *Alnus incana* ssp. *rugosa* at disturbed sites only recently available for colonization by alder in Maine (Holman and Schwintzer, 1987), on *Alnus glutinosa* in the young dune region in the Netherlands (van Dijk, 1984), and on *Alnus incana* ssp. *incana* invading the experimental field in Finland (Weber, 1986). It may also contribute to the great preponderance of sp^- nodules in *Comptonia peregrina* that occurs primarily on recently disturbed sites (Schwintzer, 1989).

In contrast to sp^- frankiae, sp^+ frankiae may be maintained on a given site primarily by growth within nodules, where they produce numerous spores (Houwers and Akkermans, 1981; van Dijk, 1984; van Dijk *et al.*, 1988). The spores are then released to the soil by nodule decay and may germinate and infect roots growing in their vicinity (van Dijk, 1979). Since there is no evidence that sp^+ strains make significant growth in soil, their persistence at a given site would depend on continued presence of suitable hosts (Holman and Schwintzer, 1987). Thus under natural conditions sp^+ nodules would be expected primarily at sites where the host has been present for extended periods. This mechanism may explain the relatively frequent occurrence of sp^+ nodules on *Alnus incana* ssp. *rugosa* growing at streamsides and lakeshores in Maine, where the host has presumably been present for centuries (Holman and Schwintzer, 1987), on *Myrica gale* in southern and eastern Maine (Kashanski and Schwintzer, 1987), and on *Alnus glutinosa* in the old dune region in the Netherlands (van Dijk, 1984).

Soil pH may also influence the distribution of nodule spore type. Even when other factors favor sp^- nodules, sp^+ nodules become increasingly common at low soil pH (Holman and Schwintzer, 1987; Kashanski and Schwintzer, 1987). The mechanisms by which pH helps to determine nodule type cannot be identified at the present time but may be related to reduction of saprophytic growth of sp^- frankiae at low pH. Evidence for reduced growth of sp^- frankiae at low pH comes from growth in pure culture (Burggraaf and Shipton, 1982; Shipton and Burggraaf, 1983; Smolander *et al.*, 1988) and greater survival of frankiae introduced into soils at pH > 6.0 than at pH < 4.5 (Smolander *et al.*, 1988). Reduced saprophytic growth of sp^- frankiae would favor sp^+ strains whose resistant spores can presumably tolerate harsher conditions (Holman and Schwintzer, 1987). Soil pH may be responsible for the increased occurrence of sp^+ nodules at low pH on *Alnus incana* ssp. *rugosa* and on *Myrica gale* at some habitats in Maine and possibly on *Alnus* spp. in Finland (Holman and Schwintzer, 1987; Kashanski and Schwintzer, 1987).

Given the complex environments found in the field, host selection, length of time the host has been present, and soil pH almost certainly interact to determine spore type distribution. Moreover, additional factors may be involved as well. Possible involvement of additional factors is suggested by the strong positive correlations in *Myrica gale* between the proportion of sp^+ nodules and the length of the growing season (Kashanski and Schwintzer, 1987).

Acknowledgments

I thank John D. Tjepkema for stimulating discussions and my colleagues for sending preprints and sharing unpublished information. This work was supported by USDA Grant 87-CRCR-1-2440.

References

Baker, D. D. (1982). A cumulative listing of isolated frankiae, the symbiotic nitrogen-fixing actinomycetes. *Actinomycetales* **17,** 35–42.

Baker, D. D. (1987). Relationship among pure cultured strains of *Frankia* based on host specificity. *Physiol. Plant.* **70,** 245–248.

Burggraaf, A. J. P., and Shipton, W. A. (1982). Estimates of *Frankia* growth under various pH and temperature regimes. *Plant Soil* **69,** 135–147.

Burggraaf, A. J. P., Quispel, A., Tak, T., and Valstar, J. (1981). Methods of isolation and cultivation of *Frankia* species from actinorhizas. *Plant Soil* **61,** 157–168.

Daniere, C., Capellano, A., and Moiroud, A. (1986). Dynamique de l'azote dans un peuplement naturel d'*Alnus incana* (L.) Moench. *Acta Oecol./Oecol. Plant.* **7,** 165–175.

Domenach, A. M., Kurdali, F., Daniere, C., and Bardin, R. (1988). Determination de l'identite isotopique de l'azote fixe par le *Frankia* associe au genre *Alnus*. *Can. J. Bot.* **66,** 1241–1247.

Hahn, D., Starrenburg, M. J. C., and Akkermans, A. D. L. (1988). Variable compatibility of cloned *Alnus glutinosa* ecotypes against ineffective *Frankia* strains. *Plant Soil* **107,** 233–243.

Hall, R. B., McNabb, H. S., Jr., Maynard, C. A., and Green, T. L. (1979). Toward development of optimal *Alnus glutinosa* symbioses. *Bot. Gaz. (Chicago), Suppl.* **140,** S120–S126.

Holman, R. M., and Schwintzer, C. R. (1987). Distribution of spore-positive and spore-negative nodules of *Alnus incana* ssp. *rugosa* in Maine, USA. *Plant Soil* **104,** 103–111.

Hooker, J. E., and Wheeler, C. T. (1987). The effectivity of *Frankia* for nodulation and nitrogen fixation in *Alnus rubra* and *A. glutinosa*. *Physiol. Plant.* **70,** 333–341.

Houwers, A., and Akkermans, A. D. L. (1981). Influence of inoculation on the yield of *Alnus glutinosa* in The Netherlands. *Plant Soil* **61,** 189–202.

Kashanski, C. R. (1986). "The Distribution of Spore-Positive and Spore-Negative Nodules of *Myrica gale* L. in Maine," M.S. thesis. Univ. of Maine, Orono, Maine.

Kashanski, C. R., and Schwintzer, C. R. (1987). Distribution of spore-positive and spore-negative nodules of *Myrica gale* in Maine, USA. *Plant Soil* **104,** 113–120.

Kurdali, F., Capellano, A., Moiroud, A., and Domenach, A. M. (1989). Study of the contribution of the shoot and/or root of *Alnus* sp. in the compatibility between the host and a Sp^+ *Frankia* strain using a grafting technique. *Plant Soil* **113,** 101–109.

Lalonde, M., Simon, L., Bousquet, J., and Seguin, A. (1988). Advances in the taxonomy of *Frankia*: Recognition of species *alni* and *elaeagni* and novel subspecies *Pommerii* and *Vandijkii*. *In* "Nitrogen Fixation: Hundred Years After" (H. Bothe, F. J. de Bruijn, and W. E. Newton, eds.), pp. 671–680. Fischer, Stuttgart.

Malcolm, D. C., Hooker, J. E., and Wheeler, C. T. (1985). *Frankia* symbiosis as a source of nitrogen in forestry: A case study of symbiotic nitrogen-fixation in a mixed Alnus–Picea plantation in Scotland. *Proc. R. Soc. Edinburgh, Sect. B* **85,** 263–282.

Monz, C. A., and Schwintzer, C. R. (1989). The physiology of spore-negative and spore-positive nodules of *Myrica gale*. *Plant Soil* **118,** 75–87.

Noel, A. R. A. (1964). A staining and mounting combination for sections of plant tissues. *Stain Technol.* **39,** 324–325.

Normand, P., and Lalonde, M. (1982). Evaluation of *Frankia* strains isolated from provenances of two *Alnus* species. *Can. J. Microbiol.* **28,** 1133–1142.

Schwintzer, C. R. (1989). All field-collected actinorhizae examined in *Comptonia peregrina* and *Myrica pensylvanica* in Maine are spore negative. *Can. J. Bot.* **67,** 1460–1464.

Schwintzer, C. R., Berry, A. M., and Disney, L. D. (1982). Seasonal patterns of root nodule growth, endophyte morphology, nitrogenase activity, and shoot development in *Myrica gale*. *Can. J. Bot.* **60,** 746–757.

Sellstedt, A. (1988). Nitrogenase activity, hydrogen evolution and biomass production in different Casuarina symbioses. *Plant Soil* **105,** 33–40.

Sellstedt, A., Huss-Danell, K., and Ahlqvist, A.-S. (1986). Nitrogen fixation and biomass production in symbioses between *Alnus incana* and *Frankia* strains with different hydrogen metabolism. *Physiol. Plant.* **66,** 99–107.

Sheppard, L. J., Hooker, J. E., Wheeler, C. T., and Smith, R. I. (1988). Glasshouse evaluation of the growth of *Alnus rubra* and *Alnus glutinosa* on peat and brown earth soils when inoculated with four sources of *Frankia*. *Plant Soil* **110,** 187–198.

Shipton, W. A., and Burggraaf, A. J. P. (1983). Aspects of cultural behavior of *Frankia* and possible ecological indications. *Can. J. Bot.* **61,** 2783–2792.

Silvester, W. B., Whitbeck, J., Silvester, J. K., and Torrey, J. G. (1988a). Growth, nodule morphology and nitrogenase activity of *Myrica gale* with roots grown at various oxygen levels. *Can. J. Bot.* **66,** 1762–1771.

Silvester, W. B., Silvester, J. K., and Torrey, J. G. (1988b). Adaptation of nitrogenase to varying oxygen tension in root nodules of *Alnus incana* ssp. *rugosa*. *Can. J. Bot.* **66,** 1772–1779.

Simon, L., Stein, A., Cote, S., and Lalonde, M. (1985). Performace of *in vitro* propagated *Alnus glutinosa* (L.) Gaertn. clones inoculated with *Frankiae*. *Plant Soil* **87,** 125–133.

Smolander, A., and Sundman, V. (1987). *Frankia* in acid soils of forests devoid of actinorhizal plants. *Physiol. Plant.* **70,** 297–303.

Smolander, A., van Dijk, C., and Sundman, V. (1988). Survival of *Frankia* strains introduced into soil. *Plant Soil* **106,** 65–72.

Tjepkema, J. D., and Winship, L. J. (1980). Energy requirements for nitrogen fixation in actinorhizal and legume root nodules. *Science* **209,** 279–281.

Tjepkema, J. D., Schwintzer, C. R., and Benson, D. R. (1986). Physiology of actinorhizal nodules. *Annu. Rev. Plant Physiol.* **37,** 209–232.

Tjepkema, J. D., Schwintzer, C. R., and Monz, C. A. (1988). Time course of acetylene reduction in nodules of five actinorhizal genera. *Plant Physiol.* **86,** 581–583.

Torrey, J. G. (1987). Endophyte sporulation in root nodules of actinorhizal plants. *Physiol. Plant.* **70,** 279–288.

VandenBosch, K. A., and Torrey, J. G. (1983). Host–endophyte interactions in effective and ineffective nodules induced by the endophyte of *Myrica gale. Can. J. Bot.* **61,** 2898–2909.

VandenBosch, K. A., and Torrey, J. G. (1984). Consequences of sporangial development for nodule function in root nodules of *Comptonia peregrina* and *Myrica gale. Plant Physiol.* **76,** 556–560.

VandenBosch, K. A., and Torrey, J. G. (1985). Development of endophytic *Frankia* sporangia in field- and laboratory-grown nodules of *Comptonia peregrina* and *Myrica gale. Am. J. Bot.* **72,** 99–108.

van Dijk, C. (1978). Spore formation and endophyte diversity in root nodules of *Alnus glutinosa* (L.) Vill. *New Phytol.* **81,** 601–615.

van Dijk, C. (1979). Endophyte distribution in the soil. *In* "Symbiotic Nitrogen Fixation in the Management of Temperate Forests" (J. C. Gordon, C. T. Wheeler, and D. A. Perry, eds.), pp. 84–94. For. Res. Lab., Oregon State Univ., Corvallis, Oregon.

van Dijk, C. (1984). "Ecological Aspects of Spore Formation in the Frankia–Alnus Symbiosis," Ph.D. thesis. State Univ., Leiden, The Netherlands.

van Dijk, C., and Merkus, E. (1976). A microscopical study of the development of a spore-like stage in the life cycle of the root-nodule endophyte of *Alnus glutinosa* (L.) Gaertn. *New Phytol.* **77,** 73–91.

van Dijk, C., Sluimer, A., and Weber, A. (1988). Host range differentiation of spore-positive and spore-negative strain types of *Frankia* in stands of *Alnus glutinosa* and *Alnus incana* in Finland. *Physiol. Plant.* **72,** 349–358.

Weber, A. (1986). Distribution of spore-positive and spore-negative nodules in stands of *Alnus glutinosa* and *Alnus incana* in Finland. *Plant Soil* **96,** 205–213.

Weber, A., Nurmiaho-Lassila, E.-L., and Sundman, V. (1987). Features of the intrageneric *Alnus–Frankia* specificity. *Physiol. Plant.* **70,** 289–296.

Wheeler, C. T., McLaughlin, M. E., and Steele, P. (1981). A comparison of symbiotic nitrogen fixation in Scotland in *Alnus glutinosa* and *Alnus rubra. Plant Soil* **61,** 169–188.

Wheeler, C. T., Hooker, J. E., Crowe, A., and Berrie, A. M. M. (1986). The improvement and utilization in forestry of nitrogen fixation by actinorhizal plants with special reference to *Alnus* in Scotland. *Plant Soil* **90,** 393–406.

10

The Molecular Genetics of *Frankia*

Beth C. Mullin
Department of Botany
The University of Tennessee
Knoxville, Tennessee, U.S.A.

Chung Sun An
Department of Botany
Seoul National University
Seoul, Korea

Our understanding of the molecular genetics of *Frankia* is in its infancy. Although the last decade led to significant advances in the understanding of many aspects of actinorhizal symbioses, for those interested in the molecular genetics of *Frankia*, much of this decade has been spent working on the most basic aspects of *Frankia* genetics. The goal of most studies directed toward the genetics of *Frankia* is to understand, on the molecular level, the essential elements of the nitrogen-fixing symbioses in which it participates. These elements include the status of *Frankia* in the soil and its response to the presence of host and nonhost plants, the contribution of *Frankia* to the initiation and development of the nodule, and the role of *Frankia* in nodule metabolism. Also of interest is the extent of genetic diversity observed among isolates. In this chapter, we will present what is known about *Frankia* genetics and discuss some of the problems encountered in working with this actinomycete symbiont. The reader is also encouraged to see two other recent reviews on this subject: those by Normand and Lalonde (1986) and by Simonet *et al.* (1990).

The Biology of *Frankia* and Actinorhizal Plants

I. *Frankia* as an Actinomycete

Several years before the successful isolation of *Frankia* strain CpI1 (Callaham *et al.*, 1978), electron microscopic studies of actinorhizal nodules revealed evidence of the actinomycetous character of nodule endophytes (Becking *et al.*, 1974). This observation was confirmed by the morphological examination of *in vitro*-cultured *Frankia* strain CpI1 (Newcomb *et al.*, 1979). Over the past decade, several hundred isolates have been cultured from a wide range of host plants. These cultured *Frankia* have a relatively uniform morphology that, although actinomycetous in character, differs significantly from that of other actinomycetes.

A. Physical Properties of the *Frankia* Genome

The availability of pure cultures of *Frankia* has made it possible to examine the physical properties of the *Frankia* genome and to compare its genome to that of other actinomycetes. The base composition of the DNA of several *Frankia* isolates as determined by thermal denaturation, buoyant density, and nucleoside analysis falls within the range of 68–72% guanine plus cytosine (An *et al.*, 1983). These values are well within the range of values determined for other actinomycetes of the same cell wall type (54–75%) (An *et al.*, 1987) and above the values reported for other nitrogen-fixing bacteria (40–65%) (Buchanan and Gibbons, 1974).

Direct comparison of the genomes of *Frankia* strains ArI4 and EuI1 to 11 non-*Frankia* actinomycetes of the same cell wall type by DNA–DNA hybridization failed to reveal any significant relatedness between *Frankia* and any of the other actinomycetes tested (An *et al.*, 1987). However, analysis of the 16 S ribosomal RNA genes of two *Frankia* strains revealed a distinct phylogenetic relationship between *Frankia* and actinomycetes in the genus *Geodermatophilus* (Fox and Stackebrandt, 1987; Hahn *et al.*, 1989), a genus of soil actinomycetes with multilocular sporangia.

The genome molecular weight of *Frankia* isolates ArI4 and EuI1 has been determined by reassociation kinetic analysis to be 8.3×10^9 and 6.0×10^9, respectively (An *et al.*, 1985a). These values are between two and two and a half times the size of the *Escherichia coli* genome and at the high end of the range of genome sizes reported for other actinomycetes (Bradley *et al.*, 1978). In the absence of highly repeated sequences in the DNA, the genome size can provide insight into the coding capacity of an organism. A genome molecular weight of 8.3×10^9 is equivalent to a genome with approximately 12×10^6 base pairs. If the average gene has a length of 2000 bp, *Frankia* strain ArI4 has the capacity to code for over 6000 average-size proteins, more than twice the number *E. coli* is capable of making. It is not difficult to understand the need for an increased coding capacity in *Frankia*. *Escherichia coli*, a morphologically simple intestinal bacterium, has evolved in a relatively stable environment rich in nutrients. *Frankia*, on the other hand, is a soil microorganism that is structurally more complex than *E. coli*. It has branching filaments, vesicles, sporangia, and spores and lives in the soil as well as in nodule tissue. Although there are no estimates of the percentage of the coding capacity of *Frankia* that is unique to its symbiotic role, it is likely to be substantial.

Analysis of the DNA reassociation curves for a second more rapidly reassociating component did not reveal the presence of high or moderately repeated sequences in the genomes of ArI4 and EuI1 (An *et al.*, 1985a). The thermal denaturation curves used to estimate the G + C content of 12 *Frankia* isolates were sharp, indicating a homogeneous distribution of bases in the DNA. Likewise, the photoelectric scans of DNA from the same 12 *Frankia* strains centrifuged to equilibrium in analytical CsCl gradients revealed symmetrical DNA peaks with the exception of strain $ACN1^{AG}$, which exhibited a slight shoulder, indicating some heterogeneity in the distribution of bases. Strain $ACN1^{AG}$ has subsequently been shown to harbor a plasmid (Simonet *et al.*, 1985) and it is reasonable to suggest that the plasmid may have a different base composition than the genomic DNA or may be supercoiled and thus behave as a satellite band in the CsCl gradient. Analysis of nucleoside composition of DNA from four *Frankia* strains by HPLC revealed no detectable modified or unusual bases (An *et al.*, 1983). If such bases are present, they comprise less than 4% of the total bases.

B. Genetic Diversity among *Frankia* Isolates

The generally uniform morphology and DNA base composition of *Frankia* isolates examined give little indication of the diversity that exists within the genus. As more strains have become available for study, they have been subjected to nodulation (Baker, 1987) and efficiency testing (Normand and Lalonde, 1982), serological analysis (Baker *et al.*, 1981), analysis of total protein patterns (Benson and Hanna, 1983; Benson *et al.*, 1984; Gardes and Lalonde, 1987), isozyme analysis (Gardes *et al.*, 1987), carbon utilization studies (Shipton and Burggraaf, 1982; Bloom *et al.*, 1989a), lipid analysis (Lalonde *et al.*, 1988), restriction pattern analysis (An *et al.*, 1985b; Dobritza, 1985; Bloom *et al.*, 1989b), restriction fragment length polymorphism analysis (Normand *et al.*, 1988; A. Nittayajarn, D. Baker, and B. Mullin, unpublished observations; A. Hilger and D. Myrold, personal communication), and DNA homology testing

(An *et al.*, 1985a; Bloom *et al.*, 1989b; Fernandez *et al.*, 1989). All of these methods have revealed the existence of a tremendous diversity among isolated *Frankia* strains.

The depth of the genetic diversity among these strains is best measured by DNA–DNA hybridization of total cellular DNAs. The first such DNA hybridization studies included 19 *Frankia* strains isolated from 12 different host plant species located in five different countries (An *et al.*, 1985a). Hybridization values ranged from 97 to 10% and exhibited no consistent correlation with host plant species or geographical origin. All isolates tested, however, showed a high degree of sequence conservation to probe strains in at least 10% of their genome as determined by melting curves of the hybrid DNAs. Subsequent studies of *Frankia* isolated from *Myrica pensylvanica*, all growing in the state of New Jersey (U.S.A.), demonstrated the same range of genetic divergence with DNA hybridization values ranging from 99 to 12% (Bloom *et al.*, 1989b). Fernandez *et al.* (1989) have recently conducted DNA–DNA hybridization studies on a large number of *Frankia* isolates and their results confirm earlier observations of the depth of genetic diversity among isolated frankiae.

Although DNA–DNA hybridization studies can reveal the genetic distance between organisms, they cannot reveal the extent of genetic diversity among closely related organisms. Restriction fragment patterns of total genomic DNA do, however, provide a sensitive measure of diversity within populations of frankiae. For example, using restriction fragment pattern analysis it is possible to distinguish between *Frankia* isolate AvcI1 and its cultural variant WgAvcI1 (An *et al.*, 1985b; Ide, 1986). This method of analysis has been used by Dobritsa (1985), An *et al.* (1985b), and Bloom *et al.* (1989b) to compare isolated strains and to group them according to restriction patterns. In one such study (Bloom *et al.*, 1989b) nine distinct restriction patterns were found among 16 *Frankia* strains all isolated from *Myrica pensylvanica* nodules from five locations in the state of New Jersey (U.S.A.) All but one collection site had representatives from two or more restriction pattern groups. A companion study revealed that each unique restriction pattern represented strains with unique carbohydrate utilization patterns (Bloom *et al.*, 1989a).

Genetic diversity among *Frankia* isolated from a single site has been assessed by gel electrophoresis of total cellular proteins. In a study of *Frankia* isolates from an *Alnus incana* ssp. *rugosa* stand in the state of Connecticut (U.S.A.), Benson and Hanna (1983) placed 43 nodule isolates into five gel groups based on one-dimensional SDS–polyacrylamide gel electrophoresis of total cell proteins with one group predominating. Total cell protein patterns were very similar and yet were distinct. Further analysis of a subset of these isolates by 2-D gel electrophoresis resulted in their regrouping into three rather than four gel groups (Benson *et al.*, 1984). A single isolate comprising a fifth gel group was not subjected to 2-D analysis. The difference in groupings demonstrates the increased resolution of 2-D gel electrophoresis. In this case the assignment of isolates to gel groups was less influenced by major protein bands as a result of 2-D analysis.

On a yet more limited scale, diversity may exist within or surrounding a single nodule. One of seven nodules examined by Benson and Hanna (1983) for dual infections did result in the culturing of isolates representing two separate protein gel groups. Likewise, Bloom *et al.* (1989b) found that three of ten nodules examined for dual infection gave rise to distinct isolates based on restriction fragment patterns and DNA–DNA hybridization of total DNA. In one case the isolates from a single nodule showed only 36% DNA sequence similarity based on solution hybridization. It is important to note that these results do not shed light on the question of whether two *Frankia* strains can be fully functional cosymbionts within the same cells of a nodule, rather only that they can both be cultured from the same nodule.

Burggraaf and Valstar (1984) have reported what they interpret as marked heterogeneity within the single isolate *Frankia* sp. LDAgp1, an *Alnus glutinosa* isolate that without careful analysis might have been interpreted as indicating genetic instability of the isolate. The initial isolate was only marginally infective on *A. glutinosa* seedlings, resulting in the formation of only a few nodules several weeks after inoculation. Reisolates from these nodules were fully infective, forming numerous nodules within 2–3 weeks after inoculation. Further differences were seen in substrate utilization. Reisolates were not able to grow on glucose but could use Tween 80 as a sole carbon source, whereas clones of the original isolate were able to utilize glucose but not Tween 80. The ability of the original culture of LDAgp1 to grow on Tween 80 was lost after regular subculturing on propionate medium. Additional heterogeneity in growth pattern, ability to sporulate, and rates of acetylene reduction was observed among clones derived from LDAgp1 by mild sonication and high dilution of the original culture. Subcloning from representatives of these three types of clones resulted in subclones that showed no variation from the initial clones from which they were derived. This indicates a lack of heterogeneity in the individual clones. Burggraaf and Valstar (1984) have suggested that the heterogeneity present in the initial culture may be due to the presence of plasmids, heteroclones, or more than one chromosome per cell. It is also possible that the initial isolate consisted of more than one *Frankia* strain or that heterogeneity was inadvertently introduced subsequent to the initial isolation.

C. Genetic Stability of *Frankia*

The knowledge that there exists such a high degree of diversity among *Frankia* isolated from a single host plant species growing in a limited geographical area or even from a single nodule leads to questions of the genetic stability of these actinomycetes. Two lines of evidence suggest that the *Frankia* genome is highly stable. First, despite the diversity present at any one location, isolates from two distant locations may be remarkably similar. Benson *et al.* (1984) found representatives of all three 2-D gel protein groups discussed earlier among isolates from nodules of several different host plants growing hundreds of miles away from the initial study site (Storrs, CT, U.S.A.). Bloom *et al.* (1989b) found isolates with identical DNA restriction patterns in nodules of host plants many miles apart. Furthermore, isolated from opposite sides of the Atlantic ocean, *Casuarina* isolates G2 and D11 (Diem *et al.*, 1982) share 97% sequence similarity as measured by DNA–DNA homology testing (An *et al.*, 1985a). The second line of evidence suggesting that the *Frankia* genome is quite stable comes from observations that *Frankia* cultures appear to be stable over many passages, maintaining their initial cultural and infectivity characteristics. *Frankia* isolate AvcI1 (Baker and Torrey, 1980), which has been grown in our laboratory for several years and has been subcultured numerous times, has maintained the same set of DNA restriction patterns over time, indicating that no major changes have occurred in its genome.

Despite the overall observed stability of *in vitro*-cultured *Frankia*, cultural variants have been reported. Parson *et al.* (1985) observed a spontaneous alteration in growth habit in a colony arising from the mycelial mat of an *Alnus rubra* single colony isolate WEY 0131391 that had been repeatedly subcultured over a 24-month period. The variant (WEY 0131392), which was subcloned and cultured separately, maintained its altered morphology over repeated subculturings. The symbiotic properties of the variant did not appear to differ from those of the parent culture, but differences were apparent in total protein patterns and in some antigenic determinants. Normand and Lalonde (1986) reported the existence of a spontaneous variant (EUN1fS20) of *Frankia* strain EUN1f, which had a 1000-fold increase in resistance to streptomycin.

To isolate spontaneous *Frankia* mutants, Lechevalier *et al.* (1987) used homogenized cells from a cloned *Casuarina cunninghamiana* isolate LLR 02022 (R43) (Zhang *et al.*, 1984) to inoculate liquid cultures containing 33 m*M* L-arabinose as a carbon source. Although the parent culture could not grow on arabinose, single arabinose-utilizing colonies arose in the cultures at a relatively high frequency. One of these, designated LLR 02023, has been extensively characterized (Lechevalier *et al.*, 1987). This mutant, although very stable, has been shown to revert at low levels to the parent phenotype through the use of the penicillin selection method of Gorini and Kaufman (1960) (M. P. Lechevalier, personal communication). Restriction pattern analysis of the parent and mutant strain revealed no differences between them, indicating that no major changes had occurred in the genome (B. C. Mullin, unpublished observations). Using the same penicillin selection method, attempts to isolate B type infective mutants from A type noninfective *Frankia* cultures have not been successful (M. P. Lechevalier, personal communication).

Another example of a spontaneous change detected in a *Frankia* culture occurred in strain AvcI1. This strain, which was isolated from *Alnus viridus* ssp. *crispa* by Baker and Torrey (1980), was subcultured and sent to other laboratories. After subculturing in one of those laboratories, the isolate was returned to Baker for confirmation of its identity. This strain, WgAvcI1, was found to not differ in cultural or symbiotic properties, but upon comparison of restriction fragment patterns a marked difference in patterns was evident (An *et al.*, 1985b). The overall restriction fragment patterns are very similar if not identical. However, superimposed on this pattern are extra bright bands in WgAvcI1. The differences seen in the restriction patterns of these two isolates have remained stable over many subcultures. We interpret the extra bright bands present in WgAvcI1 as possible plasmid-derived bands because they are most prominent in the first extractions of DNA and are less prominent or not present at all in third and fourth subsequent extractions of cells (Ide, 1986). It is not known yet whether the extra bright bands are sequences unique to WgAvcI1 or are an amplification of sequences present at a lower copy number in AvcI1.

D. The Presence of Plasmids in *Frankia* Isolates

The search for plasmids among *Frankia* strains has been motivated by several factors. One major factor is the need for a vector for transformation of *Frankia* cultures. An indigenous plasmid could be developed into a transformation vector and a mobile plasmid with a broad host range would be of even greater value in genetic studies. A second factor has been an attempt to find convenient methods to identify, group, or track

Frankia strains in the environment. And finally, the knowledge that some *Rhizobium* species carry symbiotic genes on plasmids (Prakash and Atherly, 1986) has stimulated research directed toward defining the role of plasmids in *Frankia*.

Although plasmidlike DNAs had been reported to be present in *Frankia* vesicles found within *Alnus glutinosa* nodule tissue (Dobritsa, 1982), the first published report of the presence of plasmids in cultured frankiae appeared in 1983 (Normand *et al.*, 1983). Of 39 strains of *Frankia* isolated from *Alnus* spp., *Elaeagnus* spp., and *Ceanothus* spp., plasmids were detected in only 4 strains. Strains ArI3 and CpI1 were both found to harbor plasmids of 8 and 18 kb, which on further analysis were found to have identical restriction patterns (Simonet *et al.*, 1985). Strain EUN1f had a 14-kb plasmid and strain ARgN22d had a total of 6 plasmids ranging in size from 8 to 32 kb as demonstrated by gel electrophoresis and electron microscopy. The 8-kb plasmid in strain ARgN22d had a restriction pattern identical to the 8-kb plasmids present in CpI1 and ArI3 (Simonet *et al.*, 1985). In subsequent analyses of subcultures of *Frankia* strain EUN1f using the same cell lysis method, additional plasmids of 40 and 80 kb were occasionally found (Normand and Lalonde, 1986). In some cases no plasmid bands were visible, indicating an instability in the presence or copy number of plasmids in EUN1f.

The 8-kb plasmid (pFQ31) isolated from *Frankia* strain ArI3 and described in the preceding paragraph as being present in two additional *Frankia* strains was cloned into pBR322 to facilitate mapping and further cloning (Normand *et al.*, 1985). A derivative of this plasmid containing the *Streptomyces* plasmid pIJ702 with a thiostrepton resistance gene was used to attempt transformation of *S. lividans* protoplasts. Although tiny colonies were observed on thiostrepton media after transformation, no further growth occurred and the recombinant plasmid could not be recovered.

Screening of more than 200 *Frankia* isolates using a more gentle cell lysis procedure resulted in the identification of 11 additional plasmid-bearing *Frankia* strains (Simonet *et al.*, 1984, 1985; Normand and Lalonde, 1986). The 14 plasmids found in these strains ranged in size from 8 to 60 kb with the exception of one 190-kb plasmid carried by strain $ARgP5^{AG}$. Restriction analysis has shown that 3 of these 11 strains carry indistinguishable 55-kb plasmids despite the fact that the host plants from which they were isolated were located as far apart as 800 km. The plasmids from two physiological group A strains appeared and disappeared with subculturing (Normand and Lalonde, 1986), as had been observed with the group A strain EUN1f.

Using a different group of isolates and a different cell lysis method,

Bloom *et al.* (1989b) have demonstrated the presence of a number of plasmidlike bands in 10 of 16 isolates from *Myrica pensylvanica* growing at five sites in the state of New Jersey (U.S.A.). This lysis method enriches for plasmid DNA in the first extraction of cells, resulting in the isolation of chromosomal DNA that is relatively free of plasmid DNA in subsequent extractions (Ide, 1986). In this way, it was possible to group isolates on the basis of chromosomal DNA restriction patterns using second extraction DNA, and at the same time to compare plasmid restriction patterns using first extraction DNA. The results of this analysis showed that in general similar isolates carry similar plasmids. Isolates with identical chromosomal restriction patterns can, however, harbor different plasmids, and isolates with different chromosomal restriction patterns can harbor similar plasmids. These results indicate that analysis of an isolate for a plasmid type cannot be used in strain identification, but can only indicate the presence or absence of that plasmid. In this study, 9 *Frankia* strains carrying plasmids were isolated from *Myrica pensylvanica* growing on coastal sand dunes. The tenth plasmid-carrying isolate was from a site with very poor drainage. This is in contrast to the other two collection sites, where no strains bearing plasmids were found and host plants were not under environmental stress.

To determine the distribution of plasmid-bearing *Frankia* strains in two alder stands in Orleans, France, *Frankia* plasmid pFQ56 was cloned into pBR322, radiolabeled, and used to probe blots of total nodule DNA that had been electrophoresed through agarose and blotted onto nitrocellulose (Simonet *et al.*, 1988a). The resulting distribution patterns clearly differentiated between areas colonized by frankiae harboring pFQ56 and those colonized by frankiae not harboring this plasmid. This technique is a powerful one because it eliminates the need to isolate and culture frankiae from nodules, thus increasing manyfold the number of nodules that can be tested for the presence of the plasmid. This technique would be suitable for the study of the distribution of plasmids in nodules of host plants growing in different environments and could lead to a definition of the role of at least some *Frankia* plasmids. A study such as this of plasmids in the nodules of *Myrica pensylvanica* in New Jersey could quickly determine whether or not a correlation exists between the presence of a plasmid and a particular environmental stress, as was suggested from the limited number of isolates studied from *Myrica pensylvanica* to date (Bloom *et al.*, 1989b).

Plasmids have been found in *Frankia* isolated from the following host plant genera: *Alnus*, *Elaeagnus*, *Comptonia*, *Sheperdia*, and *Myrica*. The presence of multiple plasmids within a single isolate is not unusual and as many as six plasmids have been observed. The copy number of

Frankia plasmids can vary within a single strain and the copy number of a specific plasmid found in two separate strains can also vary. Plasmids from physiological group B strains vary neither qualitatively nor quantitatively as a result of subculturing. Plasmids from physiological group A do vary, at times being undetectable, only to return again at a higher copy number in a later subculture. Relatively few *Frankia* strains have been shown to have plasmids. Of more than 200 isolates screened from a variety of host plants, only 8 (4%) were found to harbor plasmids. This is in sharp contrast to *Myrica pensylvanica* isolates from the state of New Jersey, where 10 out of 16 (62%) isolates were shown to have plasmids. The significance of this difference is not known because frankiae from other host plants growing in the same locations were not evaluated nor were *Myrica pensylvanica* isolates from other locations studied.

To date no functions have been assigned to *Frankia* plasmids. No unique properties or characteristics have been found in *Frankia* strains harboring plasmids. A clone of strain ArI3 regenerated from protoplasts and shown to be cured of its two plasmids pFQ31 and pFQ32 exhibited no detectable differences in cultural or symbiotic characteristics (Normand *et al.*, 1987). A 190-kb plasmid (pFQ69) found in *Frankia* ARgP5Ag (ULQ0132105099) has been shown to hybridize with the *Klebsiella nif* gene probe pSA30. However, it is not known whether these sequences represent potentially functional *nif* genes since the chromosomal DNA of the host strain also appears to carry a related sequence (Simonet *et al.*, 1986). Dobritsa *et al.*, as cited by Simonet *et al.* (1989), found hybridization of pSA30 to plasmidlike DNAs isolated from vesicles from *Alnus glutinosa* nodules, but again it is not known whether this represents a functional sequence. Other *Frankia* plasmids do not hybridize to pSA30.

II. *Frankia* as a Symbiont

It is the ability of *Frankia* to nodulate plants and to reduce nitrogen within nodule tissues that has attracted many biologists to its study. Other chapters in this book have detailed many aspects of the actinorhizal symbiosis, including host recognition, the infection process, the development of nodules, and nodule physiology and efficiency. All of these components of the symbiosis have a genetic basis and depend on the interaction between the genes of the host plant and those of the symbiont. Attempts to optimize the symbiosis for a given host plant in a selected environment will be facilitated by a better understanding of the genetics of the symbiosis. Any attempts to alter host ranges or engineer other aspects of the symbiosis are absolutely dependent on an understanding of the genetics.

The *Rhizobium*/*Bradyrhizobium*-legume symbiosis provides a good working model for the initiation of genetic studies on actinorhizal symbioses because many aspects of the symbioses are at least superficially similar. To date more than 30 rhizobial genes have been found that are essential for nodule formation and function. These genes have been divided into categories reflecting their role in the symbiosis. The reader is referred to three excellent reviews on rhizobial *nif* and *nod* genes (Gussin *et al.*, 1986; Appelbaum, 1990; Long, 1989). Undoubtedly as work progresses on the genetics of actinorhizal symbioses, families of nodulation and nitrogen fixation genes will emerge as well.

A. *nif* Genes in *Frankia*

Nitrogenase is an enzyme complex that consists of component 1, which is a tetrameric MoFe protein with two nonidentical subunits ($\alpha_2\beta_2$), and component 2, which is an Fe protein dimer with identical subunits. These two components have been found in all nitrogen-fixing bacteria examined to date, including *Frankia*. The structural genes of the nitrogenase complex are referred to as *nif*D and *nif*K for the α and β subunits of the MoFe protein and *nif*H for the Fe protein. Ruvkun and Ausubel (1980) reported the hybridization of a cloned fragment of *Klebsiella* DNA carrying the nitrogenase structural genes K, D, and H to DNA from 19 diverse nitrogen-fixing bacteria. Among the bacteria tested was *Frankia* strain CpI1. The 6.0-kb *Eco*RI fragment used as a probe hybridized to 7.4- and 5.2-kb fragments of an *Eco*RI digest of CpI1 DNA. Using the same *Klebsiella nif*KDH fragment from recombinant plasmid pSA30 (Cannon *et al.*, 1979), Ligon and Nakas (1987) reported the isolation of several hybridizing cosmid clones from a genomic library of *Frankia* strain FaC1, an effective isolate from *Alnus viridis* subsp. *crispa*. All the hybridizing cosmid clones hybridized to the *nif*K and *nif*D regions of pSA30. None of the clones hybridized to the *nif*H region, suggesting that *nif*H may not be contiguous to *nif*K and D in strain FaC1. Subsequent sequence analysis (Ligon and Nakas, 1989) has revealed that at least one of these cosmid clones (pFA28) contains *Klebsiella nif* KD sequences and not *Frankia nif* sequences.

Normand *et al.* (1988) found that *nif*H was contiguous with K and D in a number of *Frankia* strains. The *nif*KDH region from *Frankia* strain ArI3, an *Alnus rubra* isolate, was isolated from a phage genomic library and its partial restriction map is shown in Fig. 1. A *nif*H gene was subcloned from this fragment and sequenced. A potential Shine–Dalgarno sequence was found 7 bp upstream from the start codon of the gene. The open reading frame that consists of 861 bp is terminated by a single stop codon and there is a hint of a possible stemmed loop structure 28 bases

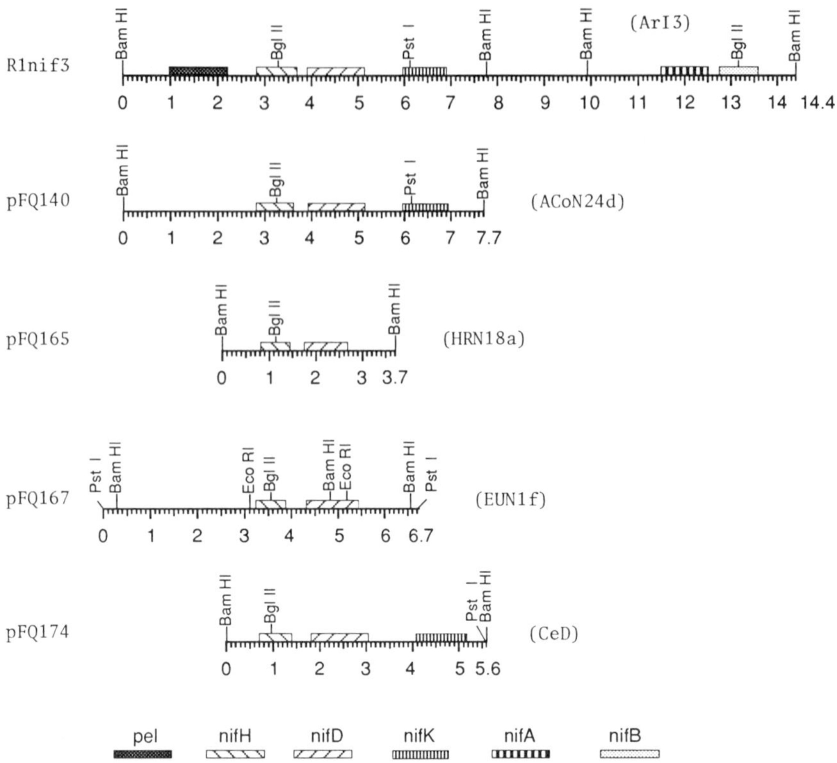

Figure 1 Partial restriction maps of cloned *Frankia* DNA containing *nif* hybridizing sequences. The shaded boxes represent the approximate location of DNA sequences that hybridize to the indicated heterologous gene probes. R1nif3 and pFQ140 are clones from strains ArI3 and ACoN24d, *A. rubra* and *A. cordata* isolates, respectively (Simonet *et al.*, 1988b; Normand *et al.*, 1988). pFQ165 is a clone from strain HRN18a, a *Hippophae rhamnoides* isolate; pFQ167 is a clone from strain EUN1f, an *Elaeagnus umbellata* isolate; and pFQ174 is a clone from CeD, a *Casuarina equisetifolia* isolate (Normand *et al.*, 1988).

3′ to this site. The amino acid sequence derived from nucleotide sequence analysis indicates that *nif*H is highly conserved. Percentage amino acid similarity values between ArI3 and *B. japonicum*, *K. pneumoniae*, *A. vinelandii*, and *Anabaena* strain 7120 are 74%, 75%, 76%, and 78%, respectively. Codon usage in *nif*H is highly skewed toward G + C with a 94% occurrence of G or C in the third position. Normand and Bousquet (1989) have sequenced *nif*H from *Frankia* strain HRN18a, a *Hippophae rhamnoides* isolate. The *nif*H from this isolate has 93% nucleotide sequence similarity and 96% derived amino acid sequence similarity with *nif*H from ArI3, setting an estimated time of divergence of these isolates at 100 million years. Nucleotide sequence comparisons such as these will undoubtedly provide valuable insight into the evolution of various groups of *Frankia* strains. Comparison of *nif*H gene sequences from *Frankia* and other nitrogen-fixing bacteria shows that the *Frankia* sequences are more similar to those from *Anabaena* than to other gram-positive nitrogen-fixing species, which supports the hypothesis that *nif* genes were transmitted horizontally among bacteria rather than vertically from a common ancestor.

The diversity present among *Frankia* isolates is reflected in the sequence divergence of *nif*K, D, and H as revealed by restriction fragment length polymorphism analysis. In three *Frankia* strains tested by Normand *et al.* (1988), *nif*K, D, and H were clustered within a 5-kb region (Fig. 1). In *Frankia* strain ULQ0132105009, *nif*K, D, and H hybridizing sequences were found in plasmid as well as chromosomal DNA. These sequences were present on different restriction fragments, indicating differences in the nucleotide sequences surrounding the clusters. No information is available yet about the similarity of the actual gene sequence. Analysis of hybridizing restriction fragments from *Alnus*-compatible, *Elaeagnus*-compatible, and *Casuarina*-compatible *Frankia* strains has indicated a high degree of restriction site conservation among the *Alnus*-compatible strains and less among the *Elaeagnus*-compatible strains (Normand *et al.*, 1988). The *Casuarina*-compatible strains examined had similar hybridizing patterns. All the strains examined, except two, had a 0.4-kb SstI fragment that mapped to the middle of *nif*H. The hypothesis made by An *et al.* (1985a) that the highly conserved portion of the *Frankia* genome detected by T_m analysis of hybrid DNAs was likely to reflect the conservation of sequences related to nitrogen fixation is supported by these observations.

In *K. pneumoniae* and in *R. meliloti*, nitrogenase expression is in part controlled by the *nif*A gene, which itself is induced by low-oxygen tension. A *nif*A probe from *R. meliloti* hybridized to cloned DNA from *Frankia* strain ArI3 and the hybridizing region was located 4.5 kb away from the *nif*HDK cluster (Fig. 1; Simonet *et al.*, 1988b). This indicates that a *nif*A-like regulatory gene may be operating in *Frankia* species. A *nif*B probe, also from *R. meliloti*, hybridized to the ArI3 clone in a position adjacent to the *nif*A-like sequence (Fig. 1). These results indicate some clustering of *nif* genes in *Frankia*, which will greatly facilitate their study.

B. *nod* Genes in *Frankia*

The ease with which several of the *nif* genes in *Frankia* have been identified is not likely to be repeated with the nodulation genes. We and others have attempted with limited success to probe *Frankia* DNA for sequences that hybridize to known nodulation genes from *Rhizobium* and *Bradyrhizobium*. Hybridization signals are weak and not reproducible enough to be used to screen genomic libraries. Reddy *et al.* (1988) have attempted to use functional complementation rather than nucleotide sequence similarity to isolate *Frankia nod* genes. Using *Frankia* DNA cloned into the broad host range vector pLAFR3 to complement mutants, several *R. meliloti* Nod^- mutant transconjugates were found to be able to nodulate alfalfa. DNA from a cosmid clone thought to contain the complementing sequences is being mapped and further complementation studies are under way (Reddy *et al.*, 1988).

All the nodulation genes of Rhizobium have been initially identified from studies of Nod^- mutants. Because such studies are not yet feasible with *Frankia*, a reasonable approach to studying the genetics of nodulation in *Frankia* is to consider carefully what enzymes might be involved in the nodulation process. Studies on the physiology, biochemistry, and fine structure of developing nodules should provide clues about what genes might be involved. Clues may also be provided by a study of the way in which certain plant pathogens interact with their host plants. Using this strategy, Séguin and Lalonde (1989) looked for genes that might be involved in plant cell wall degradation. By hybridization with a pectate lyase gene probe from *Erwinia chrysanthemia*, they found sequences with a high degree of similarity in genomic DNA from *Frankia*. It had been proposed by Lalonde and Knowles (1975) that *Frankia* degrade plant pectins during the infection process, perhaps using them as a carbon source. Simonet *et al.* (1990) have located the region of the *Frankia* genome to which the pectate lyase probe hybridizes and surprisingly it is contiguous to *nif*H (Fig. 1). Simonet *et al.* (1990) and A. Reddy, A. M. Hirsch, and J. G. Torrey (personal communication) have detected hybridization between cellulase gene probes and *Frankia* DNA.

C. Other Genes in *Frankia*

Other genes involved in nodulation or nodule physiology may also be accessible by using heterologous gene probes. For example, D. R. Benson and D. A. Rochefort (personal communication) have found that *Bradyrhizobium* genes for hydrogen uptake and *Rhizobium* genes for glutamine synthetase I and II all hybridize to *Frankia* DNA. *Frankia* strains in culture produce the plant growth regulators auxin (Wheeler *et al.*, 1984; Berry *et al.*, 1989) and cytokinin (Stevens and Berry, 1988), and these may well have an important role in nodule formation. When the cytokinin biosynthetic dimethylallyl PPi : 5′-AMP dimethylallyl transferase gene from *Agrobacterium tumefaciens* (*tzs* gene) was used to probe Southern blots of *Frankia* HFPArI3, hybridization was detected, indicating that this gene may be accessible by heterologous hybridization as well (A. M. Berry, personal communication).

III. Approaches to Studying *Frankia* Genetics

The most fruitful studies on the genetics of *Frankia* will come only after we have learned about the natural processes of genetic exchange among *Frankia* strains and have been able to develop mutation selection schemes, and mechanisms to introduce and stabilize DNA within *Frankia* cells. Until then our knowledge is limited by the use of heterologous probes and the inability to unequivocally demonstrate gene function.

Mutant *Frankia* strains are important to study because they permit both phenotypic and genotypic comparisons to be made between the wild type and mutant. Mutants provide a mechanism for determining gene function through complementation studies and can lead to information relating gene structure to gene expression. Attempts to mutagenize *Frankia* have met with limited success and only a few reports appear in the literature (Simonet *et al.*, 1990; D. R. Benson, personal communication). Because *Frankia* is a filamentous organism, the plating of single cells, which is desirable for isolating mutants, is problematic. This problem has been overcome in other actinomycete systems by the use of protoplast formation prior to plating. Faure-Raynaud *et al.* (1984) were able to produce protoplasts from pure *Frankia* strains isolated from alder using a combination of lysozyme and achromopeptidase, but no regeneration of cultures was demonstrated. Protoplast formation followed by dilution and regeneration has since been described for *Frankia* (Normand *et al.*, 1987; Tisa and Ensign, 1987), but these procedures are far from routine, partially because of the inefficiency of regeneration and the exceedingly slow growth rate of *Frankia*. To overcome these difficulties, Prakash and Cummings (1988) attempted to produce a fast-growing *Frankia* hybrid by protoplast fusion with *Streptomyces griseofuscus*. Although a fast-growing nodulating fusion product capable of fixing nitrogen was reportedly produced, no independent confirmation of this fusant has been possible (P. Normand, personal communication).

Because *Frankia* is a spore-forming actinomycete, the potential exists for using spores in mutagenesis studies. However, spores do not germinate readily, if at all, in most cultured *Frankia* strains. An exception to this is found in *Frankia* strain CeI5, where good spore germination has been observed (H. Berg, personal communication). This system has great potential to facilitate genetic studies of *Frankia* and it provides hope that other isolates will be found with spores that germinate efficiently.

Only a handful of *Frankia* mutants are available for study and fewer are available that appear to be symbiotic mutants. One such strain is AirI2, an *Alnus incana* subsp. *rugosa* isolate that has a Nif^- phenotype (Lechevalier *et al.*, 1983). Attempts to show hybridization of a *Frankia nif*HDK probe to DNA from this isolate were not successful, indicating that it is likely that the *nif* region has been deleted from this strain (Normand *et al.*, 1988). It may be possible to use this strain to develop a transformation vector using the ability to grow without reduced nitrogen as a selection mechanism. It may also be possible to use nonhost actinorhizal plants to select for the transfer of nodulation genes from one *Frankia* strain to another in a different host compatibility group. Selection for antibiotic resistance may also be feasible as a way to select cells containing recombinant DNAs. L. Simon, working with Normand and Lalonde (1986), has tested the background sensitivities of a group of *Frankia* strains to various antibiotics and has found a number of strains with very low resistance to the antibiotics gentamycin, kanamycin, neomycin, streptomycin, vancomycin, and thiostrepton.

One additional major tool that is not yet available for the study of *Frankia* genetics is a vector to introduce and replicate DNA in *Frankia* cells. The best candidates may be the *Frankia* plasmids themselves. Nothing is known yet about the role of *Frankia* plasmids, their potential to carry selectable markers, or conditions necessary to introduce them with high efficiency into *Frankia* cells. *Frankia* cells that have been cured of their plasmids, such as the plasmid-less LC2 variant of strain ArI3 (Normand *et al.*, 1987), would be excellent choices for studies on vector development and transformation. Likewise, the *Myrica pensylvanica* isolate that had the same genomic restriction pattern as a strain carrying a plasmid (Bloom *et al.*, 1989b) might be a good recipient strain for the plasmid. Actinophages may be potential vectors for transformation of *Frankia* as well. There has been only one report of a phage being present in a *Frankia* culture (M. P. Lechevalier, personal communication) and this phage has not been characterized.

IV. Concluding Remarks

The actinomycetous nature of *Frankia* and its importance in forest ecosystems have now been firmly established. We are, however, only beginning to understand the genetics of *Frankia* and the molecular biology of its symbiotic association with host plants. Fortunately, there have been extensive studies on the genetics of streptomyces and the molecular biology of the *Rhizobium*–legume symbioses. Although they are vastly different organisms from *Frankia*, we can learn a great deal from studies done on these organisms. Attention must also be focused on the ecology of *Frankia* and the molecular biology of its host plants to be able to fully utilize the nitrogen-fixing capabilities of *Frankia*.

We must keep in mind that these organisms with which it is so difficult to work and which slow our own progress as researchers are themselves tremendously successful. As a group they can nodulate a wide range of host plants and survive in the soil in the absence of known host plants. The great extent of genetic diversity among *Frankia* is a measure of their ability to change, adapt, and survive, and we as researchers must continue to search for ways to gain insight into their success.

References

An, C. S., Wills, J. W., Riggsby, W. S., and Mullin, B. C. (1983). Deoxyribonucleic acid base composition of 12 *Frankia* isolates. *Can. J. Bot.* **61,** 2859–2862.

An, C. S., Riggsby, W. S., and Mullin, B. C. (1985a). Relationships of *Frankia* isolates based on deoxyribonucleic acid homology studies. *Int. J. Syst. Bacteriol.* **35,** 140–146.

An, C. S., Riggsby, W. S., and Mullin, B. C. (1985b). Restriction pattern analysis of genomic DNA of *Frankia* isolates. *Plant Soil* **87,** 43–48.

An, C. S., Riggsby, W. S., and Mullin, B. C. (1987). DNA relatedness of *Frankia* isolates ArI4 and EuI1 to other actinomycetes of cell wall type III. *Actinomycetes* **20,** 50–59.

Appelbaum, E. (1990). The *Rhizobium/Bradyrhizobium*–legume symbiosis. *In* "The Molecular Biology of Symbiotic Nitrogen Fixation" (P. M. Gresshoff, ed.), pp. 131–158. CRC Press, Boca Raton, Florida.

Baker, D. D. (1987). Relationships among pure cultured strains of *Frankia* based on host specificity. *Physiol. Plant.* **70,** 245–248.

Baker, D., and Torrey, J. G. (1980). Characterization of an effective actinorhizal microsymbiont, *Frankia* sp. AvcI1 (Actinomycetales). *Can. J. Microbiol.* **26,** 1066–1071.

Baker D. D., Pengelly, W. L., and Torrey, J. G. (1981). Immunological analysis of relationships among the isolated frankiae (*Actinomycetales*). *Int. J. Syst. Bacteriol.* **31,** 148–151.

Becking, J. H., de Boer, W. E., and Houwink, A. L. (1964). Electron microscopy of the endophyte of *Alnus glutinosa*. *Anthonie van Leeuwenhoek* **30,** 343–376.

Benson, D. R., and Hanna, D. (1983). *Frankia* diversity in an alder stand as estimated by sodium dodecyl sulfate–polyacrylamide gel electrophoresis of whole-cell proteins. *Can. J. Bot.* **61,** 2919–2923.

Benson, D. R., Buchholz, S. E., and Hanna, D. G. (1984). Identification of *Frankia* strains by two-dimensional polyacrylamide gel electrophoresis. *Appl. Environ. Microbiol.* **47,** 489–494.

Berry, A. M., Kahn, R. K. S., and Booth, M. C. (1989). Identification of indole compounds secreted by *Frankia* HFPArI3 in defined culture medium. *Plant Soil,* **118,** 205–209.

Bloom, R. A., Lechevalier, M. P., and Tate, R. L. III, (1989a). Physiological, chemical, morphological and plant-infectivity characteristics of *Frankia* isolates from *Myrica pensylvanica*: Correlation to DNA restriction patterns. *Appl. Environ. Microbiol.* **55,** 2161–2166.

Bloom, R. A., Mullin, B. C., and Tate, R. L., III (1989b). DNA restriction patterns and solution hybridization studies of *Frankia* isolates from *Myrica pensylvanica* (bayberry). *Appl. Environ. Microbiol.* **55,** 2155–2160.

Bradley, S. G., Enquist, L. W., and Schribner, H. E., III (1978). Heterogeneity among deoxyribonucleotide sequences of Actinomycetales. *In* "Genetics of the Actinomycetales" (E. Freerksen, I. Tarnok, and J. H. Thumin, eds.), pp. 207–224. Fischer, Stuttgart.

Buchanan, R. E., and Gibbons, N. E. (1974). "Bergey's Manual of Determinative Bacteriology." Williams & Wilkins, Baltimore, Maryland.

Burggraaf, A. J. P., and Valstar, J. (1984). Heterogeneity within *Frankia* sp. LDAgp1 studied among clones and reisolates. *Plant Soil* **78,** 29–43.

Callaham, D., Del Tredici, P., and Torrey, J. G. (1978). Isolation and cultivation *in vitro* of the actinomycete causing root nodulation in *Comptonia*. *Science* **199,** 899–902.

Cannon, F. C., Riedel, G. E., and Ausubel, F. M. (1979). Overlapping sequences of *Klebsiella pneumoniae nif* DNA cloned and characterized. *Mol. Gen. Genet.* **174,** 59–66.

Diem, H. G., Gauthier, D., and Dommergues, Y. R. (1982). Isolation of *Frankia* from nodules of *Casuarina equisetifolia*. *Can. J. Microbiol.* **28,** 526–530.

Dobritsa, S. V. (1982). Extrachromosomal circular DNAs in endosymbiont vesicles from *Alnus glutinosa* nodules. *FEMS Microbiol. Lett.* **15,** 87–91.

Dobritsa, S. V. (1985). Restriction analysis of the *Frankia* spp. genome. *FEMS Microbiol. Lett.* **29,** 123–128.

Faure-Raynaud, M., Bonnefoy, M.-A., Perradin, Y., Simonet, P., and Moiroud, A. (1984). Protoplast formation from *Frankia* strains. *Microbios* **41,** 159–166.

Fernandez, M. P., Meugnier, H., Grimont, P. A. D., and Bardin, R. (1989). Deoxyribonucleic acid relatedness among members of the genus *Frankia*. *Int. J. Syst. Bacteriol.* **39,** 424–429.

Fox, G. E., and Stackebrandt, E. (1987). The application of 16S rRNA cataloguing and 5S RNA sequencing in bacterial systematics. *Methods Microbiol.* **19,** 405–458.

Gardes, M., and Lalonde, M. (1987). Identification and subgrouping of *Frankia* strains using sodium dodecyl sulfate polyacrylamide gel electrophoresis. *Physiol. Plant.* **70,** 237–244.

Gardes, M., Bousquet, J., and Lalonde, M. (1987). Isozyme variation among 40 *Frankia* strains. *Appl. Environ. Microbiol.* **53,** 1596–1603.

Gorini, L., and Kaufman, H. (1960). Selecting bacterial mutants by the penicillin method. *Science* **131,** 604–605.

Gussin, G. N., Ronson, C. W., and Ausubel, F. M. (1986). Regulation of nitrogen fixation genes. *Annu. Rev. Genet.* **20,** 567–591.

Hahn, D., Lechevalier, M. P., Fisher, A., and Stackebrandt, E. (1989). Evidence for a close phylogenic relationship between members of the genera *Frankia*, *Geodermatophilus*, and "*Blastococcus*" and emdenation of the family Frankiaceae. *System. Appl. Microbiol.* **11,** 236–242.

Ide, P. I. (1986). "Optimization of Conditions for Restriction Fragment Length Polymorphism Analysis in the Identification of *Frankia* Strains," M.S. thesis. Univ. of Tennessee, Knoxville, Tennessee.

Lalonde, M., and Knowles, R. (1975). Ultrastructure, composition, and biogenesis of the encapsulation material surrounding the endophyte in *Alnus crispa* var. *mollis* root nodules. *Can. J. Bot.* **53,** 1951–1971.

Lalonde, M., Simon, L., Bousquet, J., and Séguin, A. (1988). Advances in the taxonomy of *Frankia*: Recognition of species *alni* and *elaeagni* and novel subspecies *pommerii* and *vandijkii*. *In* "Nitrogen Fixation: Hundred Years After" (H. Bothe, F. J. Bruijn, and W. E. Newton, eds.), pp. 671–680. Fischer, Stuttgart.

Lechevalier, M. P., Baker, D., and Horrière, F. (1983). Physiology, chemistry, serology and infectivity of two *Frankia* isolates from *Alnus incana* subsp. *rugosa*. *Can. J. Bot.* **61,** 2826–2833.

Lechevalier, M. P., Labeda, D. P., and Ruan, J.-S. (1987). Studies on *Frankia* sp. LLR 02022 from *Casuarina cunninghamiana* and its mutant LLR 02023. *Physiol. Plant.* **70,** 249–254.

Ligon, J. M., and Nakas, J. P. (1987). Isolation and characterization of *Frankia* sp. strain FaC1 genes involved in nitrogen fixation. *Appl. Environ. Microbiol.* **53,** 2321–2327.

Ligon, J. M., and Nakas, J. P. (1989). Nucleotide and deduced amino acid sequences of *nif*K and partial sequences *nif*D from *Frankia* sp. FaC1. *Nucleic Acids Research,* **16,** 11843.

Long, S. R. (1989). *Rhizobium*–legume nodulation: Life together in the underground. *Cell (Cambridge, Mass.)* **56,** 203–214.

Newcomb, W., Callaham, D., Torrey, J. G., and Peterson, R. L. (1979). Morphogenesis and fine structure of the actinomycetous endophyte of nitrogen fixing root nodules of *Comptonia peregrina*. *Bot. Gaz. (Chicago), Suppl.* **140,** S522–S534.

Normand, P., and Bousquet, J. (1989). Phylogeny of nitrogenase sequences in *Frankia* and other nitrogen-fixing microorganisms. *J. Mol. Evol.* **29,** 436–447.

Normand, P., and Lalonde, M. (1982). Evaluation of *Frankia* strains isolated from provenances of two *Alnus* species. *Can. J. Microbiol.* **28,** 1133–1142.

Normand, P., and Lalonde, M. (1986). The genetics of actinorhizal *Frankia*: A review. *Plant Soil* **90,** 429–453.

Normand, P., Simonet, P., Butour, J. L., Rosenberg, C., Moiroud, A., and Lalonde, M. (1983). Plasmids in *Frankia* sp. *J. Bacteriol.* **155,** 32–35.

Normand, P., Downie, J. A., Johnston, A. W. B., Kieser, T., and Lalonde, M. (1985). Cloning of a multicopy plasmid from the actinorhizal nitrogen-fixing bacterium *Frankia* sp. and determination of its restriction map. *Gene* **34,** 367–370.

Normand, P., Simonet, P., Prin, Y., and Moiroud, A. (1987). Formation and regeneration of *Frankia* protoplasts. *Physiol. Plant.* **70,** 259–266.

Normand, P., Simonet, P., and Bardin, R. (1988). Conservation of *nif* sequences in *Frankia*. *Mol. Gen. Genet.* **213,** 238–246.

Parson, W. L., Robertson, L. R., and Carpenter, C. V. (1985). Characterization and infectivity of a spontaneous variant isolated from *Frankia* sp. WEY 0131391. *Plant Soil* **87,** 31–42.

Prakash, R. K., and Atherly, A. G. (1986). Plasmids of *Rhizobium* and their role in symbiotic nitrogen fixation. *Int. Rev. Cytol.* **104,** 1–24.

Prakash, R. K., and Cummings, B. (1988). Creation of novel nitrogen-fixing actinomycetes by protoplast fusion of *Frankia* with streptomyces. *Plant Mol. Biol.* **10,** 281–289.

Reddy, A., Torrey, J. G., and Hirsch, A. M. (1988). Isolation of *Frankia* strain HFPCcI3 nodulation genes by cross-species complementation of *Rhizobium meliloti*. *In* "Molecular Genetics of Plant–Microbe Interactions. Proceedings of the 4th International Symposium on Molecular Genetics of Plant–Microbe Interactions" (R. Palacios and D. P. Verma, eds.), pp. 213–214. APS Press, St. Paul, Minnesota.

Ruvkun, G. B., and Ausubel, F. M. (1980). Interspecies homology of nitrogenase genes. *Proc. Natl. Acad. Sci. U.S.A.* **77,** 191–195.

Séguin, A., and Lalonde, M. (1989). Detection of pectolytic activity and *pel* homologous sequences in *Frankia*. *Plant Soil* **118,** 221–229.

Shipton, W. A., and Burggraaf, A. J. P. (1982). A comparison of the requirements for various carbon and nitrogen sources and vitamins in some *Frankia* isolates. *Plant Soil* **69,** 149–161.

Simonet, P., Capellano, A., Navarro, E., Bardin, R., and Moiroud, A. (1984). An improved method for lysis of *Frankia* with achromopeptidase allows detection of new plasmids. *Can. J. Microbiol.* **30,** 1292–1295.

Simonet, P., Normand, P., Moiroud, A., and Lalonde, M. (1985). Restriction enzyme digestion patterns of *Frankia* plasmids. *Plant Soil* **87,** 49–60.

Simonet, P., Haurat, J., Normand, P., Bardin, R., and Moiroud, A. (1986). Localization of *nif* genes on a large plasmid in *Frankia* sp. strain ULQ0132105009. *Mol. Gen. Genet.* **204,** 492–495.

Simonet, P., Le, N. T., du Cros, E. T., and Bardin, R. (1988a). Identification of *Frankia* strains by direct DNA hybridization of crushed nodules. *Appl. Environ. Microbiol.* **54,** 2500–2503.

Simonet, P., Normand, P., and Bardin, R. (1988b). Heterologous hybridization of *Frankia* DNA to *Rhizobium meliloti* and *Klebsiella pneumoniae nif* genes. *FEMS Microbiol. Lett.* **55,** 141–146.

Simonet, P., Normand, P., Hirsch, A. M., and Akkermans, A. D. L. (1990). The genetics of the *Frankia* actinorhizal symbiosis. *In* "The Molecular Biology of Symbiotic Nitrogen Fixation" (P. M. Gresshoff, ed.), pp. 77–109. CRC Press, Boca Raton, Florida.

Stevens, G. A., and Berry, A. M. (1988). Cytokinin secretion by *Frankia* sp. HFPArI3 in defined medium. *Plant Physiol.* **87,** 15–16.

Tisa, L. S., and Ensign, J. C. (1987). Formation and regeneration of protoplasts of the actinorhizal nitrogen-fixing actinomycete *Frankia*. *Appl. Environ. Microbiol.* **53,** 53–56.

Wheeler, C. T., Crozier, A., and Sandberg, G. (1984). The biosynthesis of indole-3-acetic acid by *Frankia*. *Plant Soil* **78,** 99–104.

Zhang, Z., Lopez, M. G., and Torrey, J. G. (1984). A comparison of cultural characteristics and infectivity of *Frankia* isolates from root nodules of *Casuarina* species. *Plant Soil* **78,** 79–90.

11

Micropropagation, Tissue Culture, and Genetic Transformation of Actinorhizal Plants and *Betula*

Armand Séguin and Maurice Lalonde
Centre de Recherche en Biologie Forestière
Faculté de Foresterie et de Géomatique
Université Laval
Ste-Foy, Québec, Canada

Genetic improvement programs for actinorhizal species could take advantage of the wide geographic, ecologic, and taxonomic diversity of actinorhizal species but the long generation time of forest trees makes traditional breeding strategies prohibitively long processes. Recent progress in tissue culture and other aspects of plant biotechnology (e.g., somatic hybridization, haploid cultures, clonal propagation, and genetic transformation) of actinorhizal plants should significantly contribute to their genetic manipulation by providing plant materials with higher multiplication rates and shorter generation times. Maynard (1986) has already presented a critical and realistic point of view on the implication and limitation of *in vitro* techniques in breeding strategies for forest trees. Micropropagation, plant tissue culture, and genetic transformation techniques constitute important tools for specific investigations and manipulations of the physiologic and genetic processes involved in symbiotic actinorhizal interactions. To date, most of the work has concerned the microsymbiont, involving its isolation and growth in culture and the characterization of its morphology, physiology, and genetics (see previous chapters). Micropropagation of clonal plantlets will undoubtedly be useful in studying the host since there is great genetic diversity within actinorhizal plant genera and their provenances (Bousquet and Lalonde, Chapter 12). Great variation between *Alnus* spp. clones has been reported in relation to symbiotic performance (Simon *et al.*, 1985;

Mackay *et al.*, 1987) and the use of clonal plant material has been important in studying effective and ineffective *Frankia* strains (Nesme *et al.*, 1985; Hahn *et al.*, 1988). As already proposed by Hall *et al.* (1979), these examples demonstrate that clonal material of the host plant and pure cultures of the microsymbiont must be used for the selection and evaluation of genotypes and their combinations.

Various techniques and specific problems related to *in vitro* culture and propagation of forest trees have already been extensively reviewed in recent books edited by Bajaj (1986), Bonga and Durzan (1987), and Hanover and Keathley (1988). This review summarizes the successes obtained with actinorhizal species and the use of these techniques for research on the actinorhizal symbiosis and genetic improvement.

Unlike *Rhizobium* symbioses, in which all host species belong to the same family (Leguminosae, with the exception of *Parasponia* spp.), *Frankia* host plants are found in various plant families (see Chapter 1). Among these, the Betulaceae represents an interesting model system to study the symbiotic interactions because it includes actinorhizal (*Alnus*) and nonactinorhizal (e.g., *Betula, Corylus, Ostrya, Carpinus*) genera. Birch (*Betula* spp.) is phylogenetically closely related to some alder species (Bousquet *et al.*, 1989) and its tissue culture and genetic transformation have emphasized its importance as a model for the study and transfer of genes involved in the extension of the symbiosis to new hosts. For these reasons, progress in gene transfer and cell and tissue culture of birch is included in this review.

I. Vegetative Propagation

A. Conventional Techniques for Propagation

Conventional vegetative propagation of actinorhizal plants through cuttings is possible using standard horticultural techniques (Lundquist and Torrey, 1984; reviewed by Tremblay *et al.*, 1986). These propagation methods have been used to produce small numbers of clonal plants (Bajuk *et al.*, 1978; Gordon and Weeler, 1978; Dawson and Gordon, 1979; Hall and Maynard, 1979; Monaco *et al.*, 1979; Huss-Danell, 1980; Dawson and Sun, 1981; Hennessey *et al.*, 1987). Large-scale propagation of superior genotypes requires nursery facilities for the establishment of clonal orchards to maintain sufficient stock-plants as sources of cuttings. This strategy was recently used for a genetic improvement program based on interspecific hybridization of *Alnus* spp. (Prat, 1988).

B. *In Vitro* Techniques for Micropropagation

Micropropagation using different types of explants (e.g., shoot buds, seedlings, female inflorescences) can provide large numbers of genetically identical and disease-free plants (Brown and Sommer, 1982). This approach provides improved clones for practical use in forestry as well as for the study of symbiotic interactions. Since the first report of *in vitro* propagation of *Alnus glutinosa* (Garton *et al.*, 1981), many additional *Alnus* spp. have been micropropagated (Tremblay *et al.*, 1986), as well as members of the family Elaeagnaceae and the genus *Casuarina* (Table I).

Full description of the methods used for the establishment and manipulation of well-characterized cell culture systems has been provided by Dixon (1985), Bonga and Durzan (1987), and Pierik (1987). Figure 1 shows an outline of the different stages and objectives of cell and tissue culture of Betulaceae.

1. Explant Sources

The term "explant" refers to the initial piece of plant cultivated *in vitro*. Special care should be taken in explant selection to obtain successful *in vitro* initiation. *Alnus* spp. have been initiated from buds of aseptically germinated seedlings (Périnet and Lalonde, 1983a), 2- to 8-month-old plants grown in the greenhouse (Garton *et al.*, 1981; Tremblay *et al.*, 1984), and field plants of 2 and 25 years of age (Tremblay and Lalonde, 1984; Périnet *et al.*, 1988). Mature trees with particular traits of interest are most attractive for mass propagation. However, they are generally much more difficult to micropropagate and different strategies have to be tried. Recently, shoot cultures from mature trees of *A. incana* (syn. *A. incana* ssp. *incana*) and *A. cordata* were successfully initiated. For *A. incana*, grafting of mature scions on juvenile rootstocks was necessary before explant excision for *in vitro* culture (Périnet *et al.*, 1988). In the case of *A. cordata*, aseptic cultures were established from axillary bud explants of actively growing shoots of greenhouse cuttings (Barghchi, 1988). *In vitro* initiation from explants (stem tips, immature female inflorescences) from mature trees of *Casuarina* spp. was reported without the need of rejuvenation treatments (Duhoux *et al.*, 1986; AboEl-Nil, 1987). However, in most cases cultures were initiated with relatively young plant material (Table I).

Successful *in vitro* initiation can also depend on the provenance of the plant material. For example, *Hippophae rhamnoides* seedlings from North America were successfully propagated but not those from Europe (Montpetit and Lalonde, 1988). Variation among genotypes from the

Table I Micropropagated Actinorhizal Plants

		Multiplication medium			Rooting medium			
Species	Explant source[a] (age in months)	Salts[b]	Plant growth regulators (μM)[c]	Sugars[d] (% w/v)	Salts	Plant growth regulators (μM)	Sugars (% w/v)	Reference[e]
Alnus cordata	FI,GH-shoot tip (96)	MS	BAP (1.1–4.4)	G6	1/2 MS	IBA (1.2–4.9)	S3	1
A. crispa	GH-shoot tip (24)	WPM	IBA (0.1–0.5)	S2	—			2
A. crispa	GH-shoot tip (8)	MS	BAP (5)	S6 or G3	1/2 MS	IBA (1)	G1.5–3 or S3	3
A. crispa	FI-shoot tip (24)	MS	BAP (2.5)	G3	1/2 MS	IBA (1)	G1.5	4
A. glutinosa	GH-shoot tip (2)	WPM	BAP (1)	S2	—			5
A. glutinosa	AG-seedling (3–6)	MS or WPM	BAP (0.9–2.2)	G2-3	1/2 WPM	IBA (0.5–1)	G2–3	6
A. glutinosa	AG-seedling (1)	MS	BAP (0.5–5)	S3	1/2 MS	IBA (1)	S3	7
A. glutinosa	GH-shoot tip (24)	WPM	BAP (0.1–1)	S2	—			2
A. glutinosa	FI-shoot tip (24)	MS	BAP (2.5)	S3	1/2 MS	IBA (1)	S3–6 or G1.5–3	4
Alnus hybrids[f]	FI-shoot tip (60)	WPM[g]	IBA (0.5) + BAP (2.5)	G1.5 or F1.5	WPM	IBA (0.02–2)	G1.5–F1.5	8
A. incana	FI-shoot tip (240–300)	MS	BAP (2.5)	G1.5	1/2 MS	IBA (1)	G3	9
A. incana	AG-seedling (3–6)	MS or WPM	BAP (0.9–2.2)	G2–3	1/2 WPM	IBA (0.5–1)	G2–3	7
A. incana	FI-shoot tip (24)	MS	BAP (1–2.5)	G1.5	1/2 MS	IBA (1)	S3–6 or G1.5–3	4
A. japonica	FI-shoot tip (24)	MS	BAP (1)	G1.5	1/2 MS	IBA (1)	G3	4
A. rubra	GH-shoot tip (24)	WPM	BAP (0.1–1)	S2	—			2

A. rubra	FI-shoot tip (24)	MS	BAP (2.5)	G1.5	1/2 MS	IBA (1)	S3–6 or G1.5–3	4
A. sinuata	FI-shoot tip (24)	MS	BAP (2.5)	G3	1/2 MS	IBA (1)	G1.5	4
A. viridis	AG-seedling (3–6)	MS or WPM	BAP (0.9–2.2)	G2–3	1/2 WPM	IBA (0.5–1)	G2–3	7
A. viridis	FI-shoot tip (24)	MS	BAP (2.5)	G3	1/2 MS	IBA (1)	G1.5	4
Casuarina equisetifolia	FI-IFT (120–280)	MS	BAP (11.1) + NAA (0.05)	S3	MS[g]	NAA (0.5–1)	S3	10
Casuarina spp.	AG-seed, FI-shoot tip (48)	MS	NAA (0.005) + 2IP (5)	S3	MS[g]	IBA (6)	S3	11
Cowania mexicana	GH-shoot tip	MS	BAP (8.9) + IBA (0.5)	S4	—			12
Elaeagnus angustifolia	GH-apical bud (0.5–6)	MS[h]	Kin (5)	S3.5	1/2 MS		S3.5	13
Hippophae rhamnoides	AG-seedling (1)	MS	BAP (1)	S3	1/4 MS		S1	14
Shepherdia canadensis	FI-shoot tip (24)	SH[h]	2iP (5) + Kin (5)	S1.8 or 3.5	—			15

[a] AG, aseptically germinated; FI, field; IFT, immature female inflorescences; GH, greenhouse.

[b] MS, Murashige and Skoog high-salt medium; SH, Schenk and Hildebrandt medium; WPM, woody plant medium.

[c] 2iP, N^6-(Δ^2-isopentyenyl)-adenine; BAP, 6-benzylaminopurine; IAA, 3-indoleacetic acid; IBA, indolebutyric acid; Kin, kinetin; NAA, 1-naphthaleneacetic acid.

[d] F, fructose; G, glucose; S, sucrose.

[e] References: (1) Barghchi (1988); (2) Read *et al.* (1982); (3) Tremblay *et al.* (1984); (4) Tremblay and Lalonde (1984); (5) Garton *et al.* (1981); (6) Chalupa (1983); (7) Périnet and Lalonde (1983a); (8) Sbay *et al.* (1988); (9) Périnet *et al.* (1988); (10) Duhoux *et al.* (1986); (11) AboEl-Nil (1987); (12) Kyle and Righetti (1988); (13) Bertrand and Lalonde (1985); (14) Montpetit and Lalonde (1988); (15) Bertrand (1986).

[f] Hybrids: *A. glutinosa* × *A. incana*, *A. rubra* × *A. glutinosa*, *A. cordata* × *A. glutinosa*, *A. cordata* × *A. incana*.

[g] Activated charcoal (5–20g/liter).

[h] Polyvinylpyrollidone (0.1%).

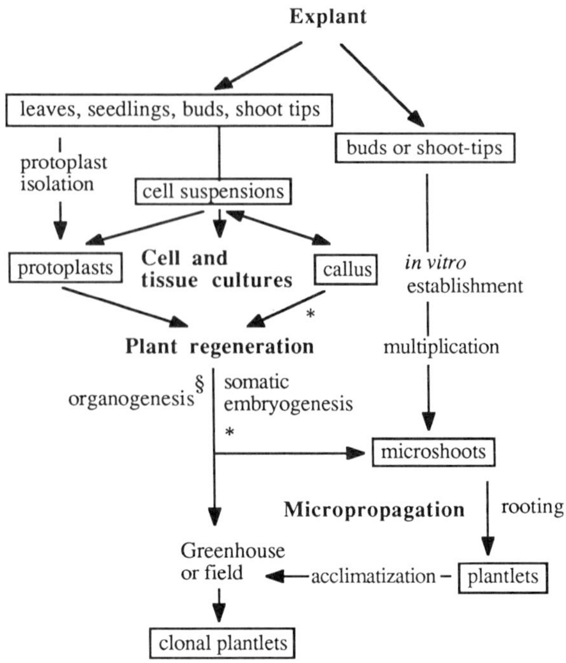

Figure 1 Flowchart showing different approaches for the *in vitro* propagation of plant genotypes. §Note that shoot regeneration has been observed with subculturable callus lines of *Betula pendula* (Huhtinen and Yahyaoglu, 1974; Simola, 1985). *This process has not yet been accomplished in actinorhizal plants.

same provenance may also occur. This was observed with mature trees of *A. incana,* where 36% of the explanted buds of two genotypes were successfully established as axenic shoot cultures, while three other genotypes had success rates ranging from 0 to 6% (Périnet *et al.*, 1988).

Finally, contamination does not seem to be a major problem for *in vitro* initiation of actinorhizal plants. No universal procedure has been adopted but tissues such as shoot tips and apical buds have usually been disinfected using the procedures described by Tremblay *et al.* (1984).

2. Culture Conditions for Shoot Multiplication and Elongation

Cytokinins are required for the development of axillary shoots during the multiplication step (Fig. 2A). Different types and concentrations of cytokinins were tested to optimize shoot proliferation of various species of actinorhizal plants. For many *Alnus* spp., 6-benzylaminopurine (BAP)

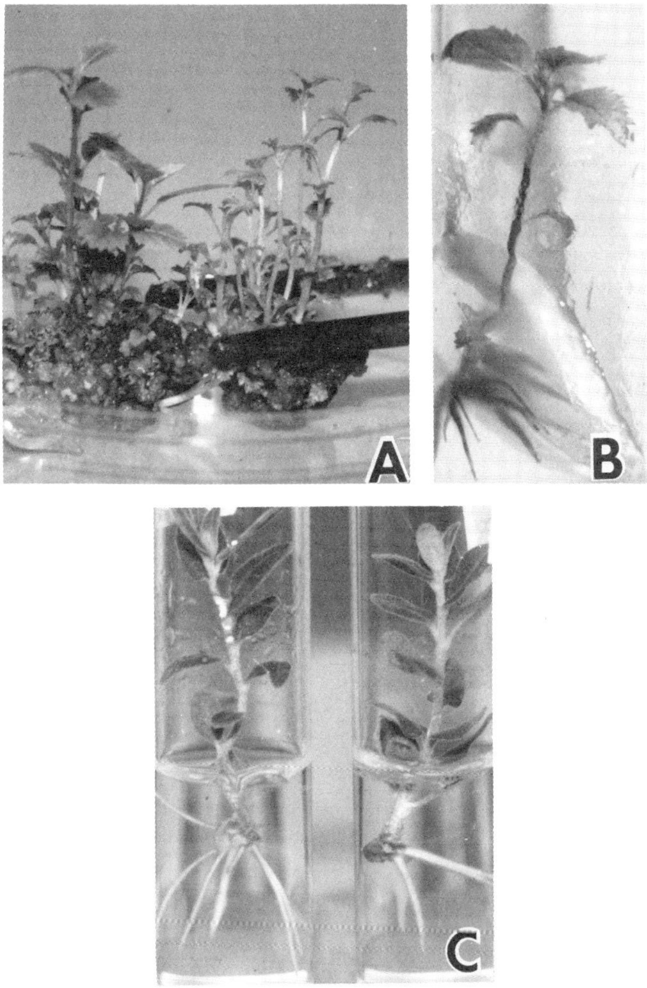

Figure 2 (A) Micropropagation of *Alnus glutinosa*. (B) Root development on micropropagated plantlet of *A. glutinosa* in agar medium. (C) Root development on micropropagated plantlets of *Elaeagnus angustifolia* in liquid medium (L. J. Bertrand and M. Lalonde, unpublished observations).

yielded the maximum number of shoots when used at concentrations ranging from 0.1 to 5 μM (Table I).

Mineral salt formulations, carbohydrate sources, and vitamins are fundamental components of culture media and special attention should be given to constituents and concentrations. MS salt formulation

(Murashige and Skoog, 1962) was used for most species of actinorhizal plants and, in other cases, SH (Schenk and Hildebrandt, 1972) or WPM (Loyd and McCown, 1980) medium was used with success (see Table I). Among the different types and concentrations of carbohydrate sources tested, sucrose and glucose generally gave good results (Table I). Variability in responses could be observed between clones of the same species (Tremblay *et al.*, 1986). In one case, shoot elongation and weight increment were stimulated by the addition of activated charcoal (Sbay *et al.*, 1988).

3. Rooting and Acclimatization

Rooting of actinorhizal plants has been reported using mist beds in the greenhouse as well as *in vitro* with specific conditions (Figs. 2B and 2C). The possibility of rooting microshoots directly into a conventional medium represents considerable commercial advantage by eliminating further costly transfers in axenic culture. Rooting conditions have been reviewed for *Alnus* spp. by Tremblay *et al.* (1986). Périnet *et al.* (1988) reported successful rooting (*in vitro* and in the greenhouse) of micropropagated mature trees of *A. incana.* For most actinorhizal plants, *in vitro* rooting was achieved by lowering MS salt concentration to one-half or one-fourth strength and by the addition of 3-indolebutyric acid (IBA). Addition of activated charcoal (5–20 g/liter) was used for rooting of *Casuarina* spp. (Table I).

Acclimatization to reduced humidity is necessary when micropropagated plantlets are transferred from *in vitro* to greenhouse or field conditions. Conditions for acclimatization have been described for *Alnus* spp. by Tremblay *et al.* (1986). These methods involve a progressive decrease of the relative humidity in mist beds or special containers and they can be applied to many species of woody plants. In most reported cases, 100% survival in soil was obtained for the various species tested.

C. Cell and Tissue Culture

In early work on tissue culture of *Alnus* spp., cambial tissue, leaves, stem sections (see Tremblay *et al.*, 1986), and root nodules (Becking, 1965) were used for callus induction. Since then, cell suspension culture and protoplast isolation and culture have been described. Successes with *Populus* spp. should be mentioned because this genus represents a model system for genetic transformation and vegetative regeneration *in vitro.* Further advances are imperative for the utilization of new approaches to the genetic manipulation of woody plants, such as haploid cultures (Chen, 1987; Bonga *et al.*, 1988), somaclonal selection, somatic hybridization, and genetic transformation (Riemenschneider *et al.*, 1988). It is now realistic to anticipate that the combination of *in vitro* techniques and genetic transformation technology can contribute to the advancement of research on nitrogen fixation and on genetic improvement of Betulaceae.

1. Callus and Cell Suspension Cultures of Betulaceae

Callus culture may be initiated using various types of explants. A callus consists of proliferating cells that can be serially subcultured and grown for extended periods. Callus may be induced on medium containing high concentrations of auxin, particularly the analogue 2,4-dichlorophenoxyacetic acid (2,4-D). For *Alnus crispa* (syn. *A. viridis* ssp. *crispa*), *A. glutinosa*, and *A. incana*, friable and fast-growing calli (Fig. 3A) were produced from leaves (Tremblay *et al.*, 1986). These calli have been maintained in culture for more than 5 years without visible change in growth rate and pigmentation. Transfer of friable callus to liquid medium was successful in initiating cell suspension culture of *Alnus* spp. (Tremblay *et al.*, 1986; see Fig. 3C).

Similarity between restriction patterns of DNA isolated from callus and cell suspension cultures suggests that these cultures are genetically stable (Giasson and Lalonde, 1987). Callus and cell suspension cultures (Figs. 3B and 3D) were also developed for *B. papyrifera* (Tremblay, 1988).

2. Protoplast Isolation and Culture

Protoplasts can be isolated from leaves, cell suspensions, or callus cultures by enzymatic digestion of cell walls. The development of protoplast technology has been crucial for the application of techniques such as somatic hybridization and direct gene transfer. Huhtinen *et al.* (1982) isolated protoplasts from leaves of *A. glutinosa* and *A. incana* seedlings and reported formation of cell colonies following their culture. Protoplasts were also isolated using cell suspension cultures of *A. incana* (Tremblay *et al.*, 1985) and *B. papyrifera* (Fig. 4A; Tremblay, 1988). Methodology using the agarose bead technique (Shillito *et al.*, 1985) was described for the culture of these protoplasts, leading to regeneration of cell walls, cell division, and microcallus formation (Tremblay *et al.*, 1985). An alternative method for high yields of protoplasts from leaves of micropropagated *Populus* could probably be used for Betulaceae (Russell and McCown, 1986).

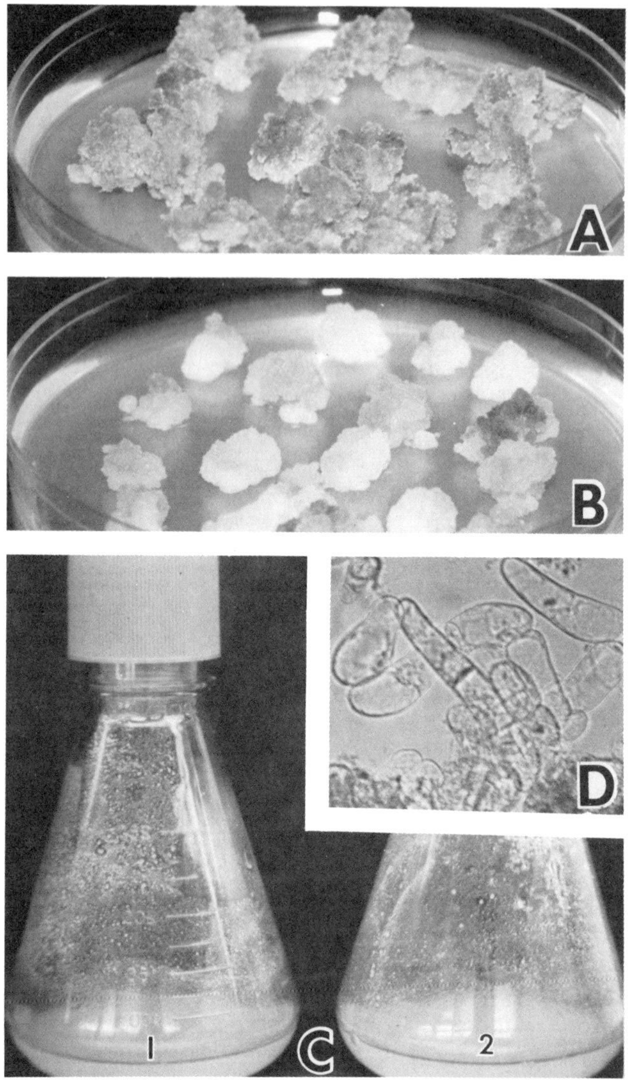

Figure 3 Subculturable callus line from leaf of *Alnus incana* (A) and *Betula papyrifera* (B). Fast-growing cell suspension cultures (C) of *B. papyrifera* (1) and *A. incana* (2) with microscopic view of cultured cells of *A. incana* (D).

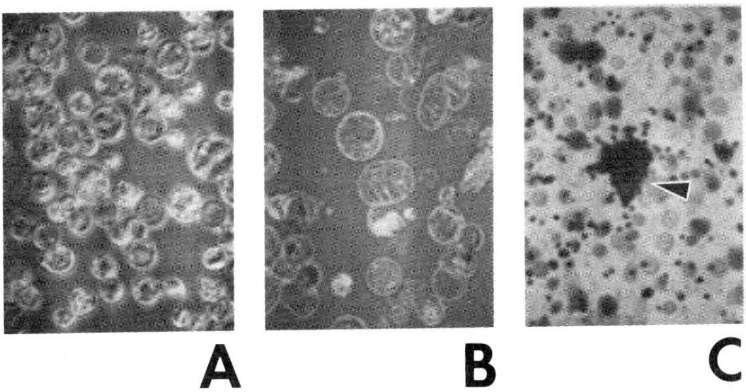

Figure 4 (A) Freshly isolated *Betula papyrifera* protoplasts prepared according to the procedure of Tremblay (1988), mean diameter of protoplasts being 50–55 μm. (B) Protoplasts of *B. papyrifera* after electroporation treatment as described by Séguin and Lalonde (1988) to introduce the NPT-II gene. (C) Microcallus (arrow) derived from stably transformed *B. papyrifera* protoplasts after 8-week growth in medium containing the antibiotic G418. These are digitized images.

3. Prospects in Plant Regeneration

Although the concerns of this review are focused on actinorhizal plants, it is important to also consider progress in regeneration of other forest trees. Haissig (1989) presented a comprehensive list of forest trees regenerated from tissues of nonmeristematic origins. At the present time, few forest tree species can be regenerated from nonvascularized tissue such as callus, cell suspensions, and protoplasts. The regeneration of whole plants from cell and tissue culture is a key step for genetic improvement through breeding or recombinant DNA technology. Adventitious shoots or embryos may arise directly on the explant or indirectly through an intermediary callus.

Attempts to achieve plant regeneration from callus cultures of *Alnus* spp. have not been successful and organogenic potential has not yet been identified in other actinorhizal plants. Complete plantlets were regenerated from adventitious buds developed from leaf laminae of *A. crispa* and *A. glutinosa* (F. M. Tremblay, unpublished). Other members of the Betulaceae family have shown organogenic potential (McCown, 1989). Huhtinen and Yahyaoglu (1974) first reported the regeneration of whole plants from *Betula pendula* callus. Since then, plantlets were regenerated from callus of *B. papyrifera* (Mackay *et al.*, 1988) and *B. pendula* (Srivastava and Steinhauer, 1981; Simola, 1985). Minocha *et al.* (1986) reported a low frequency of shoot regeneration from leaf and stem callus of *B. papyrifera*. Transfer of the callus into liquid medium failed to establish cell suspension cultures. *Betula* spp. provide good examples of the difficulty in making generalizations concerning *in vitro* regeneration potential because major differences can be observed between genotypes (McCown, 1989).

Somatic embryogenesis, even though it has been achieved with only a limited number of forest tree species, is a promising alternative (Boulay, 1987; Tulecke, 1987; Durzan, 1988). Progress with certain coniferous species has been rapid over the last few years. The advantages of somatic embryogenesis over organogenesis include the large numbers of somatic embryos that can be maintained in small volumes of liquid culture. In some cases, somatic embryos would be an appropriate alternative to seeds because somatic embryos can be encapsulated.

II. Genetic Transformation Systems for Betulaceae

Genetic transformation is a multistep process that uses both molecular and cellular techniques. First, a gene or group of genes encoding for a specific character or desired trait must be isolated and cloned. The functional gene(s) should include regulatory sequences that ensure proper expression after introduction into the plant, particularly if bacterial marker gene(s) are used. Second, the DNA fragment of interest must be introduced into a plant cell using a direct transfer or an *Agrobacterium*-mediated method. In successful transformations, DNA sequences are integrated into the plant genome and they are expressed and maintained through subsequent cell divisions. Finally, the transformed plant cells must be regenerated into whole plants (Fig. 5).

Simple model systems that use easily detected genes are usually utilized when developing transformation systems. Such markers have been used to monitor DNA transfer into *Alnus* and *Betula* spp. by *Agrobacterium*-mediated methods and by electroporation (Mackay *et al.*, 1988; Séguin and Lalonde, 1988). There has been little work on other actinorhizal species but success in genera closely related to *Alnus* spp. is very encouraging.

A. DNA Transfer Using *Agrobacterium tumefaciens*

Agrobacterium tumefaciens is a soil bacterium that naturally infects many dicotyledonous and gymnospermous plants and subsequently induces tumorous plant growth commonly called crown gall disease (Klee *et al.*, 1987). Wounding of plant tissue promotes *Agrobacterium* infection. Dur-

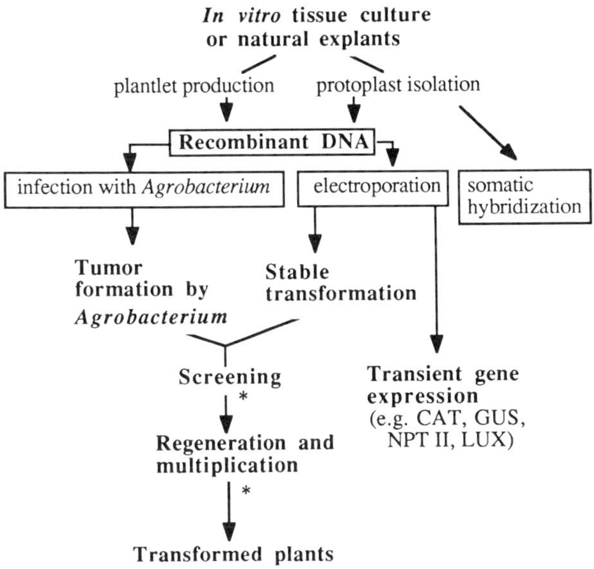

Figure 5 Flowchart showing approaches to genetic transformation of plants. *This process has not yet been accomplished in actinorhizal plants.

ing the natural infection cycle, a large bacterial plasmid called the tumor-inducing (Ti) plasmid is transferred into plant cells and this process is dependent on the bacterium. Integration of a small region of the Ti plasmid, referred to as the T-DNA, into the host plant genome leads to tumor induction (Chilton *et al.*, 1980). The T-DNA region is flanked by inversely repeated sequences of 25 base pairs known as the left and right borders. These sequences are required for T-DNA transfer and they are the recognition sequences for a site-specific endonuclease. A part of the T-DNA is responsible for tumor development by overproduction of phytohormones and another part for the synthesis of compounds called opines, which are metabolic substrates for the bacterium. Virulence genes occur in a region on the Ti plasmid (*vir* region) outside the T-DNA and they are important for the transfer of the T-DNA. Successful transformation of plant tissue can be confirmed by the detection of opines and by hybridization experiments that demonstrate the integration of the T-DNA.

Only recently has successful transformation of forest trees using *Agrobacterium* been reported (see Table II). Among members of the Betulaceae, nine *in vitro*-propagated clones of the genera *Alnus* and *Betula* were transformed by two wild-type strains of *A. tumefaciens* (Mackay *et*

Table II Genetic Transformation of Woody Plants

Species	Plant material	Transformation method: *Agrobacterium*[a]	Transformation method: Electroporation[b]	Regeneration potential	Reference
Alnus glutinosa	*In vitro* shoot segment	wt Ti		Callus	Mackay *et al.* (1988)
A. incana	*In vitro* shoot segment	wt Ti		Callus	Mackay *et al.* (1988)
A. incana	Protoplast from cell susp.		GUS	Microcallus	Séguin and Lalonde (1988)
Betula papyrifera	*In vitro* shoot segment	wt Ti		Shoot	Mackay *et al.* (1988)
B. payrifera	Protoplast from cell susp.		GUS	Microcallus	Unpublished observations
Casuarina equisetifolia	Root	Ri		—	M. Phelep and E. Duhoux (personal communication)
Populus					
alba × *grandidenta*	*In vitro* leaf segment	bin Ti + NPT-II, *aroA*		Shoot	Fillatti *et al.* (1987)
trichocarpa × *deltoides*	*In vitro* shoot segment	wt Ti		Callus	Parsons *et al.* (1986)
trichocarpa × *deltoides*	*In vitro* shoot segment	wt Ri + NPT-II		Shoot	Pythoud *et al.* (1987)
Juglans regia	Somatic embryo	bin Ti		Plantlet	McGranahan *et al.* (1988)
Pseudotsuga menziesii	*In vitro* shoot and seedling	bin Ti + APH (3′) II		Callus	Dandekar *et al.* (1987)
Pinus taeda	Seedling	wt Tti		Callus	Sederoff *et al.* (1986)
P. menziesii, P. taeda	Protoplast, embryon. susp.		LUC	Embryo	Gupta *et al.* (1988)

[a] wt, wild type strain; Ti, tumor-inducing plasmid; Ri, root-inducing plasmid; bin, binary vector; NPT-II, neomycin phosphotransferase; *aroA* confers tolerance to the herbicide glyphosate (Comai *et al.*, 1983); APH(3′) II, aminoglycoside phosphotransferase.

[b] Transient gene expression with GUS, β-glucuronidase; LUC, firefly luciferase.

al., 1988). Preconditioning treatment of shoots, on a rooting medium containing 1 m*M* of indole-3-butyric acid (IBA), was necessary to obtain high incidence of tumorigenesis after inoculation. Strain Ach5 was found to be generally more effective in inducing tumors than C58. The tumors were excised from plantlets and grown in calli in the absence of phytohormones. All tumors induced by the two strains synthesized opines and integration of T-DNA into plant genomes was confirmed.

Morphogenic differentiation into leaves and shoots was observed in tumorous calli from one of four clones of *B. papyrifera* (Mackay *et al.*, 1988), indicating an intraspecific variation in responses. Since leaves taken from these regenerated shoots did not synthesize opines and Southern blot analysis did not demonstrate T-DNA integration, regeneration had occurred from nontransformed cells. It seems that the transformed cells of the initial morphogenic tumors provided an environment suitable for shoot regeneration. Fillatti *et al.* (1987) have also observed enhanced rates of shoot regeneration from *Agrobacterium*-transformed *Populus* tissue. They suggested that plant regeneration was enhanced by the hormonal environment determined by either the oncogenes of the wild-type T-DNA or the *trans*-zeatin gene carried within the *vir* region.

However, it is generally necessary to use modified *Agrobacterium* strains, in which the oncogenic part of the T-DNA has been eliminated, to avoid production of transformed cells that are incapable of regeneration into normal plants. The so-called "disarmed" Ti plasmids have had the oncogenes of their T-DNA replaced with selectable marker genes (Bevan *et al.*, 1983; Fraley *et al.*, 1983; Herrera-Estrella *et al.*, 1983), thus allowing the selection of transformed cells on selective medium. Positive plant transformation has been obtained with disarmed Ti plasmids (Zambryski *et al.*, 1983), but their large sizes have limited molecular and genetic manipulations. This limitation was bypassed recently by using binary vectors, one plasmid containing the *vir* genes responsible for the transfer of the T-DNA and the other containing genetically engineered T-DNA. The smaller T-DNA plasmid is constructed *in vitro* with considerations for proper replication in both *Escherichia coli* and *Agrobacterium* cell systems (Hockema *et al.*, 1983; Bevan, 1984; An *et al.*, 1985; Klee *et al.*, 1985; Koncz and Schell, 1986) and for facilitating foreign DNA insertion. Preliminary experiments with this system have been unsuccessful on *Alnus* spp. and *Betula* spp. using the leaf disk infection techniques of Horsch *et al.* (1985). This is probably due to inhibitory effects of the high phenol content in these Betulaceae species (L. Savard and M. Lalonde, unpublished observations). However, certain combinations of *Agrobacterium* strains and Ri and Ti plasmids have been successful for gene transfer into woody plant cells (Table II).

Transgenic plants have only been regenerated in two woody genera, *Populus* and *Juglans*. A transgenic *Populus* hybrid has been obtained using an oncogenic strain harboring a binary vector that contains the NPT-II and 5-enolpyruvylshikimate 3-phosphate (EPS) synthase chimeric gene fusion (Fillatti *et al.*, 1987). Furthermore, a modified "supervirulent" strain of *A. rhizogenes* has also successfully transformed a *Populus* hybrid (Pythoud *et al.*, 1987). For *Juglans regia*, a nononcogenic binary strain containing APH(3')II (aminoglycoside phosphotransferase for kanamycin resistance) was used for genetic transformation of somatic embryos (McGranahan *et al.*, 1988).

B. DNA Transfer by Electroporation

A direct gene transfer method has the advantage of not requiring extra DNA sequences other than a reporter and/or a selectable marker gene. Many methods for direct transformation of plant cells have been developed, such as chemically mediated DNA uptake (Paszkowski *et al.*, 1984), electroporation (Fromm *et al.*, 1986), microinjection (Greisbach, 1985), and particle acceleration (Klein *et al.*, 1988). Among these new techniques, enhanced transformation frequencies have been reported for electroporation. Electroporation was first used to introduce DNA into mammalian cell lines (Neumann *et al.*, 1982; Potter *et al.*, 1984), and with appropriate modifications this technique was found to be applicable to plant protoplasts (Fromm *et al.*, 1985). Electroporation involves submitting a solution containing protoplasts and cloned DNA to a high-voltage electrical pulse, which apparently creates temporary pores in the plasma membrane to allow passage of DNA.

Electroporation can be used for studying both transient and stable gene expression (Fig. 5). In fact, transient gene expression experiments are rapid tests for the expression potential of specific genes under the control of selected promotors because these experiments do not require waiting for cell colony formation or plant regeneration. Using this approach, Séguin and Lalonde (1988) were able to demonstrate that the GUS gene (Jefferson *et al.*, 1987) can be transferred into protoplasts of *A. incana*, with expression being detected 20 hr after electroporation. With the recent availability of viable *Betula* protoplasts (Tremblay, 1988), conditions for successful transformation of *B. papyrifera* protoplasts were evaluated (Fig. 4). As with *A. incana*, GUS activity in transformed *B. papyrifera* increased with higher concentrations of DNA, higher voltages for the electric pulse, and the presence of PEG in the electroporation medium.

A selectable gene marker can be used to isolate stably transformed cell lines and to study stable integration and long-term expression of introduced genes. The Tn5 neomycin phosphotransferase (NPT-II) gene confers resistance to antibiotic compounds such as kanamycin and G418 (Bevan *et al.*, 1983). A modified NPT-II gene, under the control of the 35 S promotor from cauliflower mosaic virus (CaMV), was successfully integrated into *A. incana* and *B. papyrifera* protoplasts using electroporation (unpublished observations). Since cell suspensions of *Alnus* and *Betula* spp. have shown endogenous resistance to kanamycin, transformed cells were selected as microcalli tolerant to the antibiotic G418 (Fig. 4C). The GUS gene had been introduced with the selectable marker and found to be expressed in all microcalli tested. The development of a plant regeneration system from callus remains the major obstacle to studies of the stable integration and long-term expression of introduced genes. Electroporation has been found to enhance DNA synthesis in cultured plant protoplasts (Rech *et al.*, 1988) and to increase the frequency of plant regeneration from protoplasts (Ochatt *et al.*, 1988). These findings may contribute to overcoming problems in plant regeneration in actinorhizal plants.

III. Applications of *in Vitro* Techniques

A. Study of Symbiosis

In all reports on vegetative propagation of actinorhizal plants, the capacity for nodulation was maintained. Since the publication of Lalonde *et al.* (1981) on the axenic nodulation of *A. glutinosa* seedlings, other studies using *in vitro*-propagated plantlets have confirmed that *Frankia* is able to infect root systems without any "microbial helper" (Périnet and Lalonde, 1983b). Using *in vitro*-produced clonal plantlets, Simon *et al.* (1985) confirmed the large variability in the performance observed among different *A. glutinosa* clone cuttings. Other reports using various clones of alder plants have advanced our understanding of the symbiotic process. Tremblay *et al.* (1984) were able to isolate a nonnodulating clone from a nodulating species (*A. crispa*). Studies on this micropropagated clone suggested that a significant decrease in the number of root hairs was responsible for its nodulation deficiency with many *Frankia* strains. Although this clone is not completely "nonnodulating" because it can be nodulated with the aggressive *Frankia* strain ANNI (Tremblay *et al.*, 1984), further investigations of this clone should provide an understanding of the implication of root hairs in the nodulating process. The importance of root exudates in the induction of root hair deformation has also been investigated in various clones of *A. glutinosa* (Vergnaud *et al.*, 1987). Biochemical analysis of protein and sugar content of these plant exudates revealed variation between provenances but further characterization is needed to elucidate their roles in actinorhizal interactions.

A comparative molecular study of the cellular mechanisms involved in the infection of root tissue and the formation of nodules will allow the identification of host plant genes required for the establishment of symbiosis. Products from the expression of these plant genes were first defined as "actinorhizins" by Tremblay *et al.* (1986). The identification of host proteins induced after infection with *Frankia* (unpublished observations) and the establishment of an *A. glutinosa* genomic library (B. C. Mullin, personal communication) are encouraging. Cell suspension cultures have also been used to investigate plant–*Frankia* interactions (Laliberté and Lalonde, 1989). Concerted efforts using these approaches should provide a better understanding of the molecular interactions between actinorhizal plants and *Frankia* spp.

B. Genetic Improvement

Genetic improvement of woody plants may be achieved in two different ways: classical breeding and biotechnological techniques. Both utilize approaches to exploit existing genetic variation by using recombination and selection for improved genotypes. Somaclonal variation in tissue cultures offers another opportunity to uncover natural variability in plants (Evans and Sharp, 1986; Ahuja, 1987).

Besides avoiding the long sexual generations of trees, biotechnological techniques may also overcome interspecific and other sexual incompatibilities that limit the development of hybrids. As suggested by Tremblay *et al.* (1986), a somatic hybrid combining the growth characteristics of *Alnus rubra* and the cold hardiness of *A. crispa* would be a good candidate for use in forestry in northern climates. Moreover, recent developments in protoplast isolation of *B. papyrifera* make possible experiments on somatic hybridization with *Alnus* spp. (Tremblay, 1988), which is an attractive avenue for the transfer of genes responsible for the nitrogen-fixing association to this nonactinorhizal species. All these possibilities have not yet been extensively explored. Although these approaches are promising, their development is presently limited by the lack of successful vegetative regeneration of actinorhizal plants (see Figs. 1 and 5).

Genetic transformation also offers the possibility of transferring genes between plant species that are not sexually compatible. Since certain actinorhizal plants can be clonally propagated, the stable transfer of a single dominant gene into a selected species could provide a useful genotype without sexual recombination. However, these technologies are dependent on research in plant genetics and biochemistry that identifies the gene(s) responsible for specific phenotypes. The study of the physiological processes involved in cold acclimatization in different *Alnus* species is a good example. Suspension cultures and micropropagated plantlets of *Alnus* spp. have already been used to study environmental factors that may be involved in the development of frost hardiness (Tremblay and Lalonde, 1987; Tremblay, 1989). Further investigation of cellular modifications during this process is needed to identify the specific genes involved.

A high priority must be given to the achievement of whole plant regeneration from protoplasts or from explants infected with modified strains of *A. tumefaciens*. In a few cases, protoplasts obtained from embryogenic suspension cultures were successfully regenerated into whole plants (Haissig, 1989). Such *in vitro* material is not yet available for *Alnus* spp. or other actinorhizal plants. The successes in protoplast isolation and plantlet regeneration from shoot cultures of *Populus* hybrids (Russell and McCown, 1986) may permit valuable studies of the regeneration processes. In the Betulaceae, the regeneration process remains the bottleneck that prevents full exploitation of all available techniques in genetic transformation. When this is overcome, utilization of plant tissue culture in combination with biotechnology techniques will permit comprehensive research on the actinorhizal symbiosis and the production of transgenic plants capable of interacting with *Frankia*.

Acknowledgments

We thank J. Bousquet, L. Savard, and L. Simon for constructive criticisms, F. M. Tremblay and K. Wong for reviewing the manuscript, B. C. Mullin for communication of current projects, and M. Pelchat for word processing of the manuscript. This work was supported in part by an NSERC of Canada operating grant (A-2920) and by a grant (G-83-2) from the Ministére Energie et Resources, Québec to M. L.

References

AboEl-Nil, M. M. (1987). Micropropagation of *Casuarina*. *In* "Cell and Tissue Culture in Forestry" (J. M. Bonga and D. J. Durzan, eds.), Vol. 3, pp. 400–410. Nijhoff, The Hague.

Ahuja, M. R. (1987). Somaclonal variation. *In* "Cell and Tissue Culture in Forestry" (J. M. Bonga and D. J. Durzan, eds.), Vol. 1, pp. 272–285. Nijhoff, The Hague.

An, G., Watson, B. D., Stachel, S., Gordon, M. P., and Nester, E. W. (1985). New cloning vehicles for transformation of higher plants. *EMBO J.* **4,** 277–284.

Bajaj, Y. P. S. (ed.) (1986). "Biotechnology in Agriculture and Forestry," Vol. 1. Springer-Verlag, Berlin and New York.

Bajuk, L. A., Gordon, J. C., and Promnitz, L. C. (1978). Greenhouse evaluation of the growth potential of *Alnus glutinosa* clones. *Iowa State J. Res.* **52,** 341–349.

Barghchi, M. (1988). Micropropagation of *Alnus cordata* (Loisel.) Loisel. *Plant Cell Tissue Organ Cult.* **15,** 233–244.

Becking, J. H. (1965). *In vitro* cultivation of alder root-nodule tissue containing the endophyte. *Nature (London)* **207,** 885–887.

Bertrand, L. J. (1986). "Propagation *in vitro* de Deux Arbustes Actinorhiziens: *Elaeagnus angustifolia* L. et *Shepherdia canadensis* (L.) Nutt.," M.S. thesis. Université Laval, Ste-Foy, Quebec, Canada.

Bertrand, L. J., and Lalonde, M. (1985). *In vitro* propagation and nodulation by *Frankia* of actinorhizal Russian olive (*Elaeagnus angustifolia* L.) *Plant Soil* **87,** 143–152.

Bevan, M. W. (1984). Binary *Agrobacterium* vectors for plant transformation. *Nucleic Acids Res.* **12,** 8711–8721.

Bevan, M. W., Flavell, R. B., and Chilton, M.-D. (1983). A chimaeric antibiotic resistance gene as a selectable marker for plant cell transformation. *Nature (London)* **304,** 184–187.

Bonga, J. M., and Durzan, D. J. (eds.) (1987). "Cell and Tissue Culture in Forestry," Vols. 1–3. Nijhoff, The Hague.

Bonga, J. M., von Aderkas, P., and James, D. (1988). Potential application of haploid cultures of tree species. *In* "Genetic Manipulation of Woody Plants" (J. W. Hanover and D. E. Keathley, eds.), pp. 57–78. Plenum, New York.

Boulay, M. (1987). *In vitro* propagation of tree species. *In* "Plant Tissue and Cell Cultures" (C. E. Green, D. A. Somers, W. P. Hackett, and D. D. Biesboer, eds.), pp. 367–382. Liss, New York.

Bousquet, J., Girouard, E., Strobeck, C., Dancik, B. P., and Lalonde, M. (1989). Restriction fragment polymorphisms in the rDNA region among seven species of *Alnus* and *Betula papyrifera*. *Plant Soil* **118,** 231–240.

Brown, C. L., and Sommer, H. E. (1982). Vegetative propagation of dicotyledonous trees. *In* "Tissue Culture in Forestry" (J. M. Bonga and D. J. Durzan, eds.), pp. 109–149. Nijhoff, The Hague.

Chalupa, V. (1983). Micropropagation of conifer and broadleaved forest trees. *Comun. Inst. For. Czech.* **13,** 7–39.

Chen, Z. (1987). Induction of androgenesis in hardwood trees. *In* "Cell and Tissue Culture in Forestry" (J. M. Bonga and D. J. Durzan, eds.), Vol. 2, pp. 247–268. Nijhoff, The Hague.

Chilton, M.-D., Saiki, R. K., Yadav, N., Gordon, M. P., and Quetier, F. (1980). T-DNA from *Agrobacterium* Ti plasmid is in the nuclear DNA fraction of crown gall tumor cells. *Proc. Natl. Acad. Sci. U.S.A.* **77,** 4060–4064.

Comai, L., Sen, L., and Stalker, D. (1983). An altered *AroA* gene product confers resistance to the herbicide glyphosate. *Science* **221,** 370–371.

Dandekar, A. M., Gupta, P. K., Durzan, D. J., and Knauf, V. (1987). Transformation and foreign gene expression in micropropagated Douglas-fir (*Pseudotsuga menziesii*). *Bio/Technology* **5,** 587–590.

Dawson, J. O., and Gordon, J. C. (1979). Nitrogen fixation in relation to photosynthesis in *Alnus glutinosa*. *Bot. Gaz. (Chicago), Suppl.* **140,** S70–S75.

Dawson, J. O., and Sun, S.-H. (1981). The effect of *Frankia* isolates from *Comptonia peregrina* and *Alnus crispa* on the growth of *Alnus glutinosa*, *A. cordata*, and *A. incana* clones. *Can. J. For. Res.* **11,** 758–762.

Dixon, R. A. (1985). "Plant Cell Culture: A Practical Approach." IRL Press, Oxford.

Duhoux, E., Sougoufara, B., and Dommergues, Y. (1986). Propagation of *Casuarina equisetifolia* through axillary buds of immature female inflorescences cultured *in vitro*. *Plant Cell Rep.* **3,** 161–164.

Durzan, D. J. (1988). Process control in somatic polyembryogenesis. *In* "Molecular Genetics of Forest Trees" (J.-E. Hällgren, ed.), Proc. Frans Kempe Symp., pp. 147–169. Inst. Skoglig Genet. Växtfysiol., Umeå, Sweden.

Evans, D. A., and Sharp, W. R. (1986). Applications of somaclonal variation. *Bio/Technology* **4,** 528–532.

Fillatti, J. J., Sellmer, J., McCown, B., Haissig, B., and Comai, L. (1987). *Agrobacterium* mediated transformation and regeneration of *Populus*. *Mol. Gen. Genet.* **206,** 192–199.

Fraley, R. T., Rogers, S. G., Horsch, R. B., Sanders, P. R., Flick, J. S., Adams, S., Bittner, M., Brand, L., Fink, C., Fry, J., Galluppi, G., Goldberg, S., Hoffmann, N., and Woo, S. (1983). Expression of bacterial genes in plant cells. *Proc. Natl. Acad. Sci. U.S.A.* **80,** 4803–4807.

Fromm, M., Taylor, L. P., and Walbot, V. (1985). Expression of genes transferred into monocat and dicot plant cells by electroporation. *Proc. Natl. Acad. Sci. U.S.A.* **82,** 5824–5828.

Fromm, M., Taylor, L. P., and Walbot, V. (1986). Stable transformation of maize after gene transfer by electroporation. *Nature (London)* **319,** 791–793.

Garton, S., Hosier, M. A., Read, P. E., and Farnham, R. S. (1981). *In vitro* propagation of *Alnus glutinosa*. Gaertn. *HortScience* **16,** 758–759.

Giasson, L., and Lalonde, M. (1987). Restriction pattern analysis of deoxyribonucleic acid isolated from callus and cell suspension of actinorhizal and non-actinorhizal Betulaceae. *Physiol. Plant.* **70,** 304–310.

Gordon, J. C., and Wheeler, C. T. (1978). Whole plant studies on photosynthesis and acetylene reduction in *Alnus glutinosa*. *New Phytol.* **80,** 179–186.

Greisbach, R. J. (1985). Advances in the microinjection of higher plant cells. *Bio/Technology* **3,** 348–350.

Gupta, P. K., Dandekar, A. M., and Durzan, D. J. (1988). Somatic proembryo formation and transient expression of a *luciferase* gene in Douglas-fir and loblolly pine protoplasts. *Plant Sci.* **58,** 85–92.

Hahn, D., Starrenburg, M. J. C., and Akkermans, A. D. L. (1988). Variable compatibility of cloned *Alnus glutinosa* ecotypes against ineffective *Frankia* strains. *Plant Soil* **107,** 233–243.

Haissig, B. E. (1989). Status of forest tree vegetative regeneration for biotechnology. *Am. Biotechnol. Lab.* **7,** 48–51

Hall, R. B., and Maynard, C. A. (1979). Considerations in the genetic improvement of alder. *In* "Symbiotic Nitrogen Fixation in the Management of Temperate Forests" (J. C. Gordon, C. T. Wheeler, and D. A. Perry, eds.), pp. 95–110. For. Res. Lab., Oregon State Univ., Corvallis, Oregon.

Hall, R. B., McNabb, H. S., Jr., Maynard, C. A., and Green, T. L. (1979). Toward development of optimal *Alnus glutinosa* symbioses. *Bot. Gaz. (Chicago), Suppl.* **140,** S120–S126.

Hanover, J. W., and Keathley, D. E. (eds.) (1988). "Genetic Manipulation of Woody Plants." Plenum, New York.

Hennessey, T. C., Blair, L. K., and McNew, R. W. (1985). Variation in response among three *Alnus* spp. clones to progressive water stress. *Plant Soil* **87,** 135–141.

Hennessey, T. C.,Lorenzi, E. M., and McNew, R. W. (1987). Stomatal conductance and growth of five *Alnus glutinosa* clones in response to controlled water stress. *Can. J. For. Res.* **18,** 421–426.

Herrera-Estrella, L., Depicker, A., Van Montagu, M., and Schell, J. (1983). Expression of chimeric genes transferred into plant cells using Ti plasmid-derived vector. *Nature (London)* **303,** 209–213.

Hockema, A., Hirsch, P. R., Hooykaas, P. J. J., and Schilperoort, R. A. (1983). A binary plant vector strategy based on separation of *vir* and T-region of the *Agrobacterium tumefaciens* Ti-plasmid. *Nature (London)* **303,** 179–180.

Horsch, R. B., Fry, J. B., Hoffmann, N. L., Wallroth, M., Eichholtz, D., Rogers, S. G., and Fragley, R. T. (1985). A simple and general method for transferring genes into plants. *Science* **227,** 1229–1231.

Huhtinen, O., and Yahyaoglu, Z. (1974). Das Frühe Blühen von aus Kalluskulturen herangezogenen Pflänzchen bei der Birke (*Betula pendula* Roth.). *Silvae Genet.* **23,** 32–34.

Huhtinen, O., Honkanen, J., and Simola, L. K. (1982). Ornithine- and putrescine-supported divisions and cell colony formation in leaf protoplasts of alders *Alnus glutinosa* and *A. incana*. *Plant Sci. Lett.* **28,** 3–9.

Huss-Dannell, K. (1980). Nitrogen fixation and biomass production in clones of *Alnus incana*. *New Phytol.* **85,** 503–511.

Jefferson, R. A., Kavanagh, T. A., and Bevan, M. W. (1987). GUS fusions: β-glucuronidase as a sensitive and versatile gene fusion marker in higher plants. *EMBO J.* **6,** 3901–3907.

Klee, H. J., Yanofsky, M., and Nester, E. W. (1985). Vectors for transformation of higher plants. *Bio/Technology* **3,** 637–642.

Klee, H. J., Horsch, R. B., and Rogers, S. G. (1987). *Agrobacterium*-mediated plant transformation and its further applications to plant biology. *Annu. Rev. Plant Physiol.* **38,** 467–486.

Klein, T. M., Fromm, M., Weissinger, A., Tomes, D., Schaaf, S., Sletten, M., and Sanford, J. (1988). Transfer of foreing genes into intact maize cells with high-velocity microprojectiles. *Proc. Natl. Acad. Sci. U.S.A.* **85,** 4305–4309.

Koncz, C., and Schell, J. (1986). The promoters of T-L-DNA gene 5 controls the tissue-specific expression of chimeric genes carried by a novel type of *Agrobacterium* binary vector. *Mol. Gen. Genet.* **204,** 383–396.

Kyle, N. E., and Righetti, T. L. (1988). Survival of intergeneric micrografted Rosaceae. *Bot. Gaz. (Chicago)* **149,** 21–24.

Laliberté, S., and Lalonde, M. (1989). Co-culture of *Frankia alni* subs. *pommerii* (strains ACN1^{AG}) with birch (*Betula papyrifera*) protoplasts cell. *Plant Soil* **118,** 189–197.

Lalonde, M., Calvert, H. E., and Pine, S. (1981). Isolation and use of *Frankia* strains in actinorhizae formation. *In* "Current Perspectives in Nitrogen Fixation" (A. H. Gibson and W. E. Newton, eds.), pp. 296–299. Aust. Acad. Sci., Canberra, Australia.

Loyd, G., and McCown, B. (1980). Commercially-feasible micropropagation of mountain laurel, *Kalmis latifolia*, by use of shoot-tip culture. *Proc. Int. Plant Prop. Soc.* **30,** 421–427.

Lundquist, R., and Torrey, J. G. (1984). The propagation of *Casuarina* species from rooted stem cuttings. *Bot. Gaz. (Chicago)* **145,** 378–384.

Mackay, J., Simon, L., and Lalonde, M. (1987). Effect of substrate nitrogen on the performance of *in vitro* propagated *Alnus glutinosa* clones inoculated with Sp^+ and Sp^- *Frankia* strains. *Plant Soil* **103,** 21–31.

Mackay, J., Séguin, A., and Lalonde, M. (1988). Genetic transformation of 9 *in vitro* clones of *Alnus* and *Betula* by *Agrobacterium tumefaciens*. *Plant Cell Rep.* **7,** 229–232.

Maynard, C. A. (1986). Population genetics of forest trees: Implication for the application of *in vitro* culture. *In Vitro Cell. Dev. Biol.* **22,** 231–233.

McCown, B. H. (1989). Birches (*Betula* spp). *In* "Biotechnology in Agriculture and Forestry" (Y. P. S. Bajaj, ed.), Vol. 2. Springer-Verlag, Berlin and New York.

McGranahan, G. H., Leslie, C. A., Uratsu, S. L., Martin, L. A., and Dandekar, A. M. (1988). *Agrobacterium*-mediated transformation of walnut somatic embryos and regeneration of transgenic plants. *Bio/Technology* **6,** 800–804.

Minocha, S. C., Noh, E. W., and Kausch, A. P. (1986). Tissue culture and genetic transformation in *Betula papyrifera* and *Populus tremuloides*. *Proc. Tech. Assoc. Pulp. Pap. Ind. Res. Dev. Conf.* pp. 89–92.

Monaco, P. A., Ching, T. M., and Ching, K. K. (1979). Cloning red alder genotypes by vegetative cuttings. *In* "Symbiotic Nitrogen Fixation in the Management of Temperate Forests" (J. C. Gordon, C. T. Wheeler, and D. A. Perry, eds.), p. 482 (abstr.) For. Res. Lab., Oregon State Univ., Corvallis, Oregon.

Montpetit, D., and Lalonde, M. (1988). *In vitro* propagation and subsequent nodulation of the actinorhizal *Hippophae rhamnoides* L. *Plant Cell Tissue Organ Cult.* **15,** 189–199.

Murashige, T., and Skoog, F. (1962). A revised medium for rapid growth and bioassays with tobacco tissue cultures. *Physiol. Plant.* **15,** 473–497.

Nesme, X., Normand, P., Tremblay, F. M., and Lalonde, M. (1985). Nodulation speed of *Frankia* sp. on *Alnus glutinosa, Alnus crispa,* and *Myrica gale*. *Can. J. Bot.* **63,** 1292–1295.

Neumann, E., Schaefer-Ridder, M., Wang, Y., and Hofschneider, P. H. (1982). Gene transfer into mouse lyoma cells by electroporation in high electric fields. *EMBO J.* **1,** 841–845.

Ochatt, S. J., Chand, P. K., Rech, E. L., Davey, M. R., and Power, J. B. (1988). Electroporation-mediated improvement of plant regeneration from Colt cherry (*Prunus avium* × *Pseudocerasus*) protoplasts. *Plant Sci. (Washington, D.C.)* **54,** 165–169.

Parsons, T. J., Sinkar, V. P., Stettler, R. F., Nester, E. W., and Gordon, M. P. (1986). Transformation of poplar by *Agrobacterium tumefaciens*. *Bio/Technology* **4,** 533–536.

Paszkowski, J., Shillito, R. D., Saul, M., Mandak, V., Hohn, T., Hohn, B., and Potrykus, I. (1984). Direct gene transfer to plants. *EMBO J.* **3,** 2717–2722.

Périnet, P., and Lalonde, M. (1983a). *In vitro* propagation and nodulation of the actinorhizal host plant *Alnus glutinosa* (L.) Gaertn. *Plant Sci. Lett.* **29,** 9–17.

Périnet, P., and Lalonde, M. (1983b). Axenic nodulation of *in vitro* propagated *Alnus glutinosa* plantlets by *Frankia* strains. *Can. J. Bot.* **61,** 2883–2888.

Périnet, P., Vallée, G., and Tremblay, F. M. (1988). *In vitro* propagation of mature trees of *Alnus incana* (L.) Moench. *Plant Cell Tissue Organ Cult.* **15,** 85–89.

Pierik, R. L. M. (1987). "*In vitro* Culture of Higher Plants." Nijhoff, The Hague.

Potter, H., Weir, L., and Leder, P. (1984). Enhancer-dependent expression of human k immunoglobulin genes introduced into mouse pre-B lymphocytes by electroporation. *Proc. Natl. Acad. Sci. U.S.A.* **81,** 7161–7165.

Prat, D. (1988). Intéret de l'hybridation interspécifique et de la multiplication végétative: La cas de l'aulne. *Acte Colloq. Sci. Ind. Bois, 2nd* **1,** 161–168.

Pythoud, F., Sinkar, V. P., Nester, E. W., and Gordon, M. P. (1987). Increased virulence of *Agrobacterium rhizogenes* conferred by the *vir* region of pTiBo542: Application to genetic engineering of poplar. *Bio/Technology* **5,** 1323–1327.

Read, P. E., Garton, S., Louis, K. A., and Zimmerman, E. S. (1982). *In vitro* propagation of species for bioenergy plantations. *In* "Plant Tissue Culture" (A. Fijiwara, ed.), pp. 757–758. Maruzen, Tokyo.

Rech, E. L., Ochatt, S. J., Chand, P. K., Davey, M. R., Mulligan, B. J., and Power, J. B. (1988). Electroporation increases DNA synthesis in cultured plant protoplasts. *Bio/Technology* **6,** 1091–1093.

Riemenschneider, D. E., Haissing, B. E., and Bingham, E. T. (1988). Integrating biotechnology into woody plant breeding programs. *In* "Genetic Manipulation of Woody Plants" (J. W. Hanover and D. E. Keathley, eds.), pp. 433–450. Plenum, New York.

Russell, J. A., and McCown, B. H. (1986). Culture and regeneration of *Populus* leaf protoplasts isolated from non-seedling tissue. *Plant Sci. (Washington, D.C.)* **46,** 133–142.

Sbay, H., Guillot, J., Danthu, P., and Prat, D. (1988). *In vitro* propagation of interspecific hybrids in *Alnus*. *Proc. Int. Symp. For. Tree Physiol.*, in press.

Schenk, R. V., and Hildebrandt, A. C. (1972). Medium and techniques for induction and growth of monocotyledonous and dicotyledonous plant cell cultures. *Can. J. Bot.* **50,** 199–204.

Sederoff, R., Stomp, A.-M., Chilton, W. S., and Moore, L. W. (1986). Gene transfer into loblolly pine by *Agrobacterium tumefaciens*. *Bio/Technology* **4,** 647–649.

Séguin, A., and Lalonde, M. (1988). Gene transfer by electroporation in Betulaceae protoplasts: *Alnus incana*. *Plant Cell Rep.* **7,** 367–370.

Shillito, R., Saul, M., Paszkowski, J., and Potrykus, I. (1985). High efficiency direct gene transfer to plants. *Bio/Technology* **3,** 1099–1103.

Simola, L. K. (1985). Propagation of plantlets from leaf callus of *Betula pendula* F. *purpurea*. *Sci. Hortic. (Amsterdam)* **26,** 77–85.

Simon, L., Stein, A., Côté, and Lalonde, M. (1985). Performance of *in vitro* propagated *Alnus glutinosa* (L.) Gaertn. clones inoculated with *Frankia*. *Plant Soil* **87,** 125–133.

Srivastava, P. S., and Steinhauer, A. (1981). Regeneration of birch plants from catkins tissue cultures. *Plant Sci. Lett.* **22,** 379–386.

Tremblay, F. M. (1988). Callus formation from protoplasts of*Betula papyrifera* Marsh. cell suspension culture. *J. Plant Physiol.* **133,** 247–251.

Tremblay, M.-F. (1989). "Étude de l'Acclimatation au Froid du Genre Alnus," Ph.D. thesis. Université Laval, Ste-Foy, Québec, Canada.

Tremblay, F. M., and Lalonde, M. (1984). Requirements for *in vitro* propagation of seven nitrogen-fixing *Alnus* species. *Plant Cell Tissue Organ Cult.* **3,** 189–199.

Tremblay, M.-F., and Lalonde, M. (1987). Effect of photoperiod and temperature on the development of frost hardiness in three *Alnus* species. *Physiol. Plant.* **70,** 327–332.

Tremblay, F. M., Nesme, X., and Lalonde, M. (1984). Selection and micropropagation of nodulating and non-nodulating clones of *Alnus crispa* (Ait.) pursh. *Plant Soil* **78,** 171–179.

Tremblay, F. M., Power, B. J., and Lalonde, M. (1985). Callus regeneration from *Alnus* incana protoplasts isolated from cell suspensions. *Plant Sci. (Washington, D.C.)* **41,** 211–216.

Tremblay, F. M., Périnet, P., and Lalonde, M. (1986). Tissue culture of *Alnus* spp. with regard to symbioses. *In* "Biotechnology in Agriculture and Forestry" (Y. P. S. Bajaj, ed.), Vol. 1, pp. 87–100. Springer-Verlag, Berlin and New York.

Tulecke, W. (1987). Somatic embryogenesis in woody perennials. *In* "Cell and Tissue Culture in Forestry" (J. M. Bonga and D. J. Durzan, eds.), Vol. 2, pp. 61–91. Nijhoff, The Hague.

Vergnaud, L., Chabound, A., and Rougier, M. (1987). Preliminary analysis of root exudates of *in vitro*-micropropagated *Alnus glutinosa* clones. *Physiol. Plant.* **70,** 319–326.

Zambryski, P., Joos, H., Genetello, C., Leemans, J., Van Montagu, and Schell, J. (1983). Ti plasmid vector for the introduction of DNA into plant cells without alteration of their normal regeneration capacity. *EMBO J.* **2,** 2143–2150.

12

The Genetics of Actinorhizal Betulaceae

Jean Bousquet and Maurice Lalonde
Centre de Recherche en Biologie Forestière
Faculté de Foresterie et de Géomatique
Université Laval
Ste-Foy, Québec, Canada

I. Introduction

The actinorhizal symbiosis may be viewed as a dynamic biological system in which both the host plant and the actinomycete *Frankia* interact in such a way that a selective advantage is conferred to the two partners. An appropriate approach to a study of this symbiosis is to look at the system as one entity and to consider individually the host and the microsymbiont, with their multiple interactions, so that a better understanding of the coadaptive significance of the symbiosis may be gained. Considerable efforts have been devoted during the last 20 years to attain a better knowledge of the biology and the genetics of the microsymbiont (see previous chapters). Efforts directed toward the genetics of the hosts have accumulated in recent years, and our purpose is to review these aspects and their consequences on the study of the symbiosis, with special regard to actinorhizal Betulaceae as a model system.

II. Diversity among Actinorhizal Plants

A. Taxonomy and Life History

Actinorhizal plants are dispersed among 8 dicotyledonous families and about 23 genera (for a complete list, see Chapter 1). The lack of

The Biology of *Frankia* and Actinorhizal Plants

taxonomic relatedness among them is striking. This is in opposition to the *Rhizobium* symbiosis in which most of the host genera are found in the Leguminoseae (Dixon and Wheeler, 1986), although the genus *Parasponia* is found in the Ulmaceae (Fig. 1). Actinorhizal genera are found in ancient as well as more recent lineages, and some families are completely actinorhizal (e.g., Elaeagnaceae) while others are not (e.g., Betulaceae) (Fig. 1). The classification presented in Fig. 1 (Cronquist, 1988) is quite different than the one presented by Dixon and Wheeler (1986, p. 26), which is based on Hutchinson's (1973) classification. In particular, in reference to Fig. 1, the Rosales are not ascribed to a position anterior to the Hamamelidales, and the Elaeagnaceae and Rhamnaceae are found in different orders and are shown to have diverged from the Rosales. The Datiscaceae are found in the Violales (Dilleniidae) and the Casuarinales are grouped together with the Fagales and the Myricales in the Hamamelidae. Pollen records show that the orders Urticales, Fagales, and Myricales can be found as early as the upper Cretaceous (ca. 70–100 My), while Casuarinales were found as early as the lower Quaternary (Paleocene, ca. 60 My), and the generally more advanced groups such as Rosales, Elaeagnales, and Rhamnales can be found in the Oligocene and Miocene (ca. 20–40 My) (Müller, 1981).

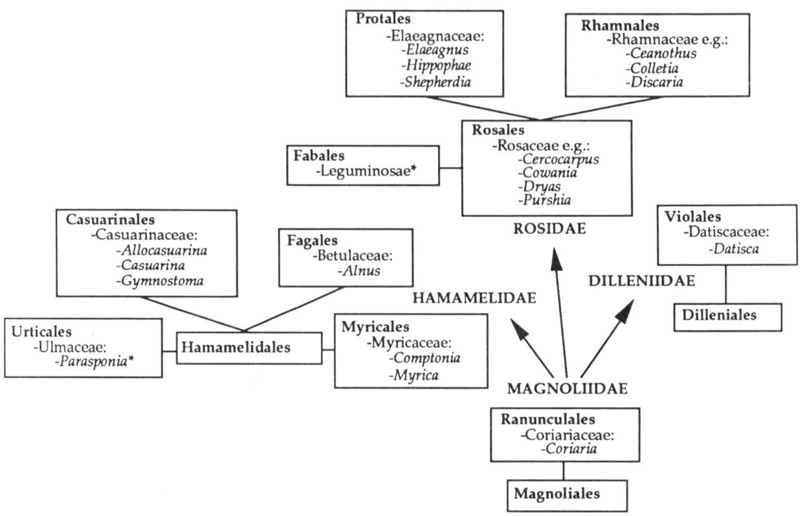

Figure 1 Taxonomic position of actinorhizal genera in the Magnoliopsida (dicotyledons) according to Cronquist (1988). Subclasses are shown in bold uppercase, orders in bold lowercase, families in lowercase, and genera in italics. *Family or genus involved in the *Rhizobium* symbiosis.

Therefore, in view of the different classifications suggested by different schools of plant taxonomy, caution must be exercised when designating certain groups as being more or less advanced.

Among the different actinorhizal genera, modes of reproduction are found to be strongly divergent, including hermaphroditic as well as monoecious and dioecious species. Aside from the fact that actinorhizal genera are dicotyledonous and that they bear nodules, other features shared by most are related to their life history: perennial nature, early-succession behavior, and generally a large reproductive effort, in terms of number of propagules produced. Therefore, most of the actinorhizal taxa could be considered as *r*-strategists (MacArthur and Wilson, 1967), invading with efficiency the open and impoverished sites where interspecific competition is usually not a limiting factor.

B. Evolutionary Trends

Two evolutionary views may be proposed to explain the lack of taxonomic relatedness among actinorhizal genera, in spite of their autecological similarities. The first view would emphasize the acquisition of the ability to associate with *Frankia* by widely divergent genera, conferring a selective advantage for specific ecological niches. For this hypothesis to hold true, the acquisition of this ability would be after the divergence of families into their respective genera, and prior to the divergence of genera into their respective species. One would also have to accept that the ability to associate with *Frankia* was passed during the speciation process. Therefore, the ability to associate with *Frankia* would be considered, among the different actinorhizal genera, as a homoplasy (see Wiley, 1981) resulting from parallel or convergent evolution, and within genera as a synapomorphy.

The second view is of an opposite dynamic process in which some of the oldest representatives of the angiosperms (e.g., Magnoliidae, lower Cretaceous) established an early association with *Frankia*. This ability would therefore be considered homologous and shared by common ancestry among the actinorhizal genera. This is assuming that the primitive angiosperms were woody forms (Cronquist, 1988) and opportunistic wind-pollinated shrubs (Tiffney, 1984; Thomas and Spicer, 1987), and that the availability of nitrogen was limited in the early Cretaceous period (Bond, 1983). Evidence of the presence of *Frankia* 100 My ago has also recently been inferred (Normand and Bousquet, 1989). Contemporary plant families and genera lost the ability to associate with *Frankia* (or gained the ability not to associate with *Frankia*), as a result of a decreasing selective advantage, likely related to an increase in nitrogen availability during the past 100 My. The ancestral ability to associate with *Frankia* was retained as a selective advantage conferred on some specialized early-successional plants of sites generally deficient in nitrogen. Results obtained from DNA hybridization studies on host plants provide support for this hypothesis (Bousquet *et al.*, 1989).

III. A Model Host Group for the Temperate Regions

A. General Considerations

For the temperate regions, *Alnus* represents the actinorhizal genus for which the largest amount of information is available concerning the variability at the infra- and interspecific level. *Alnus* is organized into several taxonomic levels, each one showing large differences in diversity (Furlow, 1979). The morphological variation and the ecological preferences have been thoroughly studied within and between species (Furlow, 1979). Cytological analyses have also been conducted on many species (e.g., Woodworth, 1929; Chiba, 1966). Studies at the DNA level have been initiated recently (see Section IV). Cell and tissue culture systems have been established (see Chapter 11). Because of the potential use of alder species in land reclamation, reforestation, biomass production, and short-rotation timber production (see Chapters 15–18), improvement of growth and other desirable phenotypic traits by means of traditional selection approaches has been achieved for several species (Hall and Maynard, 1979). *Frankia* strains have been isolated from a large number of species within *Alnus* (Lechevalier, 1986), and interspecific as well as infraspecific host–microsymbiont combinations have been tested (see Section III,E). This genus is also a member of an incompletely nodulated family, so that genetic divergence from other closely related nonnodulating genera can be examined. Clearly, alder should be considered as a model system for the temperate regions.

B. Ecogeography and Morphology

The genus *Alnus* (subclass Hamamelidae, order Fagales, family Betulaceae; Cronquist, 1988) contains from 26 to over 30 species depending on the taxonomic treatment of a number of infraspecific taxa (Table I). The early fossil record of the angiosperms indicates that *Alnus* may have originated on the Asian landmass, not later than the upper Cretaceous (70–100 My), where it was associated with other early woody angiosperms such as *Betula* and *Corylus* (Furlow, 1979). Today, alder species are found naturally mainly in the Northern Hemisphere and in a limited

Table I List of *Alnus* spp. and Chromosome Numbers[a]

Species	Chromosome number ($2n$)[b]	Species	Chromosome number ($2n$)[b]
Subgenus *Alnus*		Subgenus *Alnobetula*	
A. acuminata	—	*A. firma*	112
A. cordata	28, 42	*A. pendula*	—
A. crematogyne	—	*A. sieboldiana*	—
A. faurieri	—	*A. viridis* ssp. *crispa*	28
A. glutinosa	28, 42, 56	*maximowiczii*	—
A. incana ssp. *incana*	28	*sinuata*	28
rugosa	28	*suavendens*	—
tenuifolia	28	*viridis*	28
A. inokumae (A. hirsuta var. *microphylla)*	14	Subgenus *Clethropsis*	
A. japonica	28, 42, 56	*A. maritima*	28
A. jorullensis	—	*A. nepalensis*	—
A. hirsuta	28, 42	*A. nitida*	—
A. lanata	—		
A. matsumurae	—		
A. oblongifolia	28		
A. orientatis	28, 42		
A. rhombifolia	—		
A. rubra	28		
A. serrulata	28		
A. serrulatoides	—		
A. subcordata	28, 42, 56		
A. trabeculosa	—		

[a]From Murai (1964), Chiba (1966), Furlow (1979), and Hall and Maynard (1979).
[b]Putative basic chromosome number $x = 7$.

part of South America. They are basically early-successional colonizing species found more frequently on wet soils but also in dry habitats. Their size ranges from small shrubs to large trees.

The taxonomy of *Alnus* species has been clarified, notably by Murai (1964, 1969) and by Furlow (1979). The genus was originally considered as a single species of *Betula* (Linnaeus, 1753), whereas today these two groups are considered as separate genera based on several discontinuities, particularly differences in the morphology of the woody infructescences (Furlow, 1979). Furlow further divided the genus into three subgenera, *Alnobetula, Alnus,* and *Clethropsis*. The subgenus *Alnobetula* is typically adapted to boreal and montane regions, with most of its species characterized by a shrubby habit. The subgenus *Alnus* is considered to contain the most primitive species (Furlow, 1979). It has the largest species diversity (Table I) and generally shows a preference for temperate, mesic environments. The subgenus *Clethropsis* has been considered by some authors (e.g., Murai, 1964) as the most primitive segment of the genus and contains few species (Table I). The three subgenera are easily distinguished on the basis of bud and flower structures, leaf venation, exposure of pistillate catkins during the winter, and phenology. No naturally occurring hybridization nor successful artificial hybridization (see Sections III,H and III,I) has been observed between subgenera, suggesting strong reproductive barriers.

Within each subgenus, species are much less distinct morphologically, but nevertheless still easily distinguishable. Many species are not intrinsically reproductively isolated. Numerous artificial hybrids could be obtained within a subgenus (see Section III,I), but spatial isolation, in relation to geographical ranges, limits the amounts of interspecific gene flow (extrinsic isolation). Furlow (1979) considered as conspecific and classified as subspecies those taxa that are distinct morphologically and in habitat but that intergrade where their ranges overlap. Characters such as variation in leaf shape, margin, and pubescence or infructescence form are found to be very variable at the infraspecific level, and therefore taxa based, for instance, on extremes in leaf pubescence are no longer recognized (Furlow, 1979).

C. Reproductive Biology

Alders are allogamous, monoecious, and anemophilous. Their sexual maturity is reached early in life (generally after 3 to 6 years in natural conditions). Male sterility has been noted in a rare instance in *A. glutinosa* (Bensimon, 1985). Protogyny seems the rule rather than the exception (McVean, 1953; Furlow, 1979). High levels of self-incompatability have been reported for several species (see Hall and Maynard, 1979; Prat, 1988), and the naturally occurring outcrossing rate in *A. crispa* (syn. *A. viridis* ssp. *crispa*) is high (Bousquet *et al.*, 1987a). Therefore, we would expect, *a priori*, low-selfing rates and an open-pollinated mating system (see Section III,G).

Alders regularly produce high numbers of sexual propagules. Several million pollen grains have been estimated from a single *A. glutinosa* catkin (Erdtman, 1969, cited by Furlow, 1979) and a single individual may produce from a few hundred to a few thousand catkins annually. As well, single individuals of *A. crispa* may produce up to many thousand seeds per year. Apomixis (agamospermy) has been reported by Woodworth (1930) as a result of natural interspecific hybridization between *A. serrulata* and *A. rugosa* (syn. *A. incana* ssp. *rugosa*), but after further sampling, Furlow (1979) could not confirm these observations and concluded that such irregularity was the exception. After maturation, the seeds are progressively shed over a period of time. The seeds are usually winged and dispersed by wind or water (Furlow, 1979). Usually, the cones remain attached to the branches and still contain viable seeds for up to one year. Seed germination is epigeous and germination success is usually variable from one species to the other, depending on the crop year. Stratification may be necessary to ensure proper germination, indicating that part of a seed crop may act as a seed bank for delayed germination (Furlow, 1979). Furthermore, germination appears to be sensitive to low oxygen and moisture levels (McVean, 1953), an observation that may explain why Huenneke (1987) did not observe any seedling recruitment over a three-year period on poorly drained soil, even when *A. rugosa* thickets were producing seeds.

Naturally occurring clonal propagation may have a profound impact on plant population and subpopulation structures, resulting notably in clumped genotype distributions (Silander, 1979). Individuals of species such as *A. rugosa* or *A. sinuata* (syn. *A. viridis* ssp. *sinuata*) have been found occasionally to be interconnected by their roots (Furlow, 1979). However, Huenneke (1985, 1987) as well as Bousquet *et al.* (1987a, 1988) failed to find any evidence for such a phenomenon in the numerous sexually mature populations of *A. crispa* and *A. rugosa* they analyzed. No root suckers were noted, and the spatial distribution of individuals was found to be random with respect to their genotypes (see Huenneke, 1985). However, extensive sprouting was noted in these shrubby species, producing genetically homogeneous thickets or clumps, again based on evidence obtained from multilocus genotypic identification of individuals (Huenneke, 1985; Bousquet *et al.*, 1987a, 1988). Sprouting would enable some individuals to survive much longer than dendrometric determinations would indicate, yielding a decrease in mortality rate (Huenneke, 1987). This ability to sprout could be crucial to the survival of alder individuals colonizing periodically flooded areas such as swamps or riverbanks.

D. Ploidy Levels and Genome Expression

Alder species have been reported generally with chromosome numbers of $2n = 28$, but $2n = 56$ and $2n = 42$ have also been recorded (Table I). The latter species are not representative of a simple polyploid series and they may indicate hybridization between $2n = 28$ and $2n = 56$. Pollen quality of these putative hybrid taxa has been reported to be poor (Furlow, 1979). Woodworth (1929) first described the formation of bivalents at meiosis I for a number of alder species. The species were therefore treated as diploid with a basic chromosome number of $x = 14$ (Woodworth, 1929, 1931), until Chiba (1962) reported a diploid number $2n = 14$ in root tip cells of *A. hirsuta* var. *microphylla* (syn. *A. inokumae*), and further obtained viable artificial hybrids with $2n = 21$ by reciprocal crossing between species of $2n = 28$ and $2n = 14$ (Chiba, 1966). The basic chromosome number was therefore accepted as $x = 7$ and *Alnus* was considered polyploid.

However, evidence from allozyme data indicates that gene expression is typically diploid for all loci in species so far investigated (Huenneke, 1985; Bousquet *et al.*, 1987a, 1988). For instance, heterozygotes were never found with more than two alleles, as is observed in recent autopolyploids (Gottlieb, 1982), and no fixed heterozygosity was observed, as is often seen in recent allopolyploids (Gottlieb, 1981). At meiosis I, chromosome pairing is typical of a diploid species or at best an allotetraploid, and a recent autotetraploid origin for these species may be rejected because of the absence of quadrivalents (Woodworth, 1929). These species most likely arose from old ploidy events during the initial diversification and expansion of angiosperm flora in the Cretaceous and early Tertiary (Stebbins, 1971). This was likely followed by diploidization and gene silencing of the duplicates of many genes by mutation (Ferris and Whitt, 1977; Haufler, 1987). Therefore, several alder species, if not most of them, may be considered diploidized polyploids, with their putative diploid ancestors being extinct.

E. Variation in Host/*Frankia* Interactions at the Host Infraspecific Level

Extensive variation in specificity and effectivity of *Frankia* inoculated on several *Alnus* spp. has been regularly reported (see Chapter 5). Notable differences in rates of nitrogen fixation, nodulation speed, and growth of the host species have been shown to result from host/*Frankia* interactions using different alder species (e.g., Dawson and Sun, 1981; Nesme *et al.*, 1985). After measuring the growth performance of *A. glutinosa* clones inoculated with different *Frankia* strains, Nesme *et al.* (1985), Simon *et al.* (1985), and Prat (1989) did not find any host/*Frankia* interaction at the host infraspecific level even when the clones came from different provenances. Similarly, no host/*Frankia* interaction was noted among provenances of *A. cordata*, *A. glutinosa*, or *A. rubra* (Tessier du Cros *et al.*, 1984). However, Mackay *et al.* (1987) and Hahn *et al.* (1988) found interactions among *A. glutinosa* clones. Similar interactions have also been reported among provenances of *A. glutinosa* (Hall *et al.*, 1979) as well as among ecotypes of the same species (Hahn *et al.*, 1988). Tremblay *et al.* (1984) also reported host-dependent interaction with *Frankia* among clones of *A. crispa* relative to nodule formation. From the available information and the limited sampling with respect to infraspecific host genetic variability and number of *Frankia* strains tested, it still remains difficult to assess the extent of variation in host/*Frankia* interactions within *Alnus* spp.

The level of available combined nitrogen has been shown to have a significant influence not only on the nodulation, N_2-fixation, and general metabolism of the host (Pizelle, 1984; Sellstedt and Huss-Danell, 1986), but also on host/*Frankia* interactions (Mackay *et al.*, 1987). Other factors such as the soil type have also been found to produce similar effects (Sheppard *et al.*, 1988). These factors and several others must therefore be taken into consideration when proceeding with the evaluation of interactions and performance of the symbiosis under different cultivation systems (see Sellstedt and Huss-Danell, 1984). This apparent relative instability of the host/*Frankia* interactions induced by environmental factors may indicate that, at the host infraspecific level, the interactions are not strongly genetically controlled, and that replicated tests under various environmental conditions would be necessary to assess these trends more precisely. Large-scale studies are also needed to better assess the frequency and extent of the variation in host/*Frankia* interactions within alder species, as well as their temporal stability. These studies should include not only several *Alnus* spp. but also several provenances and many clones per provenance for every species, together with an array of pure *Frankia* strains representative of the natural habitat of each host species.

F. Infraspecific Variation in Quantitative Characters

Alnus glutinosa and *A. rubra*, because of their tree habit and their associated potential as timber species in short rotation, have attracted considerable attention from tree breeders of the Northern Hemisphere. These species have been introduced in North America and Europe (and in Africa), respectively, and tested by means of several rangewide provenance tests established in multiple locations (DeBell and Wilson, 1978; Monaco *et al.*, 1981; Hall *et al.*, 1983; DeWald and Steiner, 1986; Cannell *et al.*, 1987; Genys, 1988; Zwolinski, 1988). Unfortunately, for alder species the heritability of the most important phenotypic characters is unknown. This is due to the fact that most studies are based on provenance trials, without including half-sib (open-pollinated) or full-sib families. Estimates of general and specific combining abilities are therefore lacking and selection programs so far developed are not fully optimized. The only reported estimates of heritability were derived from a clonal test of *A. glutinosa*, with broad-sense heritability of 0.32 and 0.59 for height and diameter growth, respectively (Hall *et al.*, 1983).

Significant amounts of variation between provenances have been reported in several instances for growth traits (Hall *et al.*, 1983; Genys, 1988). However, this variation failed to follow any trend related to the geographic origin of seed sources and was therefore found to be typically ecotypic (DeBell and Wilson, 1978; DeWald and Steiner, 1986; Genys, 1988; Hahn *et al.*, 1988). Significant genotype/environment interactions have also been noted with growth traits (Hall *et al.*, 1983), and thus ranking of the provenances varied from one test site to the other. In spite of the paucity of data available on the allocation of variation within and among provenances, it appears that the variation seems to be mainly found within provenances for characters such as growth, as well as for wood density and bark thickness (DeBell and Wilson, 1978; Hall *et al.*, 1983). This trend is also observed in *Betula* spp. (Eriksson and Jonsson, 1986). Therefore, a considerable amount of variation seems to be available for selection within populations.

Phenological characters such as bud set and onset of frost hardening are strongly correlated to the origin of seed source in a clinal fashion following a latitudinal trend (DeBell and Wilson, 1978; Maynard and Hall, 1980; Hall *et al.*, 1983; DeWald and Steiner, 1986; Cannell *et al.*, 1987). Therefore, most of the variation is found among provenances for these traits, as also noted in *Betula* spp. (Eriksson and Jonsson, 1986). This is in contrast with growth traits. Even if formal estimates of heritability are not available, the phenological characters related to the cessation of growth, together with branching and specific gravity, appear to be more heritable than growth characters (DeBell and Wilson, 1978; Hall and Maynard, 1979). Budbreak generally failed to be related to the origin of seed sources. Low variation has been noted between as well as within provenances for this character (Maynard and Hall, 1980; DeWald and Steiner, 1986; Cannell *et al.*, 1987). Several other phenotypic characters have been shown to vary extensively within species (see Chapter 11; Gonnet and Danière, 1989) but sample sizes remained small and no genetic component could be derived.

G. Infraspecific Variation as Assessed by Allozyme Markers

Alders show large amounts of infraspecific genetic variation at the allozyme level (Bousquet *et al.*, 1987a, 1988). The proportion of polymorphic loci (P) and the number of alleles per locus (A) are generally larger for alder species ($P = 0.52$–0.80, $A = 1.8$–2.3) and other long-lived perennial plants such as conifers than for annuals and selfers (Hamrick *et al.*, 1979; Gottlieb, 1981; Ledig, 1986). The amounts of heterozygosity are also large and well over 10%, even when the monomorphic loci are taken into consideration (Bousquet and Lalonde, 1990). These levels of genetic variation are similar to those of plants with large chromosome numbers, long-lived perennial nature, outcrossing mating system, and anemophily (Loveless and Hamrick, 1984; Ledig, 1986). Populations of alder are also found in Hardy–Weinberg equilibrium, and thus behave essentially as panmictic units (Huenneke, 1985; Bousquet *et al.*, 1987a, 1988). Subpopulation structure and inbreeding, as evaluated by fixation indices (Nei, 1977), were minimal and high levels of recombination and extensive gene exchange among individuals within populations could be inferred (Bousquet and Lalonde, 1990). Outcrossing rates measured in populations of *A. crispa* were found to be large ($t = 0.95$), indicating an open-pollinated mating system (Bousquet *et al.*, 1987a). No subpopulation clonal structure could be detected (Huenneke, 1985; Bousquet *et al.*, 1987a, 1988).

The gene diversity found among populations of alder is very low, as opposed to the diversity found within its populations (Bousquet and Lalonde, 1990). For every alder species considered and on the average, the proportion of variation found within species is equal to or higher than 95% of the total variation. Only the residual 5% is accounted for by differentiation among populations. Generally, annuals and selfers show larger differentiation among populations than long-lived species and outcrossers (Brown, 1979; Loveless and Hamrick, 1984). The mating system could explain most of this difference, with selfing species showing very limited gene flow among populations (see Levin, 1988, for a review).

The genetic identities (Nei, 1972) found among conspecific populations of *Alnus* spp. are very large, and they are at the top level of the spectrum found for conspecific plant populations (Table II). This also indicates that conspecific populations are not differentiated to a great extent. Patterns of conditional average frequencies and related methods (Slatkin, 1985) showed that gene flow is large enough to counteract genetic drift (Bousquet *et al.*, 1987b,c, 1988, 1990b). Because alder conspecific populations are so poorly differentiated, multivariate methods such as discriminant analysis of multilocus genotypes have to be used to detect any trend in the population structure that would be associated with geography (see Bousquet *et al.*, 1987c).

H. Allozyme Divergence between Taxa

The genetic identity (Nei, 1972) deduced from allozyme data between the two subspecific parapatric taxa *A. crispa* and *A. sinuata* (Bousquet *et al.*, 1990b) was well within the range of estimates obtained between a number of conspecific populations, and this genetic identity was near the top of the range of estimates obtained between a number of subspecific taxa (Table II). However, the genetic distance (Nei, 1972) between these subspecific taxa was four to nine times greater than that evaluated between conspecific populations of alder (Table II). Introgression was noted in the sympatric zone, at the overlap of the range of the two taxa (Bousquet *et al.*, 1990b). Therefore, these taxa do not seem to be reproductively isolated intrinsically, but they are isolated geographically over much of their ranges, which agrees with their subspecific taxonomic designation (Furlow, 1979).

The hybrid zone between *A. crispa* and *A. sinuata* seemed narrow and very abrupt (Fig. 2). It has been suggested that hybrid zones may act like a bridge for a diminutive level of gene exchange, resulting in the preser-

Table II Average Genetic Identities and Average Genetic Distances[a] Found between Conspecific Populations, Subspecific Taxa, and Congeneric Species of Plants and of *Alnus*

Group	Number of species	Number of pairwise taxa comparisons	Genetic identity (±S.D.)	Genetic distance
Conspecific populations[b]	23	—	0.963 ± 0.034	0.031
Conspecific populations[c]	38	—	0.947 ± 0.041	0.054
A. crispa (juvenile pop.)[d]	1	—	0.995	0.005
A. crispa (mature pop.)[e]	1	—	0.988	0.012
A. rugosa (mature pop.)[f]	1	—	0.988	0.012
A. sinuata (juvenile pop.)[g]	1	—	0.987	0.013
Subspecific taxa[c]	—	14	0.914 ± 0.087	0.090
A. crispa–A. sinuata[g]	—	1	0.954	0.047
Congeneric species[b]	—	21	0.669 ± 0.199	0.402
Congeneric species[c]	—	37	0.671 ± 0.176	0.399
A. crispa–A. rugosa[f]	—	1	0.666	0.407

[a]Calculated according to the methods of Nei (1972).
[b]From Gottlieb (1981).
[c]From Crawford (1983).
[d]From Bousquet *et al.* (1987c).
[e]From Bousquet *et al.* (1987b).
[f]From Bousquet *et al.* (1988).
[g]From Bousquet *et al.* (1990b).

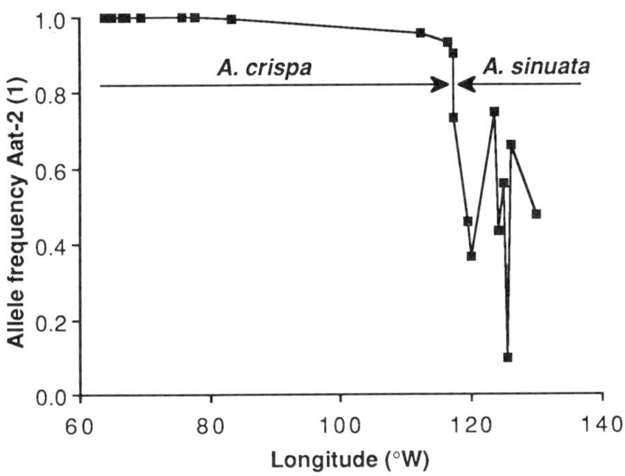

Figure 2 Frequency of the common allele at locus *Aat-2* in 20 populations of *A. crispa* and *A. sinuata* across Canada. Note the abrupt change at the overlap of species distribution.

vation of the genetic integrity of parental taxa (Barton and Hewitt, 1985; Hewitt, 1988; Bousquet *et al.*, 1990b). The lower fitness of hybrids would be caused by the insertion of foreign gene combinations into a balanced parental gene complex, with both parental complexes being more fit for their respective habitats than the hybrids. Hybrids would tend to backcross with one or the other parental species to reconstitute the parental complexes. Such introgressive hybridization has also been reported by Steele (1961) between *A. rugosa* and *A. serrulata,* using morphological characters, and Furlow (1979) similarly reported many instances of natural hybridization between subspecific taxa where their ranges overlap.

The genetic identity between two species of two different subgenera has also been obtained by comparing allozymes of *A. crispa* (subgenus *Alnobetula*) and *A. rugosa* (subgenus *Alnus*) (Bousquet *et al.*, 1988). These species have a transcontinental distribution in northern North America and are found sympatric over most of their range, but differ in their ecological microhabitat preferences. Their interspecific genetic identity was representative of the identity found between congeneric plant taxa that are reproductively isolated (Table II). The associated genetic distance between the congeneric species *A. crispa* and *A. rugosa* was about nine times larger than the distance between the subspecific taxa *A. crispa* and *A. sinuata* (Table II). No evidence for a hybrid multilocus composition was noted among over 1000 mature individuals of *A. crispa* and *A.*

rugosa analyzed, and some diagnostic alleles were fixed in one or the other species (Bousquet *et al.*, 1988).

I. Artificial Hybridization

The amplitude of interspecific genetic distances obtained from allozyme data is in agreement with artificial interspecific crosses. While no success was obtained from crosses involving different subgenera, viable hybrids were obtained from interspecific crosses within the subgenus *Alnobetula* and within the subgenus *Alnus* (Chiba, 1966; Furlow, 1979; Hall and Maynard, 1979; Prat, 1988). Generally, larger success has been obtained when subspecific taxa were crossed (Hall and Maynard, 1979). Morphological characters as well as the number of chromosomes of hybrids are usually intermediate of parental types (Chiba, 1966). Variable crossing success may also be obtained depending on the direction of the cross (Chiba, 1966).

Beyond the intrinsic value of such crosses for our understanding of the isolation barriers between *Alnus* species, these crosses have stimulated tree breeders to obtain hybrids showing vigor superior to that of the parents. Such hybrid vigor or heterosis would be caused by dominance and/or overdominance effects (for a recent account, see Zouros and Foltz, 1987). Superior growth has been obtained in some instances (Chiba, 1966; Prat, 1988), but the general view is that there is no apparent superior vigor in interspecific nor in infraspecific *Alnus* hybrids (Chiba, 1966). This view can be extended to most forest tree species (Ledig, 1986), although *Populus* may constitute an exception (Chiba, 1966; Teissier du Cros, 1984). The lack of hybrid vigor would be caused by a breakdown of coadapted gene complexes leading to developmental instability of the hybrids. In natural populations as well, hybrid zones are generally quite restricted and narrow (see Section III,H). Within conspecific alder populations, no noticeable heterozygote advantage was detected in the mature populations of *A. crispa* or *A. rugosa* analyzed for allozyme diversity, as evidenced by the observed heterozygosities, which did not deviate significantly from the proportions expected under Hardy–Weinberg equilibrium (Bousquet *et al.*, 1987a, 1988).

J. Practical Implications

Genetic variation is very large between and within alder species. In tree improvement, selection strategies should more than before, take into account the genetic variation found within populations as a potential source of genetic gain. As strongly suggested by Maynard (1986), the use of somaclonal variation, for the unique goal of producing new variants, appears to be an awkward strategy in view of the large amounts of natural variation already available and not catalogued. The largest part of the infraspecific variation is found between individuals rather than between populations and provenances. Therefore, the use of clonal material instead of open-pollinated families or provenance seedlots is imperative in any experiment where the infraspecific host variation has to be minimized for other factors to be assessed.

The genecology may not vary much from one species to the other, but no simple rule can be suggested to adjust the selection programs for species not yet studied. For instance, even if *A. crispa* and *A. rugosa* belong to different subgenera, have different ecological preferences (dry versus humid habitats), associate predominantly with different strain types of *Frankia* (Normand and Lalonde, 1982; van Dijk, 1984), diverge strongly in regard to their phenology (Furlow, 1979) and physiological status (F. M. Tremblay and M. Lalonde, unpublished observations), and show large divergence at most allozyme loci examined (Bousquet *et al.*, 1988), they have basically the same genecology in terms of diversity and distribution of that diversity. Alternately, *A. sinuata,* which does not have complete reproductive isolation with *A. crispa,* shows a larger differentiation among populations (Bousquet *et al.*, 1990b) and therefore the selection strategies may be oriented in a slightly different way for this taxa.

The significance of natural hybrid zones between several taxa must also be considered when sampling strategies are defined, and allozyme markers may be used to detect putative hybrid seedlots (Bousquet *et al.*, 1990b), and other types of markers such as chloroplast DNA restriction patterns can also be used (Szmidt *et al.*, 1988). There is also a potential for exotic species introduced into new areas, not only to disseminate their propagules (e.g., seeds) in the new environment but to spontaneously contaminate the natural gene pools of indigenous species adjacent to plantations by introgressive hybridization. These genetic markers can be used efficiently to monitor the extent of such contamination. Species with an important economic potential such as *A. glutinosa, A. rubra,* or *A. cordata,* and species of the subgenus *Clethropsis,* have to be examined at the allozyme level so the overall genetic and genecological variation within *Alnus* can be better assessed. Allozyme studies presently conducted on *A. glutinosa* (D. Prat, personal communication) and *A. maritima* (subgenus *Clethropsis*) (W. J. Elisens, personal communication) will contribute significantly to our understanding of the population and evolutionary genetics of *Alnus.*

IV. The Potential of Studies at the DNA Level

A. Current Strategies

Techniques such as restriction fragment polymorphisms (RFPs) have shown a great potential for the detection of variation at the infraspecific level, as well as for the genotyping of clonal and other pedigree material and for the establishment of linkage maps (Helentjaris *et al.*, 1985; Landry and Michelmore, 1987; Bernatsky and Tanskley, 1989). For actinorhizal species, the intensive characterization of their genomes at the DNA level is still to be done. Because of the conserved nature of certain plant DNA regions such as ribosomal DNA or chloroplast DNA (Appels and Honeycutt, 1986; Palmer and Stein, 1986; Palmer, 1987; Birky, 1988), RFPs using heterologous probes can be applied to solve many problems related to taxonomy and phylogenetics by increasing the sample of species analyzed and reducing the sampling at the infraspecific level (Fig. 3A). Comparative analysis of *Alnus* chloroplast DNA restriction patterns has also been undertaken (M. U. Stoehr, personal communication). The recent availability of alder genomic banks (B. C. Mullin, personal communication) is now providing homologous probes that may result in an increasing number of genes screened and sequenced in actinorhizal plants. Polymerase chain reaction technology (Mullis and Faloona, 1987) also holds similar promise. Recently, the genes coding for the RNA ribosomal small subunit have been amplified from a variety of *Alnus* and other actinorhizal species as well as *Betula* species, using amounts of tissue as small as a single embryo (Bousquet *et al.*, 1990a).

Until recently, other limitations included the difficulty with which perennial plant DNA obtained from mature vegetative tissues could be restricted by endonucleases. Besides the methylated nature of actinorhizal plant DNA (Giasson and Lalonde, 1987), other problems specific to these plants are the presence of secondary metabolites such as tannins and pigments that tend to bind strongly to the DNA in the course of its extraction (Roberts *et al.*, 1985). A number of extraction protocols yielding restrictable high-molecular-weight DNA have been successfully used with various actinorhizal plant tissues (Hattori and Johnson, 1985; Roberts *et al.*, 1985; Giasson and Lalonde, 1987; Hattori *et al.*, 1987; Mackay *et al.*, 1988; Bousquet *et al.*, 1989). RFPs have been reported using leghemoglobin probes hybridized to the DNA of a number of actinorhizal plants (Hattori and Johnson, 1985; Roberts *et al.*, 1985), but no infraspecific variation was noted. Giasson and Lalonde (1987), using restriction patterns of total DNA, were able to identify with certainty some alder and birch species, but no infraspecific variation was noted. The N-terminal amino acid sequences of ribulose 1,5-biphosphate carboxylase have also been reported for *A. viridis* ssp. *maximowiczii*

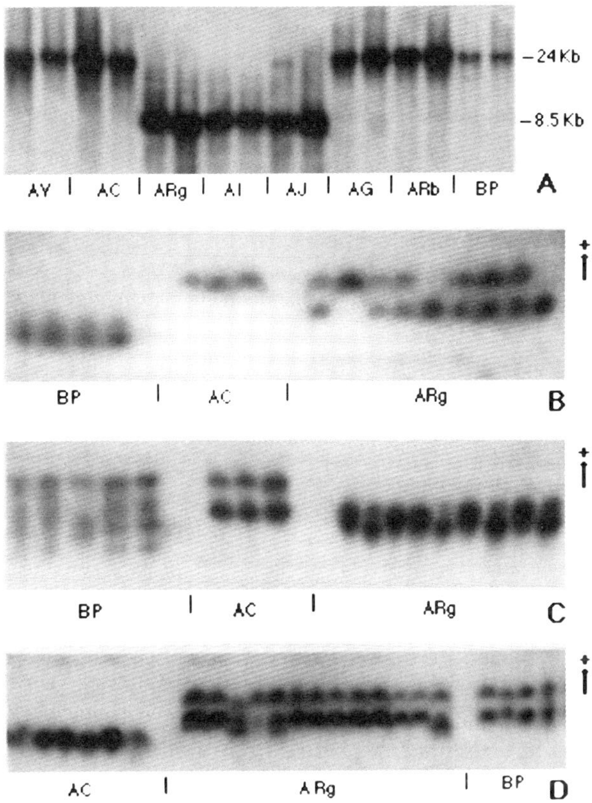

Figure 3 (A) Restriction fragment polymorphism of nuclear ribosomal DNA obtained from seven *Alnus* spp. and *Betula papyrifera* digested with *Hind*III. AV, *A. viridis* (*A. viridis* ssp. *viridis*); AC, *A. crispa;* ARg, *A. rugosa;* AI, *A. incana* (*A. incana* ssp. *incana*); AJ, *A. japonica;* AG, *A. glutinosa;* ARb, *A. rubra;* BP, *B. papyrifera.* (B,C,D) Isozyme patterns observed among different Betulaceae: (B) phosphoglucomutase (PGM): (C) phosphoglucose isomerase (PGI); (D) malate dehydrogenase (MDH). BP, *Betula papyrifera;* AC, *A. crispa;* ARg, *A. rugosa.* For leucine aminopeptidase (LAP, data not shown), no variation was observed among the species represented (digitalized plate).

and *Betula atrata,* with 5 amino acid differences out of a total of 40 amino acids compared (Martin and Dowd, 1988).

B. Interspecific Divergence in the Betulaceae

More recently, the use of ribosomal DNA heterologous probes hybridized to the DNA of several alder species and one birch species digested with many restriction endonucleases basically confirmed the taxonomic treatment of alder species based on morphometrics (Bousquet *et al.*, 1989). The sample of species considered could be divided clearly into two groups representative of the subgenera *Alnus* and *Alnobetula*. Subspecific taxa were also found close to each other. Infraspecific variation, although reduced, could also be detected. Notably, *Betula papyrifera* showed an unexpected similarity to members of the subgenus *Alnobetula*. This study indicated that *B. papyrifera* does not appear to be more divergent from alder species than do taxa representative of different *Alnus* subgenera, according to the DNA region analyzed.

Betula and *Alnus*, in spite of a differential ability to associate with *Frankia*, may not be as divergent as traditionally thought. Ecologically, *Betula* is like *Alnus* in being a successful invader of newly available habitats, especially throughout the deglaciated regions of the Northern Hemisphere (Stebbins, 1971). van Dijk (1984) and Smolander (1990) reported evidence that *Frankia* may be as abundant in the birch rhizosphere as in the alder rhizosphere. In the Betulaceae, *Betula* is found with a basic chromosome number of $x = 7$ (Eriksson and Jonsson, 1986), the same as *Alnus*, while other members such as *Carpinus* and *Ostrya* are found with $x = 8$ or $x = 9$ (Wansher, 1934). It is difficult at present to evaluate the number of gene differences between the genera *Betula* and *Alnus* relative to that number between extant species within each genus. On the basis of the analysis of floral morphology, Gottlieb (1984) found that few major gene differences would be responsible for the absence or presence of discrete characters, while quantitative differences in shape and size would be caused by the cumulative effect of a larger number of small gene differences. It is also likely that the genes involved in the ability to associate with *Frankia* may represent only a small subset of the genome, whether they are present or not, or expressed differentially. Leghemoglobin genes, first thought to be specific to legumes and subsequently to all nitrogen-fixing plants, have been shown to be also present in non-nitrogen-fixing genera such as *Betula* (Hattori and Johnson, 1985), *Celtis*, and *Trema* (Appleby *et al.*, 1988). Preliminary analysis of isozyme patterns between *B. papyrifera* and two extant taxa of the genus *Alnus* showed that as much similarity could be detected between *Alnus* and *Betula* as between *Alnus* species (Fig. 3).

These recent findings suggest that the selective loss of the ability to associate with *Frankia* in the course of angiosperm evolution may be very likely in view of the apparent close genetic relatedness between *Alnus* and *Betula*. As discussed by Furlow (1979), *Betula* and *Alnus* apparently formed one line of descent early in the history of the Betulaceae. Because all alder species examined nodulate and no birch species can do so, it is reasonable to assume that the ability to associate with *Frankia* was lost early on in *Betula*, before its diversification. The potential of *Betula* as an experimental recipient for host genes or host gene modifiers involved in the association with *Frankia* may be considered seriously, in view of the gene delivery systems already available in the Betulaceae (see Chapter 11).

Acknowledgments

We thank H. Kope, P. Normand, A. Séguin, and L. Simon for constructive discussions and for reviewing the manuscript, and W. J. Elisens, B. C. Mullin, D. Prat, and M. U. Stoehr for communication of current projects. This work was made possible by La Fondation Desjardins and NSERC scholarships to J. B., and an NSERC operating grant (A-2920) to M. L.

References

Appels, R., and Honeycutt, R. L. (1986). rDNA: Evolution over a billion years. *In* "DNA Systematics. II: Plants" (S. K. Dutta, ed.), pp. 81–135. CRC Press, Boca Raton, Florida.

Appleby, C. A., Bogusz, D., Dennis, E. S., and Peacock, W. J. (1988). A role for haemoglobin in all plant roots? *Plant Cell Environ.* **11,** 359–367.

Barton, N. H., and Hewitt, G. M. (1985). Analysis of hybrid zones. *Annu. Rev. Ecol. Syst.* **16,** 113–148.

Bensimon, L. C. (1985). Male sterility in *Alnus glutinosa* (L.) Gaertn. *Silvae Genet.* **34,** 69–72.

Bernatsky, R., and Tanskley, S. D. (1989). Restriction fragments as molecular markers for germplasm evaluation and utilisation. *In* "The Use of Plant Genetic Resources" (A. H. D. Brown, O. H. Frankel, D. R. Marshall, and J. T. Williams, eds.), pp. 353–362. Cambridge Univ. Press, Cambridge, England.

Birky, C. W., Jr. (1988). Evolution and variation in plant chloroplast and mitochondrial genomes. *In* "Plant Evolutionary Biology" (L. D. Gottlieb and S. K. Jain, eds.), pp. 23–53. Chapman & Hall, London.

Bond, G. (1983). Taxonomy and distribution of non-legume nitrogen fixing systems. *In* "Biological Nitrogen Fixation in Forest Ecosystems: Foundations and Applications" (J. C. Gordon and C. T. Wheeler, eds.), pp. 55–88. Nijhoff/W. Junk, The Hague.

Bousquet, J., and Lalonde, M. (1990). The population genetics of alder: An overview. *In* "Biochemical Markers in the Population Genetics of Forest Trees" (H. H. Hattemer and S. Fineschi, eds.), pp. 105–112. SPB Academic Publishing bv, The Hague.

Bousquet, J., Cheliak, W. M., and Lalonde, M. (1987a). Allozyme variability in natural populations of green alder (*Alnus crispa*) in Québec. *Genome* **29,** 345–352.

Bousquet, J., Cheliak, W. M., and Lalonde, M. (1987b). Genetic differentiation among 22 mature populations of green alder (*Alnus crispa*) in central Québec. *Can. J. For. Res.* **17,** 219–227.

Bousquet, J., Cheliak, W. M., and Lalonde, M. (1987c). Genetic diversity within and among juvenile populations of green alder (*Alnus crispa*) in Canada. *Physiol. Plant.* **70,** 311–318.

Bousquet, J., Cheliak, W. M., and Lalonde, M. (1988). Allozyme variation within and among mature populations of speckled alder (*Alnus rugosa*) and relationships with green alder (*A. crispa*). *Am. J. Bot.* **75,** 1678–1686.

Bousquet, J., Girouard, E., Strobeck, C., Dancik, B. P., and Lalonde, M. (1989). Restriction fragment polymorphisms in the rDNA region among seven species of *Alnus* and *Betula papyrifera*. *Plant Soil* **118,** 231–240.

Bousquet, J., Simon, L., and Lalonde, M. (1990a). DNA amplification from vegetative and sexual tissues of trees using polymerase chain reaction. *Can. J. For. Res.* **20,** in press.

Bousquet, J., Cheliak, W. M., Wang, J., and Lalonde, M. (1990b). Genetic divergence and introgressive hybridization between *Alnus sinuata* and *A. crispa* (Betulaceae). *Plant Syst. Evol.*, in press.

Brown, A. H. D. (1979). Enzyme polymorphism in plant populations. *Theor. Pop. Biol.* **15,** 1–42.

Cannell, M. G. R., Murray, M. B., and Sheppard, L. J. (1987). Frost hardiness of red alder (*Alnus rubra*) provenances in Britain. *Forestry* **60,** 57–67.

Chiba, S. (1962). Studies on the breeding of *Betula* and *Alnus* species. (1) On the differences of morphological characters and chromosome numbers between *Alnus hirsuta* and *Alnus hirsuta* var. *microphylla*. *Nippon Rin Gakkaishi* **44,** 237–243.

Chiba, S. (1966). "Studies on the Tree Improvement by Means of Artificial Hybridization and Polyploidy in *Alnus* and *Populus* Species," Bull. Oji Inst. For. Tree Improv. 1. Oji Paper Co., Japan.

Crawford, D. J. (1983). Phylogenetic and systematic inferences from electrophoretic studies. *In* "Isozymes in Plant Genetics and Breeding" (S. D. Tanksley and T. J. Orton, eds.), Part A, pp. 257–287. Elsevier, Amsterdam.

Cronquist, A. (1988). "The Evolution and Classification of Flowering Plants," 2nd ed. Allen, Lawrence, Kansas.

Dawson, J. O., and Sun, S.-H. (1981). The effect of *Frankia* isolates from *Comptonia peregrina* and *Alnus crispa* on the growth of *Alnus glutinosa, A. cordata*, and *A. incana* clones. *Can. J. For. Res.* **11,** 758–762.

DeBell, D. S., and Wilson, B. C. (1978). Natural variation in red alder. *In* "Utilization and Management of Alder" (D. G. Briggs, D. S. DeBell, and W. A. Atkinson, eds.), Gen. Tech. Rep. PNW-70, pp. 193–208. U.S. Dep. Agric., For. Serv., Washington, D.C.

DeWald, L. E., and Steiner, K. C. (1986). Phenology, height increment, and cold tolerance of *Alnus glutinosa* populations in a common environment. *Silvae Genet.* **35,** 205–211.

Dixon, R. O. D., and Wheeler, C. T. (1986). "Nitrogen Fixation in Plants." Blackie, Glasgow and London.

Eriksson, G., and Jonsson, A. (1986). A review of the genetics of *Betula*. *Scand. J. For. Res.* **1,** 421–434.

Ferris, S., and Whitt, G. (1977). Loss of duplicate gene expression after polyploidization. *Nature (London)* **265,** 258–260.

Furlow, J. J. (1979). The systematics of the American species of *Alnus* (Betulaceae). *Rhodora* **81,** 1–121 and 151–248.

Genys, J. B. (1988). Intraspecific variation among 28 different sources of black alder, *Alnus glutinosa* (Betulaceae). *Castanea* **53,** 71–79.

Giasson, L., and Lalonde, M. (1987). Restriction pattern analysis of deoxyribonucleic acid isolated from callus and cell suspension of actinorhizal and non-actinorhizal Betulaceae. *Physiol. Plant.* **70,** 304–310.

Gonnet, J.-F., and Danière, C. (1989). Individual variation in the flavonoid aglycones excreted on the leaves of *Alnus glutinosa* (Betulaceae)-I. *Biochem. Syst. Ecol.* **17,** 239–247.

Gottlieb, L. D. (1981). Electrophoretic evidence and plant populations. *Prog. Phytochem.* **7,** 1–46.

Gottlieb, L. D. (1982). Conservation and duplication of isozymes in plants. *Science* **216,** 373–380.

Gottlieb, L. D. (1984). Genetics and morphological evolution in plants. *Am. Nat.* **123,** 681–708.

Hahn, D., Starrenburg, M. J. C., and Akkermans, A. D. L. (1988). Variable compatibility of cloned *Alnus glutinosa* ecotypes against ineffective *Frankia* strains. *Plant Soil* **107,** 233–243.

Hall, R. B., and Maynard, C. A. (1979). Considerations in the genetic improvement of alder. *In* "Symbiotic Nitrogen Fixation in the Management of Temperate Forests" (J. C. Gordon, C. T. Wheeler, and D. A. Perry, eds.), pp. 95–110. For. Res. Lab., Oregon State Univ. Press, Corvallis, Oregon.

Hall, R. B., McNabb, H. S., Jr., Maynard, C. A., and Green, T. L. (1979). Toward development of optimal *Alnus glutinosa* symbioses. *Bot. Gaz. (Chicago), Suppl.* **140,** S120–S126.

Hall, R. B., Miller, G. A., Robinson, T. L., and Onokpise, O. U. (1983). Developing *Alnus* for use in intensive culture. *In* "Intensive Plantation Culture: 12 Years Research" (E. A. Hansen, ed.), Gen. Tech. Rep. NC-91, pp. 35–45. U.S. Dep. Agric., For. Serv., Washington, D.C.

Hamrick, J. L., Linhart, Y. B., and Milton, J. B. (1979). Relationships between life history characteristics and electrophoretically detectable genetic variation in plants. *Annu. Rev. Ecol. Syst.* **10,** 173–200.

Hattori, J., and Johnson, D. A. (1985). The detection of leghemoglobin-like sequences in legumes and non-legumes. *Plant Mol. Biol.* **4,** 285–292.

Hattori, J., Gottlieb-McHugh, S. G., and Johnson, D. A. (1987). The isolation of high-molecular-weight DNA from plants. *Anal. Biochem.* **165,** 70–74.

Haufler, C. H. (1987). Electrophoresis is modifying our concepts of evolution in homosporous pteridophytes. *Am. J. Bot.* **74,** 953–966.

Helentjaris, T., King, G., Slocum, M., Siedenstrang, C., and Wegman, S. (1985). Restriction fragment polymorphisms as probes for plant diversity and their development as tools for applied plant breeding. *Plant Mol. Biol.* **5,** 109–118.

Hewitt, G. M. (1988). Hybrid zones—Natural laboratories for evolutionary studies. *Trends Ecol. Evol.* **3,** 158–167.

Huenneke, L. F. (1985). Spatial distribution of genetic individuals in thickets of *Alnus incana* ssp. *rugosa,* a clonal shrub. *Am. J. Bot.* **72,** 152–158.

Huenneke, L. F. (1987). Demography of a clonal shrub, *Alnus incana* ssp. *rugosa* (Betulaceae). *Am. Midl. Nat.* **117,** 43–55.

Hutchinson, J. (1973). "The Families of Flowering Plants Arranged According to a New System of Their Probable Phylogeny," 3rd ed. Oxford Univ. Press (Clarendon), London and New York.

Landry, B. S., and Michelmore, R. W. (1987). Methods and applications of restriction fragment polymorphism analysis to plants. *In* "Tailoring Genes for Crop Improvement: An Agricultural Perspective" (G. Bruening, J. Harada, T. Kosuge, and A. Hollaender, eds.), pp. 25–44. Plenum, New York.

Lechevalier, M. P. (1986). Catalog of *Frankia* strains, 2nd ed. *Actinomycetes* **19,** 131–162.

Ledig, F. T. (1986). Heterozygosity, heterosis, and fitness in outbreeding plants. *In* "Conservation Biology: The Science of Scarcity and Diversity" (M. E. Soulé, ed.), pp. 77–104. Sinauer, Sunderland, Massachusetts.

Levin, D. A. (1988). Local differentiation and the breeding structure of plant populations. *In* "Plant Evolutionary Biology" (L. D. Gottlieb and S. K. Jain, eds.), pp. 305–337. Chapman & Hall, London.

Linnaeus, C. (1753). "Species Plantarum," tome II. Holmia, Laurentii Salvii (Facsimile ed., W. Junk, 1908).

Loveless, M. D., and Hamrick, J. L. (1984). Ecological determinants of genetic structure in plant populations. *Annu. Rev. Ecol. Syst.* **15,** 65–95.

MacArthur, R. H., and Wilson, E. O. (1967). "The Theory of Island Biogeography." Princeton Univ. Press, Princeton, New Jersey.

Mackay, J., Simon, L., and Lalonde, M. (1987). Effect of substrate nitrogen on the performance of *in vitro* propagated *Alnus glutinosa* clones inoculated with Sp^+ and Sp^- *Frankia* strains. *Plant Soil* **103,** 21–31.

Mackay, J., Séguin, A., and Lalonde, M. (1988). Genetic transformation of 9 *in vitro* clones of *Alnus* and *Betula* by *Agrobacterium tumefaciens*. *Plant Cell Rep.* **7,** 229–232.

Martin, P. G., and Dowd, J. M. (1988). A molecular evolutionary clock for angiosperms. *Taxon* **37,** 364–377.

Maynard, C. A. (1986). Population genetics of forest trees: Implications for the application of *in vitro* techniques. *In Vitro Cell. Dev. Biol.* **22,** 231–233.

Maynard, C. A., and Hall, R. B. (1980). Early results of a range-wide provenance trial of *Alnus glutinosa* (L.) Gaertn. *Proc. Northeast. For. Tree Improve. Conf.* **27,** 184–201.

McVean, D. N. (1953). *Alnus. J. Ecol.* **41,** 447–466.

Monaco, P. A., Ching, T. M., and Ching, K. K. (1981). Variation of *Alnus rubra* for nitrogen fixation capacity and biomass production. *Silvae Genet.* **30,** 46–50.

Müller, J. (1981). Fossil pollen records of extant angiosperms. *Bot. Rev.* **47,** 1–142.

Mullis, K. B., and Faloona, F. A. (1987). Specific synthesis of DNA *in vitro* via a polymerase catalyzed chain reaction. *In* "Methods in Enzymology" (R. Wu, ed.), pp. 335–350. Academic Press, Orlando, Florida.

Murai, S. (1964). Phytotaxonomical and geobotanical studies on genus *Alnus* in Japan (III): Taxonomy of whole world species and distribution of each sect. *Ringyo Shikenjo Kenkyu Hokoku* **171,** 1–107.

Murai, S. (1969). Relationships of allied species between northwestern U.S.A. and Japan on the genus *Alnus*. *In* "Biology of Alder" (J. M. Trappe, J. F. Franklin, R. F. Tarrant, and G. M. Hansen, eds.), pp. 23–36. U.S. Dep. Agric., For. Serv., Pacific Northwest For. Range Exp. St., Portland, Oregon.

Nei, M. (1972). Genetic distance between populations. *Am. Nat.* **106,** 283–292.

Nei, M. (1977). *F*-Statistics and analysis of gene diversity in subdivided populations. *Ann. Hum. Genet.* **41,** 225–233.

Nesme, X., Normand, P., Tremblay, F. M., and Lalonde, M. (1985). Nodulation speed of *Frankia* sp. on *Alnus glutinosa*, *Alnus crispa* and *Myrica gale*. *Can. J. Bot.* **63,** 1292–1295.

Normand, P., and Bousquet, J. (1989). Phylogeny of *nifH* sequences in *Frankia* and in other nitrogen micro-organisms. *J. Mol. Evol.*, **29,** 436–447.

Normand, P., and Lalonde, M. (1982). Evaluation of *Frankia* strains isolated from provenances of two *Alnus* species. *Can. J. Microbiol.* **28,** 1133–1142.

Palmer, J. D. (1987). Chloroplast DNA evolution and biosystematic uses of chloroplast DNA variation. *Am. Nat., Suppl.* **130,** S6–S29.

Palmer, J. D., and Stein, D. B. (1986). Conservation of chloroplast genome structure among vascular plants. *Curr. Genet.* **10,** 823–833.

Pizelle, G. (1984). "Contribution à l'étude de la Nodulation et la Nutrition Azotée chez le genre *Alnus*," Ph.D. thesis. Université de Nancy I, Nancy, France.

Prat, D. (1988). Intérêt de l'hybridation interspécifique et de la multiplication végétative: Le cas de l'aulne. *In* "Actes du 2e Colloque Sciencés et Industries du Bois," tome 1, pp. 161–168. Arbolor, Nancy, France.

Prat, D. (1989). Effects of some pure and mixed *Frankia* strains on seedling growth in different *Alnus* species. *Plant Soil* **113,** 31–38.

Roberts, M. P., Jafar, S., and Mullin, B. C. (1985). Leghemoglobin-like sequences in the DNA of four actinorhizal plants. *Plant Mol. Biol.* **5,** 333–337.

Sellstedt, A., and Huss-Danell, K. (1984). Growth, nitrogen fixation and relative efficiency of nitrogenase in *Alnus incana* grown in different cultivation systems. *Plant Soil* **78,** 147–158.

Sellstedt, A., and Huss-Danell, K. (1986). Biomass production and nitrogen utilization by *Alnus incana* when grown on N_2 or NH_4^+ made available at the same rate. *Planta* **167,** 387–394.

Sheppard, L. J., Hooker, J. E., Wheeler, C. T., and Smith, R. I. (1988). Glasshouse evaluation of the growth of *Alnus rubra* and *Alnus glutinosa* on peat and acid brown earth soils when inoculated with four sources of *Frankia*. *Plant Soil* **110,** 187–198.

Silander, J. A. (1979). Microevolution and clone structure in *Spartina patens*. *Science* **203,** 658–660.

Simon, L., Stein, A., Côté, S., and Lalonde, M. (1985). Performance of *in vitro* propagated *Alnus glutinosa* (L.) Gaertn. clones inoculated with *Frankiae*. *Plant Soil* **87,** 125–133.

Slatkin, M. (1985). Gene flow in natural populations. *Annu. Rev. Ecol. Syst.* **16,** 393–430.

Smolander, A. (1990). *Frankia* populations in soils under different tree species—with special emphasis on soils under *Betula pendula*. *Plant Soil* **121,** 1–10.

Stebbins, G. L. (1971). "Chromosomal Evolution in Higher Plants." Arnold, London.

Steele, F. L. (1961). Introgression of *Alnus serrulata* and *Alnus rugosa*. *Rhodora* **63,** 297–304.

Szmidt, A. E., El-Kassaby, Y. A., Sigurgeirsson, A., Aldèn T., Hällgren, J. E., and Lindgren, D. (1988). Classifying seedlots of *Picea sitchensis* and *P. glauca* in zones of introgression using restriction analysis of cpDNA. *Theor. Appl. Genet.* **76,** 841–845.

Teissier, du Cros, E. (1984). Breeding strategies with poplars in Europe. *For. Ecol. Manage.* **8,** 23–39.

Teissier, du Cros, E., Jung, G., and Bariteau, M. (1984). Alder–*Frankia* interaction and alder–poplar association for biomass production. *Plant Soil* **78,** 235–243.

Thomas B. A., and Spicer, R. A. (1987). "The Evolution and Palaeobiology of Land Plants." Croom Helm, London.

Tiffney, B. H. (1984). Seed size, dispersal syndromes, and the rise of the angiosperms: Evidence and hypothesis. *Ann. Mo. Bot. Gard.* **71,** 551–576.

Tremblay, F. M., Nesme, X., and Lalonde, M. (1984). Selection and micropropagation of nodulating and non-nodulating clones of *Alnus crispa* (Ait.) Pursh. *Plant Soil* **78,** 171–180.

van Dijk, C. (1984). "Ecological Aspects of Spore Formation in the *Frankia–Alnus* Symbiosis," Ph.D. dissertation. Rijks-Universiteit, Leiden, The Netherlands.

Wanshcr, J. H. (1934). The basic chromosome number of the higher plants. *New Phytol.* **33,** 101–126.

Wiley, E. O. (1981). "Phylogenetics: The Theory and Practice of Phylogenetic Systematics." Wiley, New York.

Woodworth, R. H. (1929). Cytological studies in the Betulaceae. II. *Corylus* and *Alnus*. *Bot. Gaz. (Chicago)* **88,** 383–399.

Woodworth, R. H. (1930). Cytological studies in the Betulaceae. III. Parthenogenesis and polyembryony in *Alnus rugosa*. *Bot. Gaz. (Chicago)* **89,** 402–409.

Woodworth, R. H. (1931). Polyploidy in the Betulaceae. *J. Arnold Arbor. Harv. Univ.* **12,** 206–217.

Zouros, E., and Foltz, D. W. (1987). The use of allelic isozyme variation for the study of heterosis. *Curr. Top. Biol. Med. Res.* **13,** 1–59.

Zwolinski, J. B. (1988). Three-year results from provenance trials of *Alnus formosana, A. rubra, Calocedrus formosana, Cunninghamia lanceolata* and *Taiwania cyrptymerioides*. *S. Afr. For. J.* **146,** 34–47.

13

Techniques for Measuring Nitrogenase Activity in *Frankia* and Actinorhizal Plants

Lawrence J. Winship
School of Natural Science
Hampshire College
Amherst, Massachusetts, U.S.A.

John D. Tjepkema
Department of Botany and Plant Pathology
University of Maine
Orono, Maine, U.S.A.

The ideal nitrogenase assay would be rapid, simple, and inexpensive, and would allow the biochemical functions of the nodule to continue undisturbed. Unfortunately, no such assay exists. However, measurement of nitrogenase activity has become much easier since the development of the acetylene reduction assay (Dilworth, 1966; Schöllhorn and Burris, 1967; Hardy *et al.*, 1968) and improved mass spectrometer designs have simplified the use of the stable nitrogen-15 isotope. Still, each nitrogenase assay method has limitations and can, under certain circumstances, lead to serious errors in the estimation of nitrogen fixation. Some limitations, such as the phenomenon called the acetylene-induced decline (Minchin *et al.*, 1983), have only recently been investigated in actinorhizal plants (Tjepkema *et al.*, 1988; Silvester and Winship, 1990). In this chapter, we will present the current options for measuring the nitrogenase activity of *Frankia* and actinorhizal nodules, emphasizing the factors that make each method particularly suited to certain types of questions and preparations. Our goal is to make clear the pitfalls revealed by research to date as a way of helping researchers choose the most suitable method for a given investigation.

The Biology of *Frankia* and Actinorhizal Plants

I. Nitrogen Balance

One of the simplest and sometimes the best method of measuring nitrogen fixation is by following the accumulation of nitrogen in the tissues of plants growing on nitrogen-free substrates. This method can also be applied to plants growing in natural ecosystems, but becomes much more complex and less sensitive because of the need to measure changes in both plant and soil nitrogen and the need to measure other sources of nitrogen gain and loss, such as leaching, denitrification, and deposition of combined nitrogen from the atmosphere. Tissue and soil nitrogen is determined by harvesting replicate groups of plants or other material at intervals followed by wet or dry combustion and ammonia determination. These methods are covered in excellent detail elsewhere (Bergersen, 1980; Knowles, 1981; Silvester, 1983; Elkan, 1987) and are equally applicable to legume and actinorhizal systems.

Advantages:

1. integrates short-term changes in activity (diurnal and seasonal) to give the true average rate of dinitrogen fixation;
2. may be the only practical direct measurement of nitrogen fixation for ecosystem-level studies;
3. can be done with automated measurements of total organic nitrogen;
4. avoids the use of alternate nitrogenase substrates.

Problems and limitations:

1. requires destructive harvest and many replicate samples to attain statistically significant results;
2. requires that large plants be excavated intact, then tissues homogenized and subsampled—not easy for mature trees;
3. uses a total nitrogen determination that can be time-consuming, requires toxic chemicals, and can be subject to large errors if done without care;
4. cannot be used for monitoring short-term changes in fixation rates, such as those due to environmental or physiological changes;
5. can only be applied where the contribution of soil nitrogen to total plant nitrogen is known or is absent;
6. requires very precise quantitation of nitrogen content and is therefore of limited sensitivity and not useful for low-nitrogenase activities.

II. Isotope Enrichment

Direct methods other than nitrogen balance make use of isotope tracer techniques. Since the only radioactive form of nitrogen (^{13}N) has a short half-life (about 10 min), it can only be used at institutions with specialized facilities. Most workers use ^{15}N, a stable isotope, with analysis of $^{15}N/^{14}N$ ratios by mass spectrometry. Over short time periods, nitrogen fixation can be measured by following the enrichment of ^{15}N in the tissues of plants, nodules, or bacteria exposed to ^{15}N-enriched dinitrogen gas. The details of isotope enrichment studies are covered in excellent detail elsewhere (Burris, 1974; Bergersen, 1980; Silvester, 1983; Elkan, 1987).

A few additional points should be noted. One is that although this is a direct method, serious errors are possible, as discussed by Silvester (1983) and in Section VII. Another is that $^{15}N_2$ can be purchased in small compressed cylinders at moderate cost from suppliers such as EG&G Mound Applied Technologies, Stable Isotope Sales, Miamisburg, Ohio. Finally, a service is now available for the analysis of $^{15}N/^{14}N$ ratios and nitrogen content of dried and ground plant material (Waikato Stable Isotope Unit, Ruakura Agricultural Research Centre, Private Bag, Hamilton, New Zealand).

Advantages:

1. provides a sensitive, direct measure of nitrogenase activity;
2. uses the normal, physiological substrate for nitrogenase.

Disadvantages and limitations:

1. uses expensive materials and analysis equipment and is time-consuming;
2. requires destructive harvest of plant tissue;
3. can only be practically applied to excised nodules and roots or small plants, thereby requiring the consideration of disturbance and excision effects, or the use of juvenile plants;
4. covers only short time periods, requiring consideration of variation in space and time of plants in the field and many samples.

III. Isotope Dilution after ^{15}N Enrichment

In this method, the ^{15}N content of the soil surrounding the roots of a nitrogen-fixing plant is artificially increased by the addition of

^{15}N-enriched fertilizer. In theory, the added ^{15}N uniformly labels the inorganic nitrogen of the soil. The effective $^{15}N/^{14}N$ ratio of the soil nitrogen source is estimated by measuring the $^{15}N/^{14}N$ ratio of the organic nitrogen in non-nitrogen-fixing plants growing in the same soil. The organic nitrogen of the nitrogen-fixing plants will contain less ^{15}N because of dilution by atmospheric dinitrogen. The amount of dinitrogen fixation can then be calculated by comparing the atom excess ^{15}N of the fixing and nonfixing systems.

Isotope dilution has been used most extensively in legume systems and is generally accepted as a satisfactory method for obtaining integrated measures of biological nitrogen fixation in field populations of legumes (Chalk, 1985; Danso, 1986; Rennie, 1986; see Witty, 1983, for a discussion of specific methods and results). Two critical problems arise in all studies to date; the choice of a nonfixing reference plant and the assurance of a uniformly labeled soil nitrogen pool. Because nitrogen isotopes can be fractionated by enzymatic and physical processes in plants and in the soil, the choice of a plant with very similar nitrogen uptake and allocation patterns is crucial. In the few studies that applied the isotope dilution method to actinorhizal plants, nonnodulated actinorhizal plants (e.g., *Casuarina* by Gauthier *et al.*, 1985) and closely related nonnodulating species (e.g., birch by Chatarpaul and Lachance, 1989) have been used as reference plants. The results in these studies showed good agreement between isotope dilution and nitrogen balance measurements of nitrogen fixation.

Advantages:

1. is an integrative technique that measures total biological nitrogen fixation over long periods of time;
2. allows estimation of the contribution of soil nitrogen to actinorhizal plant growth;
3. does not require the excavation, disturbance, and enclosure of nodulated roots;
4. uses the normal substrate for nitrogenase.

Disadvantages and limitations:

1. requires the addition of inorganic nitrogen to the soil or growth medium, which can alter the rate of nitrogen fixation or the formation of nodules;
2. requires destructive harvest;
3. does not provide short-term measures of dinitrogen fixation;
4. depends on the selection of a suitable reference plant, which is

difficult since nonnodulating isolines of actinorhizal plants are not available and *Frankia* is widespread in the soil;
5. is potentially very expensive because of the large amount of ^{15}N-enriched fertilizer required for full-scale field studies;
6. presents difficult sampling problems when applied to woody actinorhizal plants, since nitrogen is remobilized and may not be uniformly labeled throughout the plant.

IV. Natural Abundance Dilution

The natural abundance dilution method depends on the presence of natural differences between the soil and atmosphere in the abundance of ^{15}N. While it is definitely true that different natural plant and soil materials have different $^{15}N/^{14}N$ ratios, now more easily measured with modern mass spectrometers than in the past, the origin of these differences is still in doubt. However, work with woody legume species (Shearer *et al.*, 1983) and with various *Alnus* species (Côté and Camiré, 1984; Binkley *et al.*, 1985; Domenach and Kurdali, 1989) has shown that when applied very carefully, the natural abundance dilution method may be useful.

Advantages:
1. does not require the addition of expensive, possibly disruptive labeled fertilizer;
2. does not disturb the plants or ecosystem;
3. relies on normal nitrogenase activity and not alternative substrates;
4. is integrative, providing a measure of total long-term biological nitrogen fixation.

Disadvantages and limitations:
1. cannot work unless significant, measurable differences in ^{15}N abundance exist between nonfixing trees and dinitrogen-grown trees and between atmospheric dinitrogen and soil nitrogen;
2. requires careful choice of the proper reference plants, exploring the same soil volume and using the same soil nitrogen pool as the nitrogen-fixing plants;
3. must include the determination of ^{15}N abundance in control plants grown on a nitrogen-free medium to account for isotope fractionation during dinitrogen fixation;

4. relies on uniform distribution of ^{15}N throughout the plant if only one part of the plant, such as leaves, is sampled.

V. Acetylene Reduction

The discovery that nitrogenase can reduce acetylene to ethylene has revolutionized the study of nitrogen fixation, since it makes possible an assay for nitrogenase activity that is fast, sensitive, and nondestructive.

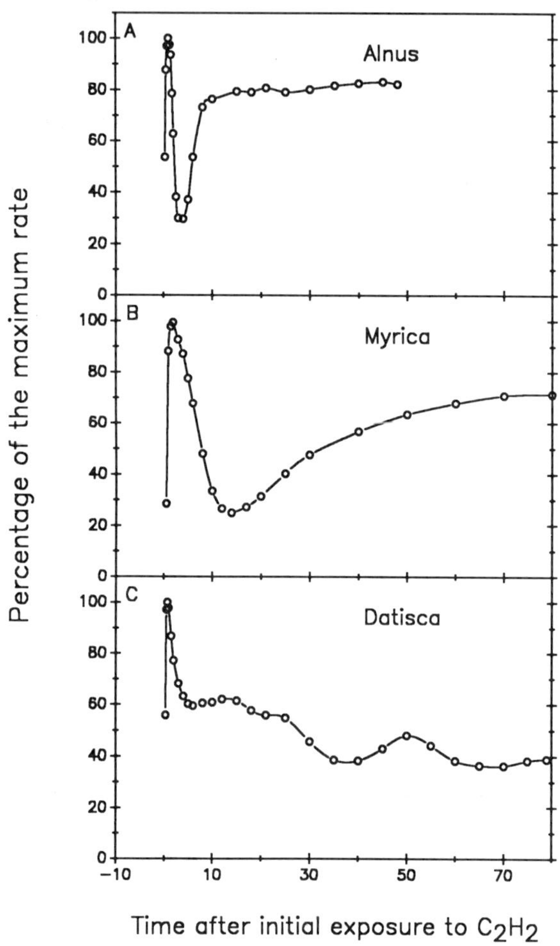

Figure 1 Effect of time in minutes on the rate of acetylene reduction. (A) *Alnus rubra;* (C) *Datisca glomerata.* Data were derived from Tjepkema *et al.* (1988). (B) *Myrica gale* (J. D. Tjepkema and C. R. Schwintzer, unpublished observations).

The acetylene reduction assay has been extensively reviewed (Hardy *et al.*, 1973; Burris, 1974; Bergersen, 1980; Silvester, 1983; Elkan, 1987), but recent work with legumes and actinorizal plants necessitates a serious reevaluation of some aspects of the use of this method.

A. Acetylene-Induced Declines and Transients

A phenomenon called the acetylene-induced decline was first noticed in legumes (Minchin *et al.*, 1983). Following exposure to 10% v/v acetylene in air (standard concentration for the acetylene reduction assay), many legume nodules show an initial steep rise in ethylene production followed by a rapid decline to levels that are less than half of the initial maximum rate. The initial rapid rise is due to the time taken by diffusion of acetylene to the site of nitrogenase followed by the diffusion of ethylene back out of the tissue. The mechanism of the decline is not known, but is hypothesized to be a result of an increase in the resistance to oxygen diffusion into the nodule. Not all legumes tested show this phenomenon, for reasons that are still unclear (Minchin and Witty, 1989).

Actinorhizal nodules show a range of responses to acetylene exposure. In some cases there is little decline or a transient decline followed by recovery to rates near or equal to the initial maximum rate (Fig. 1A; Silvester *et al.*, 1988; Tjepkema *et al.*, 1988; Silvester and Winship, 1990). In other cases there is only a limited recovery (Figs. 1B and 1C; Monz and Schwintzer, 1989).

When the initial peak rate of acetylene reduction is greater than subsequent rates, it is not clear whether it is the best measure of nitrogenase activity. Points in favor of the initial rate as the best measure of nitrogenase activity are that it minimizes possible effects on metabolism due to cessation of NH_3 formation and it is rapidly measured. On the other hand, it might represent an artificially high rate of nitrogenase activity that declines because it cannot be sustained.

B. Open versus Closed Gas Exchange Systems

Until the discovery of acetylene-induced declines and transients, most assays for actylene reduction were done in closed containers. After introduction of 10% acetylene, gas samples are withdrawn and analyzed for ethylene at intervals. The rate of acetylene reduction is calculated from the rate of ethylene accumulation.

When the acetylene reduction rate reaches a peak value within 1 min of acetylene addition and then declines after an additional minute (Fig. 1), it is clear that a closed system assay is not able to measure this initial rate accurately. Thus closed systems can only be used when the rate of acetylene reduction is relatively constant as a function of time. Whether or not this is true can only be established by use of an open system assay.

During an open system assay, the substrate gas is pumped continuously through the cuvette at a known flow rate, maintaining constant steady-state conditions around the nodule or root system. The difference between the concentrations of gas at the inlet and outlet of the cuvette is then used to estimate nitrogenase activity. Numerous examples of open gas exchange methods developed for legume systems are available in the literature (Denison *et al.*, 1983). Open gas exchange methods have also been applied to actinorhizal plants and to cultures of *Frankia* (Winship and Tjepkema, 1982; Silvester *et al.*, 1988, 1989; Tjepkema *et al.*, 1988; Monz and Schwintzer, 1989; Silvester and Winship, 1989, 1990).

A typical continuous flow assay system (Fig. 2) is made up of a gas mixing system, a variable-speed peristaltic pump, a cuvette, and a sampling and analysis system. In most cases gas mixing can be done with large syringes or rotameter flowmeters used to fill gas-tight bags. Saran or aluminized Mylar are the least permeable to acetylene and carbon dioxide but PVC beach balls also work well for a few hours. Bags made

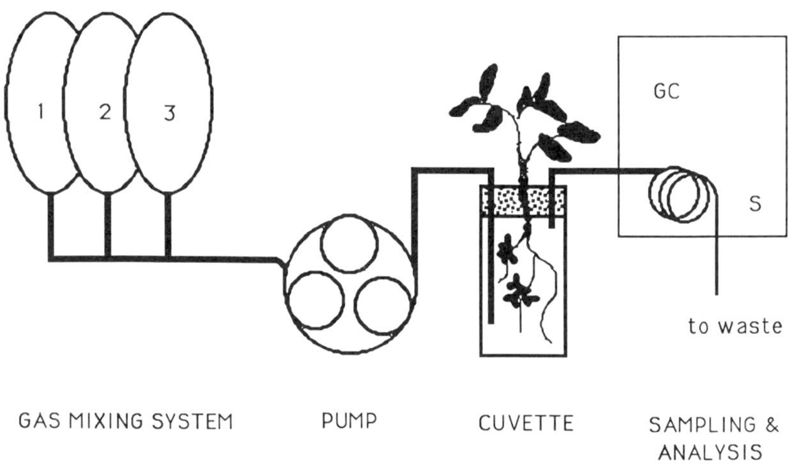

Figure 2 Schematic for a basic open (continuous flow) gas exchange system for measuring the nitrogenase (acetylene reduction) activity of intact, nodulated root systems. Gas reservoirs 1, 2, and 3 contain different premade gas mixtures and are connected to the variable-speed peristalic pump by a manifold of three-way valves. The plant is sealed into the cuvette with a split rubber stopper and adhesive mastic ("Fun-tac"). In this diagram, the outgoing gas stream is automatically analyzed by an FID gas chromatograph (GC) with a motorized gas sampling valve (S).

of polyethylene and rubber are highly permeable to acetylene and carbon dioxide and should not be used. Gas conditions can be rapidly changed by switching between premade bags of gas. More sophisticated changes in gas concentrations can be made with an electronically controlled gas mixing manifold, such as the one described by Layzell *et al.* (1989).

The flow rate of gas through the system can be adjusted for the level of activity so that gas concentrations are high enough to be accurately quantified. If the system is working correctly, the calculated nitrogenase activity will be independent of flow rate. The gas flow rate should also be matched to the speed of the phenomenon under consideration. Rapid changes in activity, such as those due to shifts in oxygen concentration or to the acetylene-induced decline (see the following), will be masked by the "buffering" effect of the cuvette volume unless the flow rate is greater than three cuvette volumes per minute.

Assay chambers (cuvettes) for open (or closed) assays can be as simple as a plastic syringe body or as complex as a specially machined brass chamber incorporating a built-in thermistor and channels for water from a circulating thermostated water bath (Winship and Tjepkema, 1982). Many workers have used plastic greenhouse pots as part of the chamber, designing special lids and bottoms to enclose the root volume of intact plants. Lucite (Plexiglas) chambers have also been used (Huss-Danell, 1978), as have simple Mylar or saran bags, or polyethylene tubs (Arnone, 1988). *Frankia* cultures have been assayed while held on moist filter paper in a Swinney filter holder (Silvester and Winship, 1989). For studies of intact, attached roots or root systems, a common problem has been sealing the stem or root into the assay chamber. Materials that have worked well include closed-cell neoprene foam plugs, Terostat putty, and a commercially available mastic for sticking picture to walls (called "Blu-tak," "Holdit," or "Fun-tak"). The assay cuvette should be held at a constant temperature, most simply by immersion in a thermostated water bath.

Sampling for analysis can be easily accomplished by inserting the exit tube from the cuvette into a beaker of water and allowing the gas to bubble out the end. Gas can then be withdrawn at intervals directly through the walls of the tubing. The bubbling in the beaker indicates that the cuvette is not leaking badly and prevents inadvertent sampling of atmospheric gas by over-quick withdrawal of the syringe sample. A simple cheap, repeating syringe as described by Silvester *et al.* (1989) greatly reduces sample variation. Alternatively, the outlet gas can be directed through the sample loop of a gas sampling valve of a gas chromatograph (Fig. 2). Sample analysis can be automated to provide a measurement of nitrogenase activity at 1- to 2-min intervals.

Open systems provide a sensitive, continuous picture of nitrogenase activity, avoiding artifacts due to changes in nodule physiology or at least making such changes evident! While open methods require one more piece of equipment, a pump, the apparatus can still be quite simple and easy to operate. Cuvettes for open systems can be simpler, since small leaks are less important because the contents are under positive pressure and thus cannot be diluted by gases leaking into the cuvette. The analysis of gas samples is the same as for closed systems.

C. Acetylene Sources and Hazards

Acetylene contains significant quantities of ethylene as an impurity, especially when obtained from compressed cylinders. This contaminating ethylene can be minimized by generating acetylene in the laboratory by adding lumps of calcium carbide to a large excess of water. However, the product contains approximately 400 μl/liter of phosphine and 100 to 200 μl/liter of H_2. Acetylene from cylinders has similar levels of phosphine, but higher levels of H_2 and ethylene (Hyman and Arp, 1987). Since phosphine is lethal to hamsters at concentrations as low as 8 μl/liter (*Merck Index*, 10th Ed.), it is highly desirable to purify acetylene from any source by passing it through a trap of sulfuric acid (Hyman and Arp, 1987).

It should also be noted that acetylene is explosive at concentrations as low as 2.5% in air. Thus the effluent gas from open system assays must not be allowed to contact sources of sparks, flames, or static electricity. Mixtures under greater than atmospheric pressure or at high concentrations of oxygen may be especially hazardous. On one occasion a saran bag containing 10% C_2H_2 and 90% O_2 exploded, apparently due to static electricity generated by vigorous flexing of the bag (personal observation!).

D. Ethylene Analysis

Ethylene is easily quantified and separated from acetylene with a simple isothermal gas chromatograph, with nitrogen carrier gas, and a flame ionization detector (FID). The following conditions permit an analysis time of as little as 50 sec per sample (W. B. Silvester, personal communication): column of $\frac{1}{8}$ inch (3 mm) outer diameter and 1.2 m length, packing of Porapak T ($\frac{60}{80}$ mesh), and column temperature of 75°C.

It is also possible to quantify ethylene using a thermal conductivity detector (TCD), using helium as a carrier gas (Winship and Tjepkema, 1982; Tjepkema, 1988). The TCD allows the measurement of carbon dioxide as well as ethylene and acetylene and makes it possible to measure respiration (as CO_2 evolution) simultaneously with acetylene reduction. TCDs with small cell volumes and small thermistors have a sensitivity for ethylene about 100-fold less than do FID detectors.

The gas analysis system may be enhanced with gas sampling valves, which allow extremely accurate automated sampling of the continuous output from an open assay system. An integrator or a computer equipped with analog to digital signal conversion hardware and signal processing software is also very useful for continuous analysis of the gas from an open assay system.

When measuring ethylene by gas chromatography, the acetylene peak usually follows the ethylene peak and can be used to measure the concentration of acetylene. Some workers have attempted to calculate the rate of acetylene reduction by using the amount of acetylene added to a closed container as an internal standard. Doing so can result in large errors if material of high moisture content (such as nodules) occupies a substantial fraction of the volume. This is because acetylene is much more soluble in water than is ethylene. Furthermore, many rubbers and plastics are much more permeable to acetylene than to ethylene. Thus storage of gas samples in containers with rubber or plastic parts can also lead to errors if acetylene is used as an internal standard.

E. Inhibition of Uptake Hydrogenase by Acetylene

Exposure of nodules to acetylene may cause effects that persist after the acetylene is withdrawn. Inhibition of uptake hydrogenase by acetylene is one such effect. Under normal circumstances, actinorhizal nodules evolve very little or no hydrogen (Chapter 7). All the hydrogen produced by nitrogenase is recycled by a very active uptake hydrogenase (Benson *et al.*, 1980). Following exposure to 10% v/v acetylene for 30 min, *Alnus rubra* nodulated by ArI3 evolved large amounts of hydrogen (Winship and Tjepkema, 1983). The same results have been observed for *Alnus incana* spp. *incana* and *Myrica gale*, while *Ceanothus* and *Datisca* nodules were unaffected (Winship *et al.*, 1987). Various species of *Casuarina* infected with *Frankia* from different sources showed varying amounts of inhibition (Sellstedt and Winship, 1987). In some cases, longer exposure was necessary to inhibit hydrogenase. In others, hydrogenase appears to be partially or completely resistant.

F. Conversion Factors

On the basis of the relative numbers of electrons needed to reduce N_2 to $2NH_3$ and C_2H_2 to C_2H_4, one would expect a ratio of 3:1 between the two activities. However, during N_2 reduction but not during C_2H_2 reduction, a variable part of the electron flow through nitrogenase is used to reduce H^+ to H_2. On this basis one would expect a ratio between N_2 and C_2H_2 reduction that is significantly greater than 3:1.

Experimentally determined ratios between N_2 and C_2H_2 reduction vary widely, but ratios of 3.5–4.0 have been considered normal (Silvester, 1983). However, these ratios were measured before the discovery of acetylene-induced declines, and thus the rate of acetylene reduction may have been underestimated.

In spite of the expense and difficulty of $^{15}N_2$ assays, it is clear that these assays should be done more frequently. At this point we do not understand all the factors that affect the ratio between N_2 and C_2H_2 reduction, so that the C_2H_2 reduction method cannot be used by itself to accurately measure the absolute rate of nitrogen fixation. Even relative measurements could sometimes be in error. On the other hand, the acetylene reduction method will continue to be used extensively because of its low cost, high speed, and nondestructive nature. Our recommendation at this point is to use $^{15}N_2$ to verify all major work done with the acetylene reduction assay. At the same time it should be realized that the $^{15}N_2$ method is also subject to serious errors, and close attention must be paid to avoiding these (Silvester, 1983).

G. Advantages and Disadvantages

Advantages:

1. the acetylene reduction method is rapid, simple, and inexpensive;
2. can be carried out in the field;
3. is nondestructive;
4. is specific for nitrogenase.

Disadvantages:

1. acetylene-induced declines and transients occur;
2. a conversion factor must be determined for each set of conditions to convert data to absolute rates of N_2 fixation.

VI. Hydrogen Analysis

Nitrogenase produces substantial amounts of hydrogen while reducing N_2 to ammonia. It is necessary to measure this hydrogen if total electron flow through nitrogenase is to be determined. However, with one known exception, most of this hydrogen is utilized by uptake hydrogenases in actinorhizal nodules (Chapter 7) and thus cannot be easily measured. Nonetheless, it is important for those working with actinorhizal plants to periodically measure the rate of hydrogen evolution by nodules to verify the presence of uptake hydrogenase under their own conditions. It would be of considerable interest to obtain additional strains of *Frankia* that lack uptake hydrogenase.

Hydrogen in the gas phase has usually been quantified using a gas chromatograph equipped with a thermal conductivity detector, an isothermal oven, a 1-m column of Molecular Sieve 5A, and argon carrier gas. Hydrogen leaves the column rather quickly, followed by very large oxygen and nitrogen peaks. The speed of this analysis is limited by the time required to flush the nitrogen and oxygen from the column, but can be as fast as 2.5 min.

Hydrogen can also be quantified continuously with a semiconductor-based, hydrogen-sensitive combustible gas analyzer (Layzell *et al.*, 1984, 1989). The sensor has high sensitivity, but its response to hydrogen is nonlinear and is affected by the ambient oxygen concentration, the presence of other combustible gases, and the concentration of argon and nitrogen. Despite these limitations, the sensor has been used very effectively with a computerized calibration system and complete analysis of the assay gas mixture (Layzell *et al.*, 1989).

Finally, hydrogen can be measured using a modified polarographic oxygen electrode (Hanus *et al.*, 1980). The system made by Hansatech is particularly handy since the control box has provisions for plating the electrode to permit hydrogen detection and for reversing the polarity of the bias voltage when measuring hydrogen. This method is well suited to studying hydrogen evolution by bacteria in culture. When dithionite is present in the assay mixture, as with measurements of nitrogenase activity in vesicle clusters or cell-free nitrogenase preparations, the coatings on the anode and cathode rapidly degrade and stable electrode performance is difficult (although possible) to maintain. Measurement of hydrogen oxidation, such as during assays for uptake hydrogenase, is easier since the electrode degrades much more slowly.

VII. Limitations and Problems Common to All Methods

A. Nodule Recovery and Disturbance Effects

Excavation of nodules from soil, or even from gravel in pots (Huss-Danell and Ahlqvist, 1984; Minchin *et al.*, 1986), can significantly reduce the nitrogenase activity of root nodules. Removal of nodules from roots can also lower activity (Wheeler *et al.*, 1978). The cause of the loss of activity is not fully known, but nodule tissue can be easily damaged by physical contact during handling. Introduction of water films on the nodule surface can also reduce activity by interfering with gas exchange. On the other hand, there could be dehydration of the nodules during the harvesting procedure. Finally, there may be exhaustion of energy reserves depending on the size of the reserves present at the time of detachment and the length of time after detachment.

Nonetheless, at least under some conditions, nodules with a short piece of attached root can be excised with little or no loss of nitrogenase activity. We have found that both water-cultured and field plants of *Myrica gale* suffer no loss of activity if the part of the root system bearing nodules is carefully cut away from the rest of the roots (Schwintzer and Tjepkema, 1983; Tjepkema *et al.*, 1988). Similarly, there was no loss of activity when nodules of *Alnus rubra* grown in nutrient mist were excised from the root system (Tjepkema and Winship, 1980).

Disturbance effects can be minimized or eliminated by assaying the intact root system by sealing off the top of the container in which the plants are growing and using the pot as an assay cuvette. The disadvantage is that for acetylene reduction in an open system a very high gas flow rate (three times the gas volume enclosed, per minute) will be needed if the initial peak rate of acetylene reduction is to be measured. Some workers have devised *in situ* methods to permit assay of intact plants under field conditions (Balandreau and Dommergues, 1973; Huss-Danell *et al.*, 1989).

B. Oxygen Shock

It has long been known that nitrogenase is inactivated by exposure to oxygen. Nodule homogenates and vesicle cluster preparations must be assayed anaerobically (Roelofsen and Akkermans, 1979) if nitrogenase activity is to survive. Intact plants and bacteria can also be affected by atmospheric oxygen. When actinorhizal plants are moved to air from growth conditions with lower than ambient oxygen concentrations around the roots (such as nonaerated water cultures), nitrogenase activity is significantly inhibited by exposure to the air (Silvester *et al.*, 1988). Further, when intact nodulated root systems or attached root nodules are challenged with step changes in oxygen concentration (Winship and Tjepkema, 1983; Rosendahl and Huss-Danell, 1988; Silvester and Winship, 1990), nitrogenase activity changes rapidly, followed by a recovery to prior activity. The speed of recovery is related to the intensity of the oxygen shock.

Plants and bacteria growing in unstirred liquid culture experience reduced oxygen concentrations due to the low solubility of oxygen in water and the reduced rate of diffusion of gases in liquid. Similarly, nodules covered with a film of liquid may have reduced internal oxygen concentrations. When the liquid is removed, as when cultures are transferred or the plants removed to a gaseous environment, a transient increase in oxygen can occur inside the nodules or *Frankia* vesicles, thus inactivating or transiently reducing nitrogenase activity. These factors must be considered in designing assay methods for intact root nodules or for *Frankia* cultures.

VIII. Conclusions

Each method for measuring nitrogenase activity has its advantages, disadvantages, and sources of error. We urge the reader to consider all of these factors in deciding which method to use. We especially urge caution in the use of excised nodules, excised root systems, and the acetylene reduction assay. When it is necessary to use excised nodules or root systems, measurements on undisturbed plants should be included as controls.

The acetylene reduction assay will continue to be used because it is much faster and less expensive than $^{15}N_2$ methods. However, the rate of acetylene reduction is not constant as a function of time under many conditions, and the factors that affect the conversion ratio between N_2 and acetylene reduction are not well understood. Thus it is essential that major results obtained with the acetylene reduction method be verified with the use of $^{15}N_2$. Measurement of the accumulation of total nitrogen is another possible method of verification, if plants are grown using nitrogen-free media.

References

Arnone, J. A. (1988). "Photosynthesis, Carbon Allocation, and Nitrogen Fixation in Red Alder (*Alnus rubra* Bong.) Seedlings: Effects of Nodulation, CO_2 Enrichment, and Substrate Nitrogen," Ph.D. dissertation, 85 pp. Yale Univ., New Haven, Connecticut.

Balandreau, J., and Dommergues, Y. (1973). Assaying nitrogenase (C_2H_2) activity in the field. *Bull. Ecol. Res. Comm., NFR* **17,** 247–254.

Benson, D. R., Arp, D. J., and Burris, R. H. (1980). Hydrogenase in actinorhizal root nodules and root nodule homogenates. *J. Bacteriol.* **142,** 138–144.

Bergersen, F. R. (ed.) (1980). "Methods for Evaluating Biological Nitrogen Fixation." Wiley, New York.

Binkley, D., Sollins, P., and McGill, W. B. (1985). Natural abundance of nitrogen-15 as a tool for tracing alder-fixed nitrogen. *Soil Sci. Soc. Am. J.* **49** 444–447.

Burris, R. H. (1974). Methodology. *In* "The Biology of Nitrogen Fixation" (A. Quispel, ed.), pp. 9–33. North-Holland Publ., Amsterdam.

Chalk, P. M. (1985). Estimation of N_2-fixation by isotope dilution: An appraisal of techniques involving ^{15}N enrichment and their application. *Soil Biol. Biochem.* **17,** 389–310.

Chatarpaul, L., and Lachance, D. (1989). Is birch a suitable reference in estimating N-fixation in alders by isotope dilution? *Plant Soil* **118,** 43–50.

Côté, B., and Camiré, C. (1984). Growth, nitrogen accumulation and symbiotic dinitrogen fixation in pure and mixed plantings of hybrid poplar and black alder. *Plant Soil* **78,** 209–220.

Danso, S. K. A. (1986). Review: Estimation of N_2-fixation by isotope dilution: An appraisal of techniques involving ^{15}N enrichment and their applications—Comments. *Soil Biol. Biochem.* **18,** 243–244.

Denison, R. F., Sinclair, T. R., Zobel, R. W., Johnson, M. N., and Drake, G. M. (1983). A non-destructive field assay for soybean nitrogen fixation by acetylene reduction. *Plant Soil* **70,** 173–182.

Dilworth, M. J. (1966). Acetylene reduction by nitrogen-fixing preparations from *Clostridium pasteurianum*. *Biochim. Biophys. Acta* **127,** 285–294.

Domenach, A. M., and Kurdali, F. (1989). Estimation of symbiotic dinitrogen fixation in alder forest by the method based on natural ^{15}N abundance. *Plant Soil* **118,** 51–59.

Elkan, G. H. (ed.) (1987). "Symbiotic Nitrogen Fixation Technology." Dekker, New York.

Gauthier, H. G., Diem, H. G., Dommergues, Y. R., and Ganry, F. (1985). Assessment of N_2 fixation by *Casuarina equisetifolia* inoculated with *Frankia* 0RS021001 using ^{15}N methods. *Soil Biol. Biochem.* **17,** 375–379.

Hanus, F. J., Carter, K. R., and Evans, H. J. (1980). Techniques for the measurement of H_2 evolution by nodules. *In* "Methods in Enzymology" (A. San Pietro, ed.), Vol. 69, pp. 731–739. Academic Press, New York.

Hardy, R. W. F., Holsten, R. D., Jackson, E. K., and Burns, R. C. (1968). The acetylene–ethylene assay for N_2 fixation: Laboratory and field evaluation. *Plant Physiol.* **43,** 1185–1207.

Hardy, R. W. F., Burns, R. C., and Holsten, R. D. (1973). Applications of the acetylene–ethylene assay for the measurement of nitrogen fixation. *Soil Biol. Biochem.* **5,** 47–81.

Huss-Danell, K. (1978). Nitrogenase activity measurements of intact plants of *Alnus incana*. *Physiol. Plant.* **43,** 372–376.

Huss-Danell, K., and Ahlqvist, A.-S. (1984). Nitrogenase activity in root nodule homogenates of *Alnus incana*. *Plant Soil* **78,** 159–170.

Huss-Danell, K., Lundquist, P.-O., and Ekblad, A. (1989). Growth and acetylene reduction activity by intact plants of *Alnus incana* under field conditions. *Plant Soil* **118,** 61–73.

Hyman, M. R., and Arp, D. J. (1987). Quantification and removal of some contaminating gases from acetylene used to study gas-utilizing enzymes and microorganisms. *Appl. Environ. Microbiol.* **53,** 298–303.

Knowles, R. (1981). The measurement of nitrogen fixation. *In* "Current Perspectives in Nitrogen Fixation" (A. H. Gibson and W. E. Newton, eds.), pp. 327–333. Aust. Acad. Sci., Canberra, Australia.

Layzell, D. B., Weagle, G. E., and Canvin, D. T. (1984). A flow-through H_2 gas analyzer for use in nitrogen fixation studies. *Plant Physiol.* **75,** 582–585.

Layzell, D. B., Hunt, S., King, B. J., Walsh, K. B., and Weagle, G. E. (1989). A multichannel system for steady-state and continuous measurements of gas exchanges from legume roots and nodules. *In* "Applications of Continuous and Steady-State Methods to Root Biology" (J. G. Torrey and L. J. Winship, eds.), pp. 1–28. Kluwer Academic, Dordrecht, The Netherlands.

Minchin, F. R., and Witty, J. F. (1989). Limitations and errors in gas exchange measurements with legume nodules. *In* Applications of Continuous and Steady-State Methods to Root Biology" (J. G. Torrey and L. J. Winship, eds.), pp. 79–96. Kluwer Academic, Dordrecht, The Netherlands.

Minchin, F. R., Witty, J. F., Sheehy, J. E., and Muller, M. (1983). A major error in the acetylene reduction assay: Decreases in nodular nitrogenase activity under assay conditions. *J. Exp. Bot.* **34,** 641–649.

Minchin, F. R., Sheehy, J. E., and Witty, J. F. (1986). Further errors in acetylene reduction assay: Effects of plant disturbance. *J. Exp. Bot.* **37,** 1581–1591.

Monz, C. A., and Schwintzer, C. R. (1989). The physiology of spore-negative and spore-positive nodules of *Myrica gale. Plant Soil* **118,** 75–87.

Rennie, R. J. (1986). Comparison of methods of enriching a soil with nitrogen-15 to estimate dinitrogen fixation by isotope dilution. *Agron. J.* **78,** 158–163.

Roelofsen, W., and Akkermans, A. D. L. (1979). Uptake and evolution of hydrogen and reduction of acetylene by root nodules and nodule homogenates of *Alnus glutinosa. Plant Soil* **52,** 571–578.

Rosendahl, L., and Huss-Danell, K. (1988). Effects of elevated oxygen tensions on acetylene reduction in *Alnus incana–Frankia* symbioses. *Physiol. Plant.* **74,** 89–94.

Schöllhorn, R., and Burris, R. H. (1967). Acetylene as a competitive inhibitor of N_2 fixation. *Proc. Natl. Acad. Sci. U.S.A.* **58,** 213–216.

Schwintzer, C. R., and Tjepkema, J. D. (1983). Seasonal pattern of energy use, respiration, and nitrogenase activity in root nodules of *Myrica gale. Can. J. Bot.* **61,** 2937–2942.

Sellstedt, A., and Winship, L. J. (1987). Hydrogen metabolism of *Casuarina* root nodules: A comparison of two inoculum sources. *Physiol. Plant.* **70,** 367–372.

Shearer, G., Kohl, D. H., Virginia, R. A., Bryan, B. A., Skeeters, J. L., Nilsen, E. T., Sharifi, M. R., and Rundel, P. W. (1983). Estimates of N_2-fixation from variation in the natural abundance of ^{15}N in Sonoran desert ecosystems. *Oecologia* **56,** 365–373.

Silvester, W. B. (1983). Analysis of nitrogen fixation. *In* "Biological Nitrogen Fixation in Forest Ecosystems: Foundations and Applications" (J. C. Gordon and C. T. Wheeler, eds.), pp. 173–212. Nihoff/Dr. W. Junk, The Hague.

Silvester, W. B., and Winship, L. J. (1989). Continuous measurements of nitrogenase activity in the study of environmental responses of *Frankia* and other bacteria. *In* "Applications of Continuous and Steady-State Methods to Root Biology" (J. G. Torrey and L. J. Winship, eds.), pp. 67–78. Kluwer Academic, Dordrecht, The Netherlands.

Silvester, W. B., and Winship, L. J. (1990). Transient response of actinorhizal nodules to changes in oxygen and acetylene. *Plant Physiol.*, in press.

Silvester, W. B., Whitbeck, J., Silvester, J. K., and Torrey, J. G. (1988). Growth, nodule morphology and nitrogenase activity of *Myrica gale* grown with roots at various oxygen levels. *Can. J. Bot.* **66,** 1762–1771.

Silvester, W. B., Parsons, R., Minchin, F. R., and Witty, J. F. (1989). Simple apparatus for growth of nodulated plants and for continuous nitrogenase assay under defined gas phase. *In* "Applications of Continuous and Steady-State Methods of Root Biology" (J. G. Torrey and L. J. Winship, eds.), pp. 55–66. Kluwer Academic, Dordrecht, The Netherlands.

Tjepkema, J. D. (1988). Oxygen regulation and energy usage in actinorhizal symbioses. *In* "Biological Nitrogen Fixation—Recent Developments" (S. Rao, ed.), pp. 311–333. Oxford Univ. Press IBH Publ., New Dehli, India.

Tjepkema, J. D., and Winship, L. J. (1980). Energy requirement for nitrogen fixation in actinorhizal and legume root nodules. *Science* **209,** 279–281.

Tjepkema, J. D., Schwintzer, C. R., and Monz, C. A. (1988). Time course of acetylene reduction in nodules of five actinorhizal genera. *Plant Physiol.* **86,** 581–583.

Wheeler, C. T., Cameron, E. M., and Gordon, J. C. (1978). Effects of handling and surgical treatments on nitrogenase activity in root nodules of "Alnus glutinosa, with special reference to the application of indole-acetic acid. *New Phytol.* **80,** 175–178.

Winship, L. J., and Tjepkema, J. D. (1982). Simultaneous measurement of acetylene reduction and respiratory gas exchange of attached root nodules. *Plant Physiol.* **70,** 361–365.

Winship, L. J., and Tjepkema, J. D. (1983). The role of diffusion in oxygen protection of nitrogenase in nodules of *Alnus rubra*. *Can. J. Bot.* **61,** 2930–2936.

Winship, L. J., Martin, K. L., and Sellstedt, A. (1987). The acetylene reduction assay inactivates root nodule uptake hydrogenase in some actinorhizal plants. *Physiol. Plant* **70,** 361–366.

Witty, J. F. (1983). Estimating N_2-fixation in the field using ^{15}N-labelled fertilizer: Some problems and solutions. *Soil Biol. Biochem.* **15,** 631–639.

14

Methods for Production and Use of Actinorhizal Plants in Forestry, Low-Maintenance Landscapes, and Revegetation

Larry F. Benoit and Alison M. Berry
Department of Environmental Horticulture
University of California, Davis, California, U.S.A.

Actinorhizal plants often serve many purposes. Common primary uses include production of biomass, fuelwood, timber, pulpwood, browse, and fruit. Actinorhizal plants are also used in amenity planting, low-maintenance landscapes, and revegetation. The adaptive ability of actinorhizal plants can greatly reduce the cost of revegetation and soil stabilization plantings.

Any use of actinorhizal plants requires appropriate plant production and inoculation techniques, which are the main subjects of this chapter. We have taken a general, holistic planning approach to the use of actinorhizal plants. Specific information for individual genera has been included mainly in tabular form for accessibility. One will no doubt note a lack of detailed information pertaining to some actinorhizal species, but we hope to have provided the reader with a reasonable approach to the use of any actinorhizal plants and to the establishment of their symbioses with *Frankia*.

I. Considerations of Site Use and Planting Goals

Deciding the primary uses of actinorhizal plants on a particular site can serve as a starting point for the process of planning for propagation, outplanting, and follow-up. The success of the planting depends on an understanding of the ecological limits of the planting site, which in turn determine the appropriate choices of actinorhizal species and the selection of propagation materials.

A. Meeting Site Objectives with Actinorhizal Plants

An early determination of the type of plantation will permit the selection of appropriate management approaches. There may be regional aspects of actinorhizal plant use to consider, as discussed in Chapters 15 through 18 in this book. On undeveloped sites, rotational plantings may be required to improve the soil for future objectives. Rotational plantings may also be useful for biomass production, forage, or revegetation. An interplanting or nurse planting needs to be compatible with the main crop or overstory species. Fast-growing actinorhizal plants may outcompete crop species such as conifers. If the actinorhizal plants will be developed as an understory, the plants chosen need to be adapted to the available light. In revegetation applications, it is important to achieve adequate coverage of the site for erosion control, watershed protection, and other objectives. With potential succession patterns in mind, a practical mix of actinorhizal and nonsymbiotic plants can be developed.

B. Physical and Vegetative Characteristics of the Site

The most basic influences on use decisions, regardless of overall site objectives, are the site climate, soil, and vegetation. An adequate evaluation of site conditions and a survey of current vegetation should be the starting point in the choice of actinorhizal species. The general habitat type or range of most actinorhizal genera is presented in Table I. Native actinorhizal plants already established on or near the site will be reasonably well adapted to climatic and other site conditions. These may provide the most effective propagative material of host and symbiont. If the site is disturbed or undeveloped, revegetation may be indicated initially, even if forestry is the ultimate goal.

II. Propagation of Actinorhizal Plants

A. Methods of Propagation

General methods of propagating woody plants are applicable to actinorhizal plants. Propagation by seed is flexible and can be used in nursery beds, for container stock, or in direct seeding on the site. There is little need of specialized propagation equipment when using seed. Where particular characteristics of source plants are desired or if seed is

Table I Habitat Type and Potential Use of Actinorhizal Plants

Genus	Habit[a]	Main use[b]
	Mesophytic and Cold-Tolerant Hosts	
Alnus	de, sh, tr	bio, wb, wp
Dryas	eg, gc	pnr, rvg, stb
Shepherdia	de, sh	rvg, shr
Myrica	de-eg, sh-tr	bank stb, wax
	Mesophytic to Xerophytic Hosts	
Casuarina, Allocasuarina, and *Gymnostoma*	eg, tr, st	fw, wb, wp
Comptonia	de, sh	rvg, bank stb
Discaria	de, sh-tr	pnr, rvg, stb
Elaeagnus	de-eg, sh-tr, st	shr
Hippophae	de, sh-tr, st	dune stb, rvg
Shepherdia	de, sh	rvg, shr
	Xerophytic Hosts[c]	
Ceanothus	de-eg, gc-tr, fr	lml, rvg
Cercocarpus	eg, sh-tr, st	fw, rvg, shr
Purshia and *Cowania*	eg, gc-sh	bws, rvg, stb

[a]Habit: deciduous (de), evergreen (eg), ground cover (gc), shrub (sh), tree (tr); tolerance—fire (fr), salinity (st).

[b]Main use: biomass (bio), browse (bws), fuelwood (fw), low-maintenance landscape (lml), pioneer (pnr), revegatation (rvg), shelter (shr), stabilization (stb), windbreak (wb), wood products (wp).

[c]Note: There is a group of drought deciduous, rhamnaceous shrubs native to South America (*Colletia, Kentrothamnus, Retanilla, Talguenea,* and *Trevoa* spp.) that may have potential use in revegetation.

not available, propagation by vegetative means may be the method of choice. Other techniques of vegetative propagation such as *in vitro* propagation and tissue culture are presented in Chapter 11 of this volume.

1. Propagation by Seed and Direct Seeding

Conditions for collection, handling, and use of seed for actinorhizal genera are listed in Table II. In general, seed is collected at the peak of maturity and stored under cool, dry conditions after some degree of cleaning or processing. Seed collections from plants growing near the site offer the advantage of good site adaptation. Commercial sources of seed are available for some actinorhizal plants, such as *Alnus* and *Elaeagnus* spp. and several genera from the western United States (*Ceanothus, Cercocarpus, Purshia,* and *Shepherdia* spp.). Commercial seed quality may be variable.

For operational planning in plant production, or if direct seeding is to be used, seed viability must be determined via germination tests.

Table II Methods for Collection, Processing, and Use of Seeds of Actinorhizal Plants[a]

Genus	Collection	Extraction	Storage conditions	Pregermination treatment	Germination conditions	Outplanting age
Alnus	Brown strobili summer–fall	Dry 27–38°C or ambient	Dry, 1–3°C sealed	Overnight soak in aerated water	20–30°C, 16hr, daylight 7–21 days	15–20 weeks
Casuarina	Strip brown cones	Dry, shaking and screening	Dry, room sealed	None	20–30°C, 16 hr, 11–40 days	12–20 weeks
Ceanothus	Bag green on plant	Rolling, and fanning mill	Dry, cool sealed	Hot water soak	25°C, 12–30 days	Up to 1 year
Cercocarpus, Cowania	Shake from plant, late summer–fall	Burlap roll or hammermill to destyle	Dry, cool	Aerated water soak 24 hr, 5 weeks to stratify	15–20°C, 1–2 weeks	16–25 weeks, 1-0, 2-0
Comptonia	Pick, golden ripe fruit		Dry, cold	Scarify + 500 ppm GA_3	15–20°C, to 8 weeks	20–30 weeks
Elaeagnus	Fall, pick or strip	Dry or macerate in water to separate pulp	Dry, sealed 1–10°C	Stratify 10–90 days if hard seed: H_2SO_4 1/2–1 hr	30°C day, 20°C night	16–52 weeks
Hippophae	Pick soft fruits, late fall to early spring	Prompt clean by water maceration to remove pulp	Dry room temp., to 1–2 years	Stratify 90 days	30°C day, 20°C night, 40 days	
Myrica	Pick, strip, or shake off plants	None, but mechanically remove wax to stratify, if waxy	Dry, with wax on, if waxy	Stratify 14–90 days 500 ppm GA_3, 24 hr	23–25°C, 16 hr, 11–150 days	12–20 weeks
Purshia	Shake dark red-gray achenes onto canvas	Dewinger, rough roll, or fanning mill	Dry, cool	Stratify 5 weeks	15–20°C, 14 days	16–20 weeks
Shepherdia	Pick, strip, or flail the yellow-red berries to canvas, prevent from heating	Macerate with water to separate pulp	Dry, sealed, at 5°C	Stratify 60–90 days hard seed: H_2SO_4 scarify, stratify	20–30°C, 21–60 days	16–25 weeks, 2-0

[a]Information is from Schopmeyer (1974), Dirr (1983), Midgley *et al*. (1981), and personal experience.

Germination tests can also provide information on the effectiveness of pregermination treatments and on timing and conditions required for germination for a particular seed lot. At least 400 seeds should be tested for accurate results in a given treatment.

There are some specific requirements for actinorhizal seed germination. Schopmeyer (1974) provides information on germination of several actinorhizal plants. In *Alnus* spp. seed germination requires light (Bormann, 1983) and is greatly enhanced by a 16- to 24-hr soak in aerated water (Berry and Torrey, 1985). Germination of *Comptonia* spp. seed is increased by treatment with 500 ppm GA_3 (Del Tredici and Torrey, 1976). *Ceanothus* spp. require boiled water soaks (ca. 90°C water poured over seed, allowed to cool to 25°C). *Cercocarpus* spp. germinate best with a 24-hr leaching in aerated water, followed by a 5-week moist, cold treatment at 0°C. Seeds of *Myrica gale* germinate best if floated on water at 5°C for several weeks and then germinated in the light (Schwintzer and Ostrofsky, 1989).

Large-scale planting of actinorhizal plants is often carried out by direct seeding on the site. *Casuarina, Cercocarpus, Cowania, Elaeagnus, Myrica, Purshia,* and *Shepherdia* spp. can all be seeded directly (Midgley *et al.*, 1981; Schopmeyer, 1974). Fall seeding may be required when cold stratification is needed for seed germination. Alternatively, seed can be stratified and then sown in spring.

2. Rooted Cuttings and Other Vegetative Propagation

Collection of cutting materials should take place at the time of year when the recommended growth stage is available for the species (Table III). Greenwood cuttings are taken when the plant is in active growth and may be called "softwood" cuttings if they are from tender developing shoots or "semihardwood" cuttings if mature and growth has stopped for the season. "Hardwood" cuttings are taken when the stock plants are dormant. Hartmann and Kester (1983) provide a good general reference for vegetative propagation.

Cutting material must be available in enough quantity to allow a reasonable chance of success in producing the desired number of plants. Rooting percentages can vary up to 100% depending on species, condition, and treatment with rooting aids. One should collect cuttings from healthy, vigorous plants. Collections are best made early in the day, when plant water potential is highest, and cuttings should be processed as soon as possible. If there is a need to delay, cutting materials should be kept cool. Hardwood and some greenwood cuttings can be held in cold storage for future use, but most greenwood cuttings should be processed for rooting (stuck) within a few days of collection.

Table III Common Vegetative Propagation Methods for Genera of Actinorhizal Plants[a]

Genus	Notes on vegetative propagation
Alnus	Semihardwood and hardwood stem cuttings, 5000 ppm IBA or powder rooting aid, root 3–6 weeks in mist, sand, or water culture
Casuarina	Softwood cuttings, 50 ppm IBA, standard mist or water culture (1/4 × Hoagland's solution), 5 weeks
Ceanothus	Softwood (terminal 4–6 inches to semihardwood cuttings, root in 4–10 weeks, 5000 ppm IBA or powder rooting aid
Cercocarpus	Semihardwood cuttings, 5000 ppm IBA, root well with bottom heat and mist, 6 weeks
Comptonia	Softwood cuttings; stem layering; root pieces covered by rooting medium
Discaria	Semihardwood cuttings
Dryas	Semihardwood cuttings; layering horizontal stems
Elaeagnus	Hardwood cuttings; layering; root cuttings
Hippophae	Hardwood cuttings; layering; root cuttings
Myrica	Semihardwood cuttings, 5000 ppm IBA; layering
Purshia	Semihardwood cuttings, shade bench only, 5000 ppm IBA or powder rooting aid, 8–10 weeks
Shepherdia	Semihardwood cuttings, 8000 ppm IBA dip

[a]Information is from Bailey and Bailey (1976), Huss-Danell (1981), Lundquist and Torrey (1984), Thomas (1986), and personal experience.

Rooting aids in common use include an aqueous dip in the potassium salt of indole-3-butyric acid (IBA) or IBA and/or naphthalene acetic acid (NAA) in a talc powder base. Cuttings may be soaked in low concentrations of IBA for up to 24 hr or more commonly given a quick dip (10–15 sec) in concentrations above 1000 ppm.

Alnus spp. are relatively easy to propagate vegetatively under standard mist bench conditions, as well as in sand or water culture (Huss-Danell, 1981). *Casuarina* and related species can be similarly propagated (Lundquist and Torrey, 1984). In our experience, *Ceanothus, Cercocarpus,* and *Purshia* are intolerant of moisture accumulation. Successful rooting of these genera may require reducing mist frequency or rooting under shade with only daily irrigation. Rooting media should be selected in relation to the water tolerance of specific cuttings. A standard mix of perlite: vermiculite (1 : 1) may be used to root *Alnus* cuttings, while 2 : 1 or 3 : 1 mixes result in better survival and rooting of *Ceanothus, Cercocarpus,* and *Purshia* cuttings (Thomas, 1986; L. F. Benoit, unpublished observations). An overview of the known cutting propagation potential for various actinorhizal plants is presented in Table III. If a particular genus is not included, one might try propagation techniques for a similar habitat type or for that of a genus in the same family.

B. Plant Production

The time of outplanting depends on seasonal climatic conditions of the site and should be selected to provide the best opportunity for successful establishment of each species. The desired planting date can then be used to schedule the production operations.

1. Container Plants and Related Planting Logistics

Containerized seedlings can be started in flats or seeded directly in the container to be used for outplanting. Sufficient seed should be sown to ensure usable plants for each container or section. Results of germination tests for the particular seed lot are used with probability tables to provide a guide to appropriate seeding rates (Balmer and Space, 1976). Additional losses may be experienced during early seedling growth as a result of poor vigor, disease, or other problems. Direct seeding in containers offers the advantage of minimizing transplant losses and root or root nodule damage. Starting seeds in flats, on the other hand, permits the selection of more uniform plants.

Choice of soil mix or potting medium can influence early seedling growth and can provide texture and moisture retention characteristics similar to those of the planting site. When available, soilless mixes such as peat-based mixes, or perlite-vermiculite, offer a useful range of soil texture and moisture-holding capacity (Hartmann and Kester, 1983).

Containerized seedlings are easy to transport and handle at the time of outplanting. Many types of liners and plant dividers are available for large-scale production. Liner materials range from paper or treated cardboard to plastic. Liner plants can be grown in large numbers per unit area and are used commonly in forest tree production (Tinus and McDonald, 1979). A standard nursery container used for smaller quantities of larger-size woody plants is the gallon can (cylinder, approx. 3.6 liters). For most actinorhizal plants, containers should be of appropriate size for a production period of about 20 weeks (or up to a year in some cases). Growing plants in containers offers some potential to choose the planting season, since container stock can be outplanted in spring or fall.

2. Bareroot Plants

Bareroot plants are inexpensive to produce in large quantities if land and appropriate equipment are available. Seed can be lined out in the nursery and the plants dug (lifted) by hand or machine. Bareroot plants can be planted directly or stored and require little space for shipping. Planting sites for bareroot plants need to have sufficient friable soil to place around the roots in the planting hole, and site conditions at planting must be suitable to ensure new root and top growth. Irrigation enhances the success of bareroot plantings by settling soil around the roots and may be required for some period after planting if soil moisture is not adequate. In addition to *Alnus* and other deciduous species, *Casuarina* spp. can be planted bareroot (Midgley *et al.*, 1981), as well as *Purshia* and *Cercocarpus* spp. (Shaw, 1984). Outplanting of bareroot stock is limited to the dormant season or early growing season.

If bareroot plants are nodulated, desiccation of nodules during handling may be a problem. Use of bareroot nodulated plants has been successful when plants were dug just prior to planting (Côté and Carmiré, 1984), but little information is available on possible handling and storage problems of nodulated stock.

3. Growth Conditions for Production of Plant Material

Nitrogen is a critical determinant for successful nodulation, plant size, and outplanting success, although adequate amounts of phosphorus and other nutrients are also necessary for optimal plant growth. Excess nitrogen availability during production can inhibit nodulation and result in tender, luxuriant shoot growth. If nitrogen is deficient, total plant growth may be reduced.

C. Maintenance of Plants to Enhance Establishment of Symbioses

1. Fertilizer Effects on Nodulation

It is well known that excess nitrogen inhibits nodulation during plant production (Dixon and Wheeler, 1983). As little as 53 ppm N (as ammonium nitrate) inhibited nodulation of *Purshia* spp. (Righetti and Munns, 1982). Nodulation was inhibited in *Casuarina cunninghamiana* with 105 ppm N (as ammonium sulfate), while 69 ppm N (as potassium nitrate) resulted in a 66% reduction in successfully nodulated plants (Zhongze and Torrey, 1985). In *Ceanothus* spp., pretreatment with nitrogen (75 ppm N, as ammonium nitrate) delayed nodulation up to 8 weeks after fertilizer treatment was stopped (Thomas and Berry, 1989). In short-term experiments, potassium nitrate inhibited nodulation in *Alnus, Casuarina,* and *Myrica* spp. at a concentration range of 21 to 28 ppm N (Kohls and Baker, 1989). The quality of irrigation water should be known, since it may already contain high levels of nitrate (up to 30 ppm in some agricultural areas of the United States). If phosphorus is limiting, nodulation may be reduced or blocked (Dommergues *et al.*, 1984; Israel, 1987). Soil pH, which can be affected by fertilizer, is also important in the nodulation process (Dixon and Wheeler, 1983).

2. Moisture Conditions

While the range of soil moisture availability tolerated by actinorhizal plants is wide, mesic conditions generally appear to enhance nodulation by *Frankia*. In flooded conditions (Schwintzer, 1985; Schwintzer and Lancelle, 1983), and in aeroponic tanks, nodulation tends to occur in the upper part of the root system rather than in the solution. In *Ceanothus* and *Cercocarpus* spp., nodulation in the greenhouse occurs under conditions of good soil drainage and adequate moisture availability, but not in extremely wet or dry conditions (Thomas, 1986; L. F. Benoit, unpublished observations). Water stress can reduce nodulation and nitrogen fixation (Dommergues *et al.*, 1984), but established nodules on *Ceanothus* spp. survived through two drought years without irrigation (L. F. Benoit, unpublished observations).

III. Approaches to Establishing Actinorhizal Symbioses

When new species are to be introduced into an area, it may be necessary to introduce any desired root symbionts by inoculation prior to outplanting (Dreyfus *et al.*, 1987). Actinorhizal nodules are perennial, persisting in some cases up to 5 years or more (Akkermans and van Dijk, 1981; Schwintzer *et al.*, 1982), and may become large in size. Nodules established by initial inoculation may thus provide nitrogen for several years. These early nodules could serve as a source of inoculum in subsequent years, provided subsequent root growth is adequate in the zone of nodulation (Wilson and Coutts, 1985).

Although inoculation is more likely to ensure the establishment of successful nodule symbioses, it may not be required during early plant production if the same actinorhizal species are present and well nodulated on or near the site. The presence and distribution of infective and effective *Frankia* on a site can be determined by growing desired actinorhizal plants in separate soil core samples (Arveby and Huss-Danell, 1988).

A. Types of Inoculum and Their Collection and Storage

Frankia strains × host species combinations can vary considerably in the degree of effectiveness in nitrogen fixation. Performance of some host–strain combinations has been evaluated (Diem *et al.*, 1988; Dreyfus *et al.*, 1987; Hooker and Wheeler, 1987; Monz and Schwintzer, 1989; Reddell *et al.*, 1988). *Frankia* isolates of known infectivity provide the most predictable source of inoculum. Other types of inoculum include: nodule inoculum from the same or cross-compatible host; rhizosphere soil associated with nodules on a cross-compatible host; and rhizosphere soil associated with the roots of an actinorhizal host with unknown nodulation.

1. *Frankia* Isolate Inoculum

Frankia isolates of known infectivity and effectivity on the particular actinorhizal host provide a measure of certainty of success. Uniform suspensions can be prepared at predictable inoculum densities. Starter cultures of selected frankiae may be requested from individual researchers or institutions. Several *Frankia* strains are on deposit in the USDA actinomycetal collection (Northern Regional Research Center, ARS, Peoria, IL) and may be requested. The propagation of *Frankia* isolates for inoculum requires specific facilities and techniques (Chapters 3 and 16).

Current research in several laboratories is directed toward methods of large-scale preservation of *Frankia* in dried form. Some dried preparations of *Frankia* in calcium alginate have been stored for 2–3 years without apparent loss of infectivity (Diem *et al.*, 1988). In *Frankia* strain HFPCcI3, spores are necessary for infectivity following air drying (S. Burleigh, unpublished observations). *Frankia* inoculum, whether fresh or dried, is not yet commercially available.

2. Nodule and Rhizosphere Soil Inoculum

Ground nodule suspensions or rhizosphere soil can be used as inocula for nodulation of actinorhizal hosts. Nodulated actinorhizal plants near the planting site are usually an excellent source of nodule or soil inoculum for the same species or cross-compatible actinorhizal plants. An evaluation of inoculum infectivity and effectiveness in nitrogen fixation should be carried out for nodule or soil sources prior to large-scale inoculation.

a. Collection of Nodule or Rhizosphere Soil Inoculum. Nodules are more likely to be found on vigorously growing plants with deep green foliage rather than on less vigorous plants in a given site. In general, the tips of active nodule lobes are rather white in appearance. Nodules often occur within 1 m of the root crown on many actinorhizal plants. However, in some situations, nodules may be at inaccessible depths (Silvester *et al.*, 1985; Virginia *et al.*, 1986; Williams *et al.*, 1986).

b. Storage of Nodule and Soil Inoculum. Inoculum infectivity can be maintained by short-term cold storage. We have stored both fresh and dry *Ceanothus* nodules in soil, near 0°C, for up to one year, with no measurable loss of nodulation potential. Low or excessive moisture in cold storage can cause desiccation of inoculum or overgrowth of other organisms. Alternatively, nodules can be freeze-dried and stored at −20°C. Infectivity of freeze-dried nodule inoculum has been maintained after 36 months in storage (Q. Liu, unpublished observations). The infectivity of nodule inoculum may be maintained in some species by drying over silica gel (e.g., *Alnus* and *Casuarina*). In other species (e.g., *Ceanothus* and *Cercocarpus*), infectivity is rapidly lost with or following this treatment (Q. Liu, unpublished observations; L. F. Benoit, unpublished observations). Preservation of infectivity with frozen nodules or nodule suspensions has also been reported (Akkermans and Houwers, 1979).

B. Inoculation Techniques

Frankia must be introduced in such a way as to ensure that actively growing roots come in contact with the inoculum. Details on preparation of inocula are given in Chapters 5 and 16. Aqueous suspensions are very commonly used to inoculate actinorhizal plants. For *Frankia* isolate inoculum, a packed cell volume (pcv) of approximately 1–5 μl per plant (after washing and centrifugation) provides adequate nodulation (Berry and Torrey, 1985; Stowers and Smith, 1985). For ground nodule suspensions, 0.5–2 mg nodule fresh weight or more per plant may be required as inoculum (Akkermans and Houwers, 1979; Périnet *et al.*, 1985). In our experience, the predictability of nodulation with rhizosphere soil may be lower than that of nodule or *Frankia* isolate inocula, thus 2–5 g or more of soil inoculum per plant may be required for adequate nodulation (Thomas, 1986).

1. Inoculation and Nodulation in Nursery Production

Inoculum suspensions can be mixed with soil or potting medium prior to planting. Inoculum has also been successfully sprayed over the surface of soil mix on a large scale (Périnet *et al.*, 1985). If small plants are to be inoculated at potting, root systems can be sprayed with or dipped in the inoculum suspension (Berry and Torrey, 1985). Dry inoculum can be rehydrated and applied as a suspension or applied dry in the potting medium or planting hole (Diem *et al.*, 1988). Pelleted seed that includes *Frankia* inoculum has not been reported, but such formulations could be feasible using a process similar to those used for rhizobial inoculation of legume seed (Dreyfus *et al.*, 1987).

2. *In Situ* Inoculations

In situ inoculation is most efficiently carried out at the time of planting. More research is needed on methods for providing inoculum in the planting hole and utilizing calcium alginate or similar preparations of *Frankia* isolates. Direct seeding may require inoculation as a drench at the time of seeding or by incorporation of inoculum into soil. Pelleted seed may be useful in direct seeding, once this technology is developed for actinorhizal plants. On sites with low phosphorus availability, it may be beneficial to include mycorrhizal inoculum as well (Dommergues *et al.*, 1984; Chapter 16).

C. Maintenance of Inoculated Plants to Enhance Symbiotic Success

The period following inoculation is a critical time for proper maintenance of actinorhizal plants. Nitrogen and other factors such as temperature, pH, and water relations can directly influence nodulation, as discussed in the following sections.

1. Low Concentrations of Available Nitrogen Required

Nodulation is highly sensitive to excess available nitrogen. As little as 6 m*M* ammonium nitrate can inhibit the nodulation of some actinorhizal plants (Righetti and Munns, 1982). It is important to stress that provision of some nitrogen enhances nodule biomass and plant productivity. In *Ceanothus*, for example, 20 ppm N resulted in an optimal combination of good plant growth and nodulation (Thomas and Berry, 1989). As previously stated, adequate phosphorus availability is required for nodulation and symbiotic nitrogen fixation. Plantings may require fertilization with phosphate. Inoculation with mycorrhizal fungi may prove to be of benefit for phosphorus nutrition.

2. Water Relations, pH, and Temperature as Symbiotic Factors

As discussed in Section II, C, 2, moderate water status may enhance the establishment of symbiosis. A soil pH in the neutral range is optimal for nodulation of most actinorhizal plants. Some hosts nodulate at lower pH, for example, *Myrica gale* nodulates best at pH 5.4–6.3 (Bond, 1951). Soil or potting mix may need to be limed in some cases, since soil pH can become extremely acid, thus limiting nodulation and plant growth. Rhizosphere temperatures between 15 and 30°C are generally required for establishing and maintaining root–nodule symbiosis (Dixon and Wheeler, 1983). Bottom heating may be useful during production of nodulated plants if temperatures are too low.

IV. Outplanting and Follow-Up

Vigorous actinorhizal plants will have the best chance of establishing successfully with appropriate planning of the outplanting operation itself. Site preparation and timing are two of the most important considerations in outplanting. Methods of site preparation are dictated by the site itself, by the overall project goals, and by economics. Some clearing or soil cultivation may be required. The planting schedule must consider factors related to ecological timing as well as the number of plants to be planted and size of crew available.

A. Soil Preparation and Planting

Soil preparation may include mechanical clearing or controlled burning of the site. The site soil may require pH adjustment, such as liming, or the addition of phosphorus and other deficient nutrients. Weed control around planting holes may be required for some period of time (one to two growing seasons) to allow for establishment of the woody host plants without excessive competition. Since the foliage of some actinorhizal plants is desirable to insects or as browse for herbivores, new plantings may require exclusion barriers for several seasons. Such protection may range from wire screen cages to fences and cattle guards. On steeply sloping sites, erosion control measures may be needed, such as cut and fill, terracing, or seeding with herbaceous plants.

B. Ecological Timing to Enhance Success of Plantings

The climate and soil moisture conditions are critical aspects of outplanting timing. In direct seeding one must take into account specific pregermination treatments required for each species. Seed that is to overwinter prior to germination must be applied at a high enough rate to allow for predation. If container stock is to be outplanted before its dormant season, a period of hardening off prior to planting will be needed. Planting during spring may also expose the plants to damaging conditions if hardening is not carried out prior to outplanting. Irrigation following outplanting can enhance the establishment of most actinorhizal species substantially. Where irrigation is not available, spring plantings should be carried out as early as possible in the moisture cycle or after danger of frost to ensure maximum growth and site adaptation in the first season. Depressions or basins around plants can also increase water availability for plant establishment.

C. Follow-up Evaluations

The first season (year) for an outplanting of perennial plants can be considered the establishment period and is the key to the success of the planting. Relatively frequent site visits may be helpful in the establishment phase, in particular to evaluate conditions of the plants and to detect problems. Even if one cannot take problem-solving action, such as irrigation in severe drought or pest management, an understanding of the causes of problems may be much more difficult to gain after the passage of several months.

V. Conclusion

We hope that the approaches to horticultural management of actinorhizal plants presented in this chapter can serve as a point of departure for future studies. We have endeavored to synthesize the information available, which is derived largely from experimental systems and based on individual species. We encourage future researchers to include details of methods and assessment of their efficacy within a horticultural context so that a greater knowledge and appreciation of the value of actinorhizal plants can be gained.

References

Akkermans, A. D. L., and Houwers, A. (1979). Symbiotic nitrogen fixers available for use in temperate forestry. *In* "Symbiotic Nitrogen Fixation in the Management of Temperate Forests" (J. C. Gordon, C. T. Wheeler, and D. A. Perry, eds.), pp. 23–35. For. Res. Lab., Oregon State Univ., Corvallis, Oregon.

Akkermans, A. D. L., and van Dijk, C. (1981). Nonleguminous root nodule symbioses with actinomycetes and rhizobium. *In* "Nitrogen Fixation: 1. Ecology" (W. J. Broughton, ed.), pp. 57–103. Oxford Univ. Press, London and New York.

Arveby, A. S., and Huss-Danell, K. (1988). Presence and dispersal of infective *Frankia* in peat and meadow soils in Sweden. *Biol. Fertil. Soils* **6,** 39–44.

Bailey, L. H., and Bailey, E. Z. (1976). "Hortus Third. A Concise Dictionary of Plants Cultivated in the United States and Canada," revised L. H. Bailey Hortorium, Cornell University. Macmillan, New York.

Balmer, W. E., and Space, J. C. (1976). "Probability Tables for Containerized Seedlings." U.S. Dep. Agric., For. Serv., Southeast. Area State Priv. For., Atlanta, Georgia.

Berry, A. M., and Torrey, J. G. (1985). Seed germination, seedling inoculation and establishment of *Alnus* spp. in containers in greenhouse trials. *Plant Soil* **87,** 161–173.

Bond, G. (1951). The fixation of nitrogen associated with the root nodules of *Myrica gale* L., with special reference to its pH relation and ecological significance. *Ann. Bot.* (*London*) **15,** 447–459.

Bormann, B. T. (1983). Ecological implications of phytochrome-mediated seed germination in red alder. *For. Sci.* **29,** 734–738.

Côté, B., and Camiré, C. (1984). Growth, nitrogen accumulation and symbiotic dinitrogen fixation in pure and mixed plantings of hybrid poplar and black alder. *Plant Soil* **78,** 209–220.

Del Tredici, P., and Torrey, J. G. (1976). On the germination of seeds of *Comptonia peregrina,* the sweet fern. *Bot. Gaz.* (*Chicago*) **137,** 262–268.

Diem, H. G., Duhoux, E., Simonet, P., and Dommergues, Y. R. (1988). Actinorhizal symbiosis biotechnology: The present and the future. *Proc. Int. Symp. Biotechnol., 8th* **2,** 984–995.

Dirr, M. A. (1983). "Manual of Woody Landscape Plants: Their Identification, Ornamental Characteristics, Culture, Propagation and Uses." Stipes, Champaign, Illinois.

Dixon, R. O. D., and Wheeler, C. T. (1983). Biochemical, physiological and environmental aspects of symbiotic nitrogen fixation. *In* "Biological Nitrogen Fixation in Forest Ecosystems: Foundations and Applications" (J. C. Gordon and C. T. Wheeler, eds.), pp. 107–171. Nijhoff/Dr. W. Junk, The Hague.

Dommergues, Y. R., Diem, H. G., Gauthier, D. L., Dreyfus, B. L., and Cornet, F. (1984). Nitrogen-fixing trees in the tropics: Potentialities and limitations. *In* "Advances in Nitrogen Fixation Research" (C. Veeger and W. E. Newton, eds.), pp. 7–13. Nijhoff, The Hague.

Dreyfus, B. L., Diem, H. G., Freire, J., Keya, S. O., and Dommergues, Y. R. (1987). Nitrogen fixation in tropical agriculture and forestry. *In* "Microbial Technology in the Developing World" (E. J. DaSilva, Y. R. Dommergues, E. J. Nyns, and C. Ralledge, eds.), pp. 7–50. Oxford Univ. Press, London and New York.

Hartmann, H. T., and Kester, D. E. (1983). "Plant Propagation Principles and Practices." Prentice-Hall, Englewood Cliffs, New Jersey.

Hooker, J. E., and Wheeler, C. T. (1987). The effectivity of *Frankia* for nodulation and nitrogen fixation in *Alnus rubra* and *A. glutinosa*. *Physiol. Plant.* **70,** 333–341.

Huss-Danell, K. (1981). Clonal differences in rooting of *Alnus incana* leafy cuttings. *Plant Soil* **59,** 193–199.

Israel, D. W. (1987). Investigation of the role of phosphorus in symbiotic dinitrogen fixation. *Plant Physiol.* **84,** 835–840.

Kohls, S. J., and Baker, D. D. (1989). Effects of substrate nitrate concentration on symbiotic nodule formation in actiorhizal plants. *Plant Soil* **118,** 171–179.

Lundquist, R., and Torrey, J. G. (1984). The propagation of *Casuarina* species from rooted stem cuttings. *Bot. Gaz. (Chicago)* **145,** 378–384.

Midgley, S. J., Turnbull, J. W., and Johnston, R. D. (eds.) (1981). "Casuarina Ecology Management and Utilization." Commonw. Sci. Ind. Res. Org., Melbourne, Australia.

Monz, C. A., and Schwintzer, C. R. (1989). The physiology of spore-negative and spore-positive nodules of *Myrica gale. Plant Soil* **118,** 75–87.

Périnet, P., Brovillette, J. G., Fortin, J. A., and Lalonde, M. (1985). Large scale inoculation of actinorhizal plants with *Frankia. Plant Soil* **87,** 175–183.

Reddell, P., Rosbrook, P. A., Bowen, G. D., and Gwaze, D. (1988). Growth responses in *Casuarina cunninghamiana* plantings to inoculations with *Frankia. Plant Soil* **108,** 79–86.

Righetti, T. L., and Munns, D. N. (1982). Nodulation and nitrogen fixation in *Purshia*: Inoculation responses and species comparisons. *Plant Soil* **65,** 383–396.

Schopmeyer, C. S. (tech. coord.) (1974). "Seeds of Woody Plants in the United States," Agric. Handb. 450. U.S. Dep. Agric., For. Serv., Washington, D.C.

Schwintzer, C. R. (1985). Effect of spring flooding on endophyte differentiation, nitrogenase activity, root growth and shoot growth in *Myrica gale. Plant Soil* **87,** 109–124.

Schwintzer, C. R., and Lancelle, S. A. (1983). Effect of watertable depth on shoot growth, root growth, and nodulation of *Myrica gale* seedlings. *J. Ecol.* **71,** 489–501.

Schwintzer, C. R., and Ostrofsky, A. (1989). Factors affecting germination of *Myrica gale* seeds. *Can. J. For. Res.* **19,** 1105–1109.

Schwintzer, C. R., Berry, A. M., and Disney, L. D. (1982). Seasonal patterns of root nodule growth, endophyte morphology, nitrogenase activity, and shoot development in *Myrica gale. Can. J. Bot.* **60,** 746–757.

Shaw, N. (1984). Producing bareroot seedlings of native shrubs. *In* "The Challenge of Producing Native Plants for the Intermountain Area" (P. M. Murphy, comp.), Gen. Tech. Rep. INT-168, pp. 6–15. U.S. Dep. Agric., For. Serv., Intermount. For. Range Exp. Stn., Ogden, Utah.

Silvester, W. B., Balboa, O., and Martinez, J. A. (1985). Nodulation and nitrogen fixation in members of the Rhamnaceae (*Colletia, Retanilla, Talguenea,* and *Trevoa*) growing in the Chilean mattorral. *Symbiosis* **1,** 29–38.

Stowers, M. D., and Smith, J. E. (1985). Inoculation and production of container-grown red alder seedlings. *Plant Soil* **87,** 153–160.

Thomas, K. A. (1986). "Vegetative Propagation and Actinorhizal Nodulation of *Ceanothus* sp.," M.S. thesis. Univ. of California, Davis, California.

Thomas, K. A., and Berry, A. M. (1989). Effects of continuous nitrogen application and nitrogen preconditioning on nodulation and growth of *Ceanothus griseus* var. *horizontalis. Plant Soil* **118,** 181–187.

Tinus, R. W., and McDonald, S. E. (1979). "How to Grow Tree Seedlings in Containers in Greenhouses," pp. 70–78. Gen. Tech. Rep. RM-60. U.S. Dep. Agric., Fort Collins, Colorado.

Virginia, R. A., Jenkins, M. B., and Jarrelll, W. M. (1986). Depth of root symbiont occurrence in soil. *Biol. Fertil Soils* **2,** 127–130.

Williams, S. E., Poth, M., and Dunn, P. H. (1986). *Ceanothus crassifolius* Torr. nodulation and nitrogen fixation across a burn chronosequence. *Agron Abstr.* **78,** 191.

Wilson, J., and Coutts, M. P. (1985). Exploiting tree crop–symbiont specificity. *In* "Attributes of Trees as Crop Plants" (M. G. R. Cannell and J. E. Jackson, eds.), pp. 359–379. Inst. Terr. Ecol., Huntingdon, England.

Zhongze, Z., and Torrey, J. G. (1985). Biological and cultural characteristics of the effective *Frankia* strain HFPCcI3 (Actinomycetales) from *Casuarina cunninghamiana* (Casuarinaceae). *Ann. Bot.* (*London*) **56,** 367–378.

15

Interactions among Actinorhizal and Associated Plant Species

Jeffrey O. Dawson
Department of Forestry
University of Illinois
Urbana, Illinois, U.S.A.

To obtain the greatest benefit from actinorhizal and associated plants, an understanding of their relationships in varying combinations and environments is necessary. Examples of such relationships exist for a diverse set of ecosystems, ranging from intensively managed crop systems to extensively managed and wild ecosystems in parks, rangelands, and forest reserves.

I. Complementary Interactions

A. The Natural Occurrence of Actinorhizal Plant Associations

Actinorhizal plants are found primarily on nitrogen-poor sites such as sandy and gravelly soils, raw mineral soils, and wet soils. They occur in forest, swamp, riparian, shrub, prairie, and desert ecosystems. In temperate and cool climates, actinorhizal plants seem to fill the ecological niche occupied by woody legumes in the tropics. Actinorhizal plants frequently occur as pioneer vegetation at early stages of plant succession following such disturbances as flooding, fires, landslides, glacial activity, and volcanic eruptions. For example, *Alnus, Hippophae,* and *Dryas* spp. were probably important in the initiation of soil formation immediately following the retreat of continental ice sheets over vast areas of Europe, Asia, and North America (Bond, 1983). Similarly *Alnus rubra* is a

The Biology of *Frankia* and Actinorhizal Plants

major colonizer of mudflows covered with ash, which resulted from the 1980 eruption of the Mount St. Helens volcano in the state of Washington, U.S.A.

Because of their ecological characteristics, exotic actinorhizal plants can become readily naturalized and sometimes weedy in the oft disturbed urban and agricultural landscape. Examples include the widespread naturalization of *Casuarina* spp. in southern Florida, U.S.A., *Elaeagnus umbellata* in the eastern United States, and *Elaeagnus angustifolia* in the western United States and Canada. *Elaeagnus umbellata* has been declared a noxious weed by weed control agencies in some areas of the eastern United States and hence is locally subject to eradication efforts and planting bans.

Many actinorhizal plants persist as understory plants in open forest stands, particularly on excessively wet or dry sites, or as a dominant species in stable plant communities. Examples include *Myrica cerifera* as an understory plant in stands of *Pinus taeda* and *Pinus elliottii* on sandy soils of the coastal plain in the southeastern United States, *Cercocarpus* spp. and *Purshia tridentata* in the understory of *Pinus ponderosa* stands in semiarid regions of western North America, and stands of *Casuarina glauca* that dominate brackish coastal swamps in southeastern Australia.

Other actinorhizal plants that occur in forest ecosystems are *Coriaria*, which precedes forest vegetation in New Zealand and is a common understory plant in exotic *Pinus radiata* plantations; *Dryas* and *Alnus*, which colonize raw glacial till in Alaska prior to succession by spruce; *Comptonia peregrina*, which occurs as an understory plant in dry open forest stands of eastern North America; *Shepherdia* and *Alnus* on poor sites with aspen and spruce in Canada; *Alnus viridis* ssp. *crispa* with jack pine on sandy soils in eastern Canada; and *Ceanothus*, an indigenous North American genus that occurs primarily on dry sites, with a variety of forest trees and other vegetation (Dawson, 1983).

In addition, actinorhizal plants are cultivated as ornamental, wildlife, timber, fuelwood, pulpwood, reclamation, nurse, and windbreak plants (Dawson, 1986). Others await culture and domestication. It is most probable that actinorhizal plants, so ubiquitous on poor soils, contribute substantially to the nitrogen economy and productivity of many ecosystems.

B. Nitrogen Contributions

One of the primary benefits of interplanting or the natural association of actinorhizal plants with other plants is their input of fixed nitrogen to soil. Nitrogen accumulation rates ranging from 60 to 320 kg per ha per year have been estimated for *Alnus* spp. (Newton *et al.*, 1968; Tarrant and Trappe, 1971). Estimates of the amounts of nitrogen that actually become available to associated plants through mineralization of organic material in stands containing actinorhizal plants range from 48 to 185 kg per ha per year for stands with alder (Daniere *et al.*, 1986; Paschke *et al.*, 1989) and up to 236 kg per ha per year for a plantation containing *Elaeagnus umbellata* (Paschke *et al.*, 1989). Estimates of nitrogen mineralization more accurately reflect the influence of actinorhizal plants on soil nitrogen fertility than estimates of total nitrogen input.

At a montane cirque lake in California, *Alnus tenuifolia* shrubs growing along the lakeshore and the banks of streams in the lake's watershed were probably a major source of nitrogen to the lake ecosystem, which included a productive trout fishery (Goldman, 1961). Increased photosynthetic activity of the lake's phytoplankton and a better development of the aquatic plant *Isoetes* were associated with the presence of *Alnus* along the shore. The cold streams that support trout and that are important for anadromous salmon spawning in the headwaters of larger rivers are often infertile. It is possible that alders commonly associated with trout and salmon waters are important sources of nitrogen to planktonic organisms at the beginning of the food chain for salmon, trout, and their fry. The importance of riparian alder vegetation to commercial salmon fisheries and sport fisheries has not been assessed.

Despite the emphasis on the benefits of nitrogen fixation to associated plants, it is important to note that actinorhizal plants can also benefit from association with others. According to W. Smith, a forestry consultant from Franklin, North Carolina, industrial experimental plots in the Pacific Northwest of the United States demonstrate that *Alnus rubra* benefits from interplanting with western red cedar (*Thuja plicata*) because of wind protection, controlled spacing, and the training effect of adjacent conifers. This results in straighter stems of greater individual *Alnus* tree diameter in contrast with many volunteer thickets of *A. rubra* in which the crowded, spindly stems are subject to bending and windthrow.

C. Recognition of Actinorhizal Plant Benefits

Many land managers are unaware of the benefits and nitrogen fixation capabilities of actinorhizal plants, notwithstanding their taxonomic diversity and widespread occurrence (Dawson, 1986). It is probable that actinorhizal plants contribute a large amount of fixed nitrogen globally, particularly to natural ecosystems, yet they are neither fully appreciated nor exploited by foresters, range managers, and agriculturists.

Recognition of the benefits of actinorhizal plants to agriculture predates the formal scientific recognition of their nitrogen fixation capabilities. Since ancient times alder (*Alnus*) has been known as a soil-improving tree in the Aomori District of northern Japan, where cutover alder sites are usually used as cropland (Yamaya, 1968).

Indigenous farmers of Central America have traditionally used alder trees in their agricultural practices. According to J. Green, a former forestry volunteer with the U.S. Peace Corps in Guatemala: "The people have, for how long I don't know, recognized that an alder left here and there in the corn (*Zea mays*) field makes the corn grow better." This practice usually involves pollarding (cutting back to the trunk to promote the growth of a dense head of small branches) of the alder trees at a height of 1 m above the ground and the scattering of cuttings on the soil surface, thereby reducing shading of the maize crop and increasing the rate of turnover of fixed nitrogen from alder tissue.

D. The Use of Actinorhizal Plantings for Forestry and Agriculture in Latin America

In South and Central America, alders are employed in highland areas for increasing food and fiber production while expanding tree cover on deforested areas. The major *Alnus* species used are *A. acuminata* and *A. jorullensis*. The name *A. jorullensis* is frequently misapplied to various Latin American alders including *A. acuminata* (Furlow, 1979). Moreover, according to Furlow (1979), *A. jorullensis* occurs only in Mexico and Central America and is more restricted to mesic sites than is *A. acuminata*, varieties of which range from Mexico to the southern Andes in South America.

In Costa Rica, upland dairy farmers plant *A. acuminata* at wide spacings in pastures combining wood production with natural nitrogen fertilization of pasture grass (Combe, 1979). In the Alto Charma region of Venezuela, Andean alder (*A. acuminata*) is the main timber species available to farmers for construction of houses (Dieudonne, 1957).

In highland areas above the city of Manizales in Colombia, farmers follow a similar practice by leaving volunteer Andean alder trees in their pastures when clearing other volunteer trees. Kikuyo grass (*Pennisetum clandestinum*) under the partial shade of alder trees continues to produce throughout the dry season while open-field kikuyo does not. Moisture inputs from condensation and drip from alder canopies during misty weather and increased nitrogen fertility are probably both responsible for improved growth of pasture grasses under alder canopies (Carlson and Dawson, 1985). Alders cultivated in pastures also provide shade and shelter for livestock while producing fuelwood and timber (Holdridge, 1951). The system increases milk production relative to systems lacking *Alnus*.

One study concludes that the potential area suitable for this *Alnus*–pasture system in Costa Rica is 50,000 ha (Canet, 1985). The study concludes that about one-third of this area already has this silvopastoral system and about half of the land so cultivated has trees of merchantable size, between 20 and 30 years old. A pasture stand aged 30 years with a density of 35 trees per hectare yielded 70 m^3 per ha of timber, 18.3 tons per ha of dry fuelwood, and 3.6 tons per ha of leaves and fine branches.

Alnus acuminata has been extensively planted in the watersheds of the cities of Manizales and Armenia in Colombia. This has resulted in seasonal moderation of streamflow and increased fog drip, thus allowing continuous water supplies during dry seasons according to P. Carlson, a former U.S. Peace Corps volunteer in the region. In neighboring cities lacking forest cover in their water catchments, streamflow ceases during the dry season. U.S. Peace Corps volunteers and local agriculture students in the Manizales region have used alder for establishing woodlots on heavily eroded sites, for direct seeding of landslide areas, and in an experimental interplanting to increase the growth of *Cedrela* trees for their valuable timber. *Alnus jorullensis* and *A. acuminata* have also been used as nitrogen-fixing shade trees for coffee and cacao plantations.

The development of silvicultural methods for growing Andean alder has resulted from several decades of alder plantation monitoring in Caldas, Colombia (Rojas *et al.*, 1978). Alder height growth of 25 m in 10 years is possible in Colombia. The native range of Andean alder is in cloud forests and moist sites of highlands with annual precipitation averaging between 2000 and 3000 mm and average daily temperature ranging between 5 and 24°C. Common tree associates of Andean alder in native highland forests of Colombia include species of *Cedrela, Juglans, Nectandra, Quercus,* and *Ocotea*. Alder heights of 35 m and diameters at breast height of 79 cm have been noted for Andean alder in Colombia.

Although Andean alder grows better in well-drained soils than in wet soils (Carlson and Dawson, 1985), it can become established and grow on boggy sites. This characteristic of Andean alder may be useful in the cool regions of the Andes, where high seasonal precipitation is coupled with low evaporative potential. In the Colombian Andes at high elevations, boggy soils can develop in open pastures next to well-drained soils in adjacent forests on the same terrain. The wet and poorly aerated soil conditions of these lands cleared of forest to create pasture may result from the loss of evapotranspirative tree cover. On these pastures mosses develop, which hold more water on the site and result in saturated soil conditions and the exclusion of many tree species and preferred pasture grasses. Since alder trees are able to tolerate these poorly drained soils, they may be able to lower soil moisture in time by increasing evapotranspiration rates as they grow. If this is the case, then planting alders may be an efficient means for converting wet sites to better-drained, more productive wood and forage sites.

E. Actinorhizal Plants Used for Forestry and Agriculture in East Asia

In Asia, *Alnus* spp. have been used in crop rotation with millet in Formosa (Silvester, 1977) and as a fuelwood crop and for land stabilization on steep, eroded lands in Nepal. *Alnus cremastogyne* is an endemic tree species in the subtropical, southwestern part of China, where it is used in mixed plantations and for land stabilization (Liu and Deng, 1987). *Casuarina oligodon* is used as a shade crop for coffee in the Papua New Guinea highlands (Ataia, 1983). During recent times in Japan, actinorhizal *Alnus, Myrica,* and *Elaeagnus* spp. have been planted in intermixtures with pines, cedar, cypress, and other trees as well as for erosion control and soil improvement (Uemera, 1964).

In Fujian Province in southeastern China, Dr. Huang Jiabin is leading research efforts to adapt actinorhizal plant culture to forest production at the Fujian Forest Research Institute in Fuzhou. Experimental plantings show much early promise for interplanting actinorhizal plants with tree crops. Examples of promising combinations include *Pinus massoniana* with *Elaeagnus oldhami* and *Myrica rubra* on poor soils in Fujian Province. Exotic *Pinus elliottii* has also exhibited improved growth when interplanted with *M. rubra* in Fujian Province. At the Research Institute's forest farm in Xiaopin, interplantings of *Castanea henreyi* with *M. rubra* are particularly promising. Another interplanting combination in the province is *Cryptomeria fortunei* with *A. cremastogyne.*

Since 1954, a vast green shelterbelt has appeared along China's southern coast. This belt of more than one million hectares is planted mainly with *Casuarina equisetifolia.* The shelterbelt plantings have controlled drifting sand dunes that were burying useful land, sheltered coastal villages from sea winds, and provided wood and fuel. Only by interplanting *C. equisetifolia* on infertile sand dunes in southeastern China is it possible to successfully grow *Eucalyptus citriodora.*

F. Mechanisms of Interaction between *Elaeagnus* and *Juglans* in Interplantings

There are numerous other examples of beneficial interactions of actinorhizal plants with associated plants, some discussed elsewhere in this book. However, our understanding of the complex biological and physical processes underlying productivity in such mixed stands of actinorhizal plants and associates is not comprehensive. The use of *Elaeagnus umbellata* and *Alnus glutinosa* as nurse crops for valuable black walnut (*Juglans nigra*) in the midwestern United States is probably the most intensively studied plantation system employing actinorhizal plants. These studies serve to illustrate the possible variety of interactions among plants and site features in mixed stands containing actinorhizal plants.

When interplanted, *Elaeagnus* dramatically increases the growth of black walnut trees in the east central United States (Table I), even when compared with benefits afforded by interplantings of walnut with *A. glutinosa* and *Robinina pseudoacacia*, a temperate climate nitrogen-fixing tree legume (Funk *et al.*, 1979; Schlesinger and Williams, 1984). One reason for *E. umbellata*'s superior growth promotion capabilities seems to be its shrubby form and spreading crown (Friedrich and Dawson, 1984). Although all the nitrogen-fixing plants employed in this interplanting study increase soil nitrogen, *E. umbellata* produces a subcanopy under the walnut, thus shading the soil, training the walnut by shade-pruning

Table I Average Height (ht), dbh or Stem Diameter at Breast Height (1.4 m), and Annual dbh Growth Rate for *Juglans nigra* Trees Interplanted with Actinorhizal Plants in Illinois, U.S.A.[a]

Interplanted species	Age, 14 years: ht (m)	Age, 14 years: dbh (cm)	Age, 18 years: ht (m)	Age, 18 years: dbh (cm)	Average annual dbh growth rate (cm per year), Age 0–14	Age 14–18	Age 0–18
Elaeagnus umbellata	10.0	15	12.2	19	1.1	1.0	1.1
Alnus glutinosa	7.6	10	10.1	14	0.7	1.1	0.8
Control	3.4	5	4.9	8	0.4	0.6	0.4

[a]Twenty-four *J. nigra* trees were spaced at 9.8 by 3.7 m with three actinorhizal plants planted between walnut trees in a row, giving an overall spacing of 2.4 by 3.7 m in replicated, bordered plots. Control plots were without interplantings. Adapted from Campbell and Dawson (1989).

lower branches (Funk *et al.*, 1979), possibly preventing the spread of spores of walnut anthracnose from litter to the walnut canopy (Kessler, 1985), and eliminating grassy and herbaceous weeds at the ground level of plantations (Friedrich and Dawson, 1984). At wider spacings of walnut, which has slow to moderate growth rates compared with many temperate, deciduous angiosperms, *Elaeagnus* shrubs have persisted for more than 20 years. Unlike *Alnus* and *Robinia, Elaeagnus* does not grow taller and faster than the walnut trees, so it does not compete as much for light, water, and nutrients. *Elaeagnus* is also more tolerant of juglone, an allelochemical produced by walnut, than *Alnus* (Rietveld, 1983), which has been eliminated from plantations with walnut owing to juglone toxicity in soils of poor internal drainage, neutral pH, and shallow depth.

At the New Zealand Forest Research Institute in Rotorua, Ian Nicholas has performed experiments with wind-screens arranged around *J. nigra* seedlings so as to block the wind without reducing light exposure. His studies show significant increase in walnut seedling growth as a result of wind protection. This is another means by which *E. umbellata* interplantings might have increased initial walnut growth in the Illinois interplantings.

In light of the many benefits of interplanting *Elaeagnus* with walnut in Illinois, foresters have debated the relative importance of nitrogen fixation in these plantations. *Elaeagnus umbellata* eliminates most weeds in the plantations, particularly grasses, which are the most competitive and detrimental weeds in local tree plantations. This fact compounds the debate because black walnut benefits more from weed control than nitrogen fertilization on the deep, rich soils most suited for production of high-quality walnut timber (von Althen, 1985; Miller *et al.*, 1987). But on less fertile soils, nitrogen deficiency can make walnut culture impractical.

To help resolve this issue, Paschke *et al.* (1989) did a study that showed annual nitrogen mineralization rates in mixed plantings of *E. umbellata* to be positively correlated with biomass of interplanted walnut in 18-year-old plantations on old field sites marginal for walnut production. The mean net mineralization rate for plots of *E. umbellata* with walnut at one location was 236 kg per ha per year in the top 20 cm of soil, which is higher than values previously reported for hardwood forest soils in North America. This value can be compared with a value of 90 kg per ha per year for walnut control plots without actinorhizal plants at the same site, a value more typical for a temperate deciduous forest. These findings suggest that nitrogen inputs from actinorhizal fixation are of considerable importance in promoting walnut growth on a marginal site, and that interplanting may enable growers to expand the range of planting sites considered suitable for walnut production.

Elaeagnus is so successful in colonizing disturbed areas and old fields that its use is prohibited in some areas. *Elaeagnus umbellata* was promoted widely in the eastern United States as a wildlife planting because its abundant fruit attracted and supported a wide variety of songbirds. The plantings succeeded, probably well beyond wildlife biologists' expectations, resulting in widespread dissemination of the seed and the weedy status of *E. umbellata*. Development of a sterile form of *E. umbellata* that could be easily propagated vegetatively would provide managers with a nurse crop, lacking the potential to become weedy, for walnut and other tree species.

In the *Elaeagnus*–walnut interplanting system, *Elaeagnus* will eventually be eliminated by shade when the walnut canopy closes. However, the effects of 20 or more years of nitrogen input and the residual effects of fixation with mineralization of organic matter added by *Elaeagnus* should prolong the benefits of nitrogen fixation to walnut growth. Since trees can conserve internal nitrogen through retranslocation, the benefits of nitrogen fertilization via actinorhizal interplantings should be further extended. An economic analysis and growth projections of the *Elaeagnus* –walnut interplantings indicated that by reducing the walnut rotation age interplanting provided benefits that could be worth U.S. $373 per ha in present value with a discount rate of 4 percent (Campbell and Dawson, 1989).

Kessler (1985) presented evidence that the shrubby subcanopy and litter of *E. umbellata* help to prevent the spread of spores of the foliar disease walnut anthracnose from infected walnut litter to the foliage of the tree canopy. His suggestion is based on experiments in which a mechanical mist-blower, commonly used to deliver pesticides, was employed to disperse fungal spores that were subsequently trapped for counting at various heights in a walnut–*Elaeagnus* interplanting. His evidence suggests a possible reason for the decrease in anthracnose symptoms observed in interplanted walnut compared with controls. It is also likely that increased nitrogen fertility in the interplanting contributed to the avoidance of this disease by walnut.

G. Actinorhizal and Associated Plant Interactions with Soil

Alnus rubra trees intermixed with Douglas-fir also seem to be able to suppress important diseases of the coniferous partner. Red alder intermixed with Douglas-fir can have the dual effects of increasing wood yields and suppressing important fungal root diseases of conifers, such as *Poria (Phellinus) weirii* (Nelson, 1975) and possibly *Fomes annosus* (personal communication with E. E. Nelson, USDA Forestry Sciences Laboratory, Corvallis, Oregon). Increased nitrification in the nitrogen-enriched soils under the alder stands increases the amount of ammonium in soil that is transformed to nitrate, which cannot be used by these fungi but which can be used by fungal antagonists, suggesting a mechanism for disease suppression (Bollen and Lu, 1968). It is also possible that phenolics and other substances produced by *A. rubra* roots and released into soil suppress these fungal root diseases. Phenolic substances produced by *A. rubra* have been found to inhibit *in vitro* growth of *P. weirii* (Li *et al.*, 1969).

There are other possible effects of actinorhizal trees and associated plants on soil. As noted previously, nitrogen increases in soil caused by actinorhizal trees and shrubs can increase the rate of nitrification, thereby lowering soil pH (Paschke *et al.*, 1989). Soil acidification can result in the increased displacement of cations from cation-exchange sites and their loss by leaching. Interplantings of tree species that do not fix nitrogen may reduce acidification and increase the accumulation and recycling of cations displaced by soil acidification.

One of the most common soil factors limiting actinorhizal nodulation and nitrogen fixation is phosphorous deficiency. Phosphorous fertilization of a *Casuarina cunninghamiana* plantation on a weathered soil in Australia proved necessary to obtain a significant growth response to *Frankia* inoculation or nitrogen fertilization (Reddell *et al.*, 1988). Huang Jiabin (personal communication) reports similar results from phosphorous fertilization of *C. equisetifolia* plantings in coastal sand dunes of Fujian Province in China, where intensive harvesting of plantation litter for fuel probably accelerates phosphorous and other nutrient depletion, resulting in second rotation growth declines. Such depletion can be particularly acute when foliage and small branches are harvested and there is no fertilization or return of the mineral-containing residues, such as ashes or animal manure.

In addition to low soil phosphorous, the absence of appropriate mycorrhizal and nitrogen-fixing symbionts (Diem and Gauthier, 1982; Reddel *et al.*, 1988), moisture stress (Kant and Narayana, 1978), low soil temperatures Reddell *et al.*, 1985), high levels of available nitrogen (Rodriguez-Barrueco, 1972), deficiencies of nutrients other than phosphorous and nitrogen (Robson and Loneragan, 1978), and low soil pH (Bond, 1957; Coyne, 1973) may inhibit field nodulation and nitrogen fixation by *Frankia* and *Rhizobium*. Salt concentration can also differentially affect the survival of actinorhizal host plants (El-Lakany and Luard, 1982) and *Frankia* strains (Dawson and Gibson, 1987). Excess soil moisture can also inhibit actinorhizal plant growth (McVean, 1956; Carlson and Dawson, 1985) and consequent capacity for symbiotic nitrogen fixation.

II. Competitive Interactions

A. Stand Dynamics

Nitrogen-fixing trees and shrubs may grow faster than, slower than, or at the same rate as interplanted or associated trees. The species composition, spacing, density, stand age, silvicultural treatments, and site characteristics interact to determine the survival, growth, form, and yield of trees in the stand. In interplantings, root and crown forms, shade tolerance, nutrient needs, and stature of interplanted species must be matched to achieve harmonious growth. When planted together, trees and shrubs will compete for light, water, and nutrients when crown and root systems are close or overlap.

Most actinorhizal trees and shrubs are shade-intolerant colonizers of disturbed sites. There are numerous examples of woody, nitrogen-fixing trees and shrubs occurring in natural ecosystems as pioneer plants on disturbed sites, where they improve soil nitrogen fertility before giving way in ecological succession to longer-lived species. In the Pacific Northwest of the United States, *A. rubra* and *Ceanothus velutinus* often colonize logged and burned areas, competing with Douglas-fir (*Pseudotsuga menziesii*) regeneration. In many cases the competition of these nitrogen fixers with the establishment of Douglas-fir after logging or fire is so severe that it negates the economic benefits afforded by nitrogen fixation by delaying the early growth and establishment and increasing the rotation age for the emerging conifer stand. The use of fire in site preparation for the regeneration of Douglas-fir stimulates *Ceanothus* seed germination and, hence, increases its potential for competing with Douglas-fir regeneration. The use of herbicides to suppress actinorhizal and other broad-leaved plants may be necessary to release shade-intolerant conifers.

In forestry the species composition, spacing, arrangement, and site characteristics necessary to secure compatible growth rates and the desired stem form for both nitrogen fixers and interplanted associates over a given rotation age must usually be determined experimentally. Results will often depend on the availability of soil nitrogen together with the increase afforded by actinorhizal plants on a given site (Binkley, 1984). For example, rapidly growing *Populus* species interplanted with *A. rubra* and *A. glutinosa* on fertile soils in North America grow more rapidly than the nitrogen-fixing alders, quickly suppressing them (Hansen and Dawson, 1982; Heilman and Stettler, 1983). However, some plantations of *Populus* trees on a low-nitrogen, mudflow substrate resulting from the volcanic eruption of Mount St. Helens in Washington State were rapidly dominated by *A. rubra* volunteering from nearby stands (personal communication with P. Heilman, Washington State University). In another case both the growth rate and the form of black walnut have been improved in Illinois by interplanting with shrubby *E. umbellata* and *A. glutinosa* trees, which grew at a lesser or comparable rate to the walnut trees. In contrast, *R. pseudoacacia* trees interplanted with walnut were cut at age 7 because they were overtopping and suppressing the walnut (Friedrich and Dawson, 1984). Thus both beneficial and competitive interactions must be considered in prescribing mixed plantings or managing mixed stands of trees on a given site.

B. Allelopathy

Nitrogen-fixing trees and shrubs can be killed or suppressed by allelochemicals from associated woody plants. Black walnut produces the allelochemical juglone (5-hydroxy-1,4-naphthoquinone), which is a potent inhibitor of mitochondrial respiration (Koeppe, 1972). Juglone at concentrations ranging from 10^{-6} to 10^{-4} *M* can reduce *Frankia* growth (Vogel and Dawson, 1986), host-plant nodulation (Vogel and Dawson, 1985), and respiration and nitrogenase activity of *A. glutinosa* roots and nodules (Neave and Dawson, 1989). Rietveld *et al.* (1983) reported high mortality of *A. glutinosa* interplanted with *J. nigra* at one location in Illinois. *Alnus glutinosa* declined and died suddenly, apparently as a result of juglone toxicity, after 8 to 13 years in four mixed plots with black walnut. Mortality of alder occurred in soils of pH 5.3 to 7.0 with poor internal drainage, but not in identical interplantings replicated on better-drained soils with pH values from 3.9 to 4.5 in the same region of southern Illinois.

Juglone toxicity is highest on poorly drained, hence less aerobic, soils in which the heterotrophic bacteria that break down juglone would be inhibited (Fisher, 1978; Schmidt, 1988). Experiments in our laboratory by Lixia Liu have shown that lowering pH to values of 4.0 and 5.5 in soils or nutrient-solution culture will reduce the toxicity of juglone to *A. glutinosa* germinants, probably because juglone is less soluble in acidic than in alkaline aqueous solutions. *Elaeagnus umbellata* is more tolerant than alder to juglone in solution culture (Rietveld, 1983). Consistent with this finding, *E. umbellata* showed no ill effects in mixed plantings with walnut even when growing adjacent to plots in which *A. glutinosa* suffered high mortality.

Variation also exists among *Frankia* strains in juglone tolerance (Vogel and Dawson, 1986). Alder tissue is high in phenolic substances (Li *et al.*, 1969). In the case of *Alnus viridis* ssp. *crispa* in Canada, high levels of the phenolic pinosylvin methyl ether seem to be an adaptation to prevent browsing by the snowshoe hare (Bryant *et al.*, 1983). *Frankia* isolates from this *Alnus* species, which is especially high in phenolic materials, were found to be more tolerant than other *Frankia* isolates to juglone and several common plant phenolic acids (Vogel and Dawson, 1986).

Thus the planting site, the actinorhizal host plant, and, perhaps, the strain of microsymbiont should be carefully matched to minimize allelopathic effects of walnut. Those employing similar mixtures of plants should be alert for allelopathic effects that may occur.

Phenolics and polyphenols are known to occur in *Populus* spp. (Dormaar, 1970; Olsen *et al.*, 1971) and may have been responsible for reduced nitrogenase activities of *Alnus incana* ssp. *rugosa* growing among *P. tremuloides* trees in northern Wisconsin (Younger and Kaptuska, 1983). Water leachates of *P. balsamifera* inhibited seed germination, growth, nodulation by *Frankia*, and nitrogenase activity of *Alnus viridis* ssp. *crispa* (Jobidan and Thibault, 1981, 1982), perhaps due to phenolics leached from poplar (*Populus*) tissues. Since intermixtures of poplars and alders are used in forestry, allelopathic influence of poplars on alders and *Frankia* could reduce productivity of these plantations.

C. Agroforestry

One factor limiting the use of actinorhizal plants in agricultural systems is their competition with intercrops. Rows and mixtures of trees or shrubs, be they actinorhizal or other woody plants, shade and compete with adjacent crops for water and nutrients. Hence the use of pollarding, mentioned earlier in this chapter, is employed in mixtures of alder with maize on steep croplands in the highlands of Guatemala. Where actinorhizal plants are used as nurse and shade crops for coffee, benefits are realized because the woody understory crop grows best at some level of shading, responds to increased nitrogen fertility, and can tolerate root competition. However, these characteristics of forest understory plants do not seem to be shared by actinorhizal plants (Côté *et al.*, 1988). This limits their utility as underplantings in forestry.

D. Physiological Processes Underlying Plant Interactions

In actinorhizal plants nitrogen fixation activity is closely related to photosynthetic activity (Gordon and Wheeler, 1978; Dawson and Gordon, 1979), hence shading in underplantings, or other environmental stresses that affect photosynthesis, would tend to reduce the amount of nitrogen fixed proportionally to its reduction of photosynthesis. Only a few actinorhizal plant species have been characterized photosynthetically (Côté *et al.*, 1988; Neave *et al.*, 1989), but the *Alnus* and *Elaeagnus* spp. so characterized have light saturation points, maximum net photosynthetic rates, and dark respiration rates typical of early to midsuccessional plants. Such characteristics suggest that these actinorhizal plants are not suitable for planting in the shady understory of other woody plants. It also provides a possible explanation for the fact that many actinorhizal species do not persist beyond early stages of plant succession following a disturbance, unless successional plants have open canopies affording only limited shading.

Elaeagnus spp., which exhibit a shrubby growth form that limits self-shading, have higher concentrations of leaf nitrogen and higher specific rates of photosynthesis than alders (Côté *et al.*, 1988), many of which are trees with greater total leaf areas. Hence the crown architecture and photosynthetic characteristics are important determinants of performance of actinorhizal plants in association with other plants. In light of the dependency of nitrogen fixation on the host plant's ability to provide photosynthate, strategies for improving symbiotic performance and production of wood and fiber in mixed stands with actinorhizal plants should focus on selection and development of photosynthetically efficient host plants of appropriate structure, growth rate, and suitability for a given environment.

One unusual aspect of alder physiology is the genus's apparent delay in autumnal foliar senescence, which is accompanied by (1) retention of green foliage well past the time of leaf drop of associated temperate deciduous trees, (2) prolonged autumnal photosynthesis (Neave *et al.*, 1989), (3) limited net autumnal resorption of foliar nitrogen (Côté and Dawson, 1986; Côté *et al.*, 1989), and (4) consequent drop of abscised leaves with high concentrations of foliar nitrogen. Hence, alders seem to shed internal nitrogen more rapidly than other woody species, accelerating the rate of nitrogen addition to soil.

III. Conclusions

Interplantings or mixed stands that include actinorhizal plants are complex systems for which there is often little understanding of the mechanisms underlying stand dynamics and productivity. To influence stand growth and structure so that growth rate and form of each woody component is optimized, the species must be compatible with one another and with management goals on a given site. We must seek to better understand both the complementary and competitive aspects of actinorhizal plants and their associates together with their site adaptabilities if we are to best realize the benefits that can be afforded by the earth's diversity of actinorhizal plants.

References

Ataia, A. (1983). *Casuarina oligodon* in the Eastern Highlands Province—Papua New Guinea. *In* "Casuarina Ecology, Management and Utilization" (S. J. Midgley, J. W. Turnbull, and R. D. Johnston, eds.), pp. 80–88. Commonw. Sci. Ind. Res. Org., Melbourne, Australia.

Binkley, D. (1984). Importance of size–density relationships in mixed stands of Douglas-fir and red alder. *For. Ecol. Manage.* **9,** 81–85.

Bollen, W. B., and Lu, K. C. (1968). Nitrogen transformations in soils beneath red alder and conifers. *In* "Biology of Alder" (J. M. Trappe, J. F. Franklin, R. F. Tarrant, and G. M. Hansen, eds.), pp. 141–148. U.S. Dep. Agric., For. Serv., Pacific Northwest For. Range Exp. Stn., Portland, Oregon.

Bond, G. (1957). The development and significance of the root nodules of *Casuarina. Ann Bot.* (*London*) **21,** 373–380.

Bond, G. (1983). Taxonomy and distribution of non-legume nitrogen-fixing systems. *In* "Biological Nitrogen Fixation in Forest Ecosystems: Foundations and Applications" (J. C. Gordon and C. T. Wheeler, eds.), pp. 55–87. Nijhoff/Dr. W. Junk, The Hague.

Bryant, J. P., Wieland, G. D., Reichardt, P. B., Lewis, V. E., and McCarthy, M. C. (1983). Pinosylvin methyl ether deters snowshoe hare feeding on green alder. *Science* **222,** 1023–1025.

Campbell, G. E., and Dawson, J. O. (1989). Growth, yield and value projections for a black walnut interplanting with black alder and autumn olive. *North. J. Appl. For.* **6,** 129–132.

Canet, G. C. (1985). Caracteristicas del sistema silvo-pastoril jaul (*Alnus acuminata*) con lecheria de altura en Costa Rica. *In* "Tecnicas de Produccion de Lena en Fincas Pequenas: Actas de los Simposios" (R. Salazar, ed.), pp. 241–250. CATIE, Turrialba, Costa Rica.

Carlson, P. J., and Dawson, J. O. (1985). Soil nitrogen changes, early growth, and response to soil internal drainage of a plantation of *Alnus jorullensis* in the Colombian highlands. *Turrialba* **35,** 141–150.

Combe, J. (1979). *Alnus acuminata* con pastoreo y con pasto de corte: Las Nubes de Coronado, Costa Rica. *In* "Taller sobre Sistemas Agroforestales en America Tropical," pp. 205–207. CATIE, Turrialba, Costa Rica.

Côté, B., and Dawson, J. O. (1986). Autumnal changes in total nitrogen, salt-extractable proteins and amino acids in leaves and adjacent bark of black alder, eastern cottonwood and white basswood. *Physiol. Plant.* **67,** 102–108.

Côté, B., Carlson, R. W., and Dawson, J. O (1988). Leaf photosynthetic characteristics of seedlings of actinorhizal *Alnus* spp. and *Elaeagnus* spp. *Photosynth. Res.* **16,** 211–218.

Côté, B., Vogel, C. S., and Dawson, J. O (1989). Autumnal changes in tissue nitrogen of autumn olive, black alder and eastern cottonwood. *Plant Soil* **118,** 21–32.

Coyne, P. D. (1973). "Some Aspects of the Autecology of Casuarinas, with Particular Reference to Nitrogen Fixation," Ph.D. thesis. Aust. Nat. Univ., Canberra, Australia.

Daniere, C., Capellano, A., and Moiroud, A. (1986). Dynamique de l'azote dans un peuplement naturel d'*Alnus incana* (L.) Moench. *Acta Oecol./Oecol. Plant.* **7,** 165–175.

Dawson, J. O. (1983). Dinitrogen fixation in forest ecosystems. *Can. J. Microbiol.* **29,** 979–992.

Dawson, J. O. (1986). Actinorhizal plants: Their use in forestry and agriculture. *Outlook Agric.* **15,** 202–208.

Dawson, J. O., and Gibson, A. H. (1987). Sensitivity of selected *Frankia* isolates from *Casuarina, Allocasuarina* and North American host plants to sodium chloride. *Physiol. Plant.* **70,** 272–278.

Dawson, J. O., and Gordon, J. C. (1979). Nitrogen fixation in relation to photosynthesis in *Alnus glutinosa. Bot. Gaz. (Chicago), Suppl.* **140,** S70–S75.

Diem, H. G., and Gauthier, D. L. (1982). Effet de l'infection endomycorhizienne (*Glomus mosseae*) sur la nodulation et la croissance de *Casuarina equisetifolia. C. R. Hebd. Seances Acad. Sci., Ser. C* **294,** 215–218.

Dieudonne, R. L. (1957). "Estudio Ecologico-silvicultural sobre el Aliso en la Region del Alto Charma," M.S. thesis, 57 pp. Fac. For. Sci., Univ. of the Andes, Merida, Venezuela.

Dormaar, J. F. (1970). Seasonal patterns of water-soluble constituents from leaves of *Populus* X "Northwest" (Hort.). *J. Soil Sci.* **21,** 105–110.

El-Lakany, M. H., and Luard, E. J. (1982). Comparative salt tolerance of selected Casuarina species. *Aust. For. Res.* **13,** 11–20.

Fisher, R. F. (1978). Juglone inhibits pine growth under certain moisture regimes . *Soil Sci. Soc. Am. J.* **42,** 801–803.

Friedrich, J. M., and Dawson, J. O. (1984). Soil nitrogen concentration and *Juglans nigra* growth in mixed plots with nitrogen-fixing *Alnus, Elaeagnus, Lespedeza,* and *Robinia* species. *Can. J. For. Res.* **14,** 864–868.

Funk, D. T., Schlesinger, R. T., and Ponder, F., Jr. (1979). Autumn-olive as a nurse plant for black walnut. *Bot. Gaz. (Chicago), Suppl.* **140,** 5110–S5114.

Furlow, J. J. (1979). The systematics of the American species of *Alnus* (Betulaceae). *Rhodora* **81**(825), 1–121 and **81**(826), 151–248.

Goldman, C. R. (1961). The contribution of alder trees (*Alnus tenuifolia*) to the primary productivity of Castle Lake, California. *Ecology* **42,** 282–288.

Gordon, J. C., and Wheeler, C. T. (1978). Whole plant studies on photosynthesis and acetylene reduction in *Alnus glutinosa. New Phytol.* **80,** 179–186.

Hansen, E. A., and Dawson, J. O. (1982). Effect of *Alnus glutinosa* on hybrid *Populus* height growth in a short-rotation, intensively cultured plantation. *For. Sci.* **28,** 49–59.

Heilman, P., and Stettler, R. F. (1983). Phytomass production in young mixed plantations of *Alnus rubra* Bong. and cottonwood in western Washington. *Can. J. Microbiol.* **29,** 1007–1013.

Holdridge, L. R. (1951). The alder, *Alnus acuminata,* as a farm timber tree in Costa Rica. *Caribb. For.* **12,** 47–53.

Jobidan, R., and Thibault, J. R. (1981). Allelopathic effects of balsam poplar on green alder germination. *Bull. Torrey Bot. Club* **108,** 413–418.

Jobidan, R., and Thibault, J. R. (1982). Allelopathic growth inhibition of nodulated and unnodulated *Alnus crispa* seedlings by *Populus balsamifera. Am. J. Bot.* **69,** 1213–1223.

Kant, S., and Narayana, H. S. (1978). Effect of water stress on growth, nodulation and nitrogen fixation in *Casuarina equisetifolia. Ann. Arid Zone* **17,** 216–221.

Kessler, K. J., Jr. (1985). Companion planting of black walnut with autumn-olive to control *Mycosphaerella* leaf spot of walnut. *In* "Proceedings of the Fifth Central Hardwood Forest Conference" (J. O. Dawson and K. A. Majerus, eds.), pp. 285–288. Dept. For., Univ. of Illinois, Urbana, Illinois.

Koeppe, D. E. (1972). Some reactions of isolated corn mitochondria influenced by juglone. *Physiol. Plant* **27,** 89–94.

Li, C. Y., Lu, K. C., Nelson, E. E., Bollen, W. B., and Trappe, J. M. (1969). Effect of phenolic and other related compounds on growth of *Poria weirii in vitro*. *Microbios* **3,** 305–311.

Liu, G., and Deng, T. (1987). A study of nodulation and nitrogen fixation of alder on the purplish soils of China. *Plant Soil* **99,** 285–290.

McVean, D. N. (1956). Ecology of *Alnus glutinosa* (L.) Gaertn. III. Seedling establishment. *J. Ecol.* **44,** 195–218.

Miller, L. A., Wray, P. H., and Mize, C. W. (1987). Response of stagnated black walnut to chemical release. *North. J. Appl. For.* **4,** 93–95.

Neave, I. A., and Dawson, J. O. (1989). Juglone reduces growth, nitrogenase activity, and root respiration of actinorhizal black alder seedlings. *J. Chem. Ecol.* **15,** 1823–1836.

Neave, I. A., Dawson, J. O., and DeLucia, E. H. (1989). Autumnal photosynthesis is extended in nitrogen-fixing European black alder compared with white basswood: Possible adaptive significance. *Can. J. For. Res.* **19,** 12–17.

Nelson, E. E. (1975). Survival of *Poria weirii* on paired plots in alder and conifer stands. *Microbios* **12,** 155–158.

Newton, M., El Hassan, B. A., and Zavitkovski, J. (1968). Role of red alder in western Oregon forest succession. *In* Biology of Alder" (J. M. Trappe, J. F. Franklin, R. F. Tarrant, and G. M. Hansen, eds.). pp. 73–84. U.S. Dep. Agric., For. Serv., Pacific Northwest For. Range Exp. Stn., Portland, Oregon.

Olsen, R. A., Odham, G., and Lindberg, G. (1971). Aromatic substances in leaves of *Populus tremula* as inhibitors of mycorrhizal fungi. *Physiol. Plant.* **25,** 122–129.

Paschke, M. W., Dawson, J. O., and David, M. B. (1989). Soil nitrogen mineralization in plantations of *Juglans nigra* interplanted with actinorhizal *Elaeagnus umbellata* or *Alnus glutinosa*. *Plant Soil* **118,** 33–42.

Reddell, P., Bowen, G. D., and Robson, A. D. (1985). The effects of soil temperature on plant growth, nodulation and nitrogen fixation in *Casuarina cunninghamiana* Miq. *New Phytol.* **101,** 441–450.

Reddell, P., Rosbrook, P. A., Bowen, G. D., and Gwaze, D. (1988). Growth responses in *Casuarina cunninghamiana* plantings to inoculation with *Frankia*. *Plant Soil* **108,** 79–86.

Rietveld, W. J. (1983). Allelopathic effects of juglone on germination and growth of several herbaceous and woody species. *J. Chem. Ecol.* **9,** 295–308.

Rietveld, W. J., Schlesinger, R. C., and Kessler, K. J., Jr. (1983). Allelopathic effects of black walnut on European black alder coplanted as a nurse species. *J. Chem. Ecol.* **9,** 1119–1133.

Robson, A. D., and Loneragan, J. F. (1978). Responses of pasture grasses to soil chemical factors others than nitrogen and phosphorous, with particular emphasis on the legume symbiosis. *In* "Plant Relations in Pastures" (J. R. Wilson, ed.), Commonw. Sci. Ind. Res. Org., Melbourne, Australia.

Rodriguez-Barrueco, G. (1972). Efecto del N-amonico sobre la fijacion de N-atmosferico en nodulos de *Casuarina*. *Anal. Edafol. Agrob.* **31,** 905–916.

Rojas, H., Macia, F., and Lastra, J. A. (1978). Monografia del *Alnus jorullensis* H. B. K. *Colomb. For.* **1**(1), 5–22.

Schlesinger, R. C., and Williams, R. D. (1984). Growth response of black walnut to interplanted trees. *For. Ecol. Manage.* **9,** 235–243.

Schmidt, S. K. (1988). Degradation of juglone by soil bacteria. *J. Chem. Ecol.* **9,** 1175–1183.

Silvester, W. B. (1977). Dinitrogen fixation by plant associations excluding legumes. *In* "A Treatise on Dinitrogen Fixation" (R. W. F. Hardy and A. H. Gibson, eds.), Sect. IV, pp. 141–190. Wiley, New York.

Tarrant, R. F., and Trappe, J. M. (1971). The role of *Alnus* in improving the forest environment. *Plant Soil, Spec. Vol.* pp. 335–348.

Uemera, S. (1964). Isolation and properties of microorganisms from root nodules of non-leguminous plants. A review with extensive bibliography. *Ringyo Shikenjo Kenkyu Hokoku* **167,** 59–91.

Vogel, C. S., and Dawson, J. O. (1985). Effect of juglone on growth *in vitro* of *Frankia* isolates and nodulation of *Alnus glutinosa* in soil. *Plant Soil* **87,** 79–89.

Vogel, C. S., and Dawson, J. O. (1986). *In vitro* growth of five *Frankia* isolates in the presence of four phenolic acids and juglone. *Soil Biol. Biochem.* **18,** 227–231.

von Althen, F. W. (1985). Revitalizing a black walnut plantation through weed control and fertilization. *For. Chron.* **61,** 71–74.

Yamaya, K. (1968). On the influence of alder (*Alnus inokumae*) on soil properties in northern Japan. *In* "Biology of Alder" (J. M. Trappe, J. F. Franklin, R. F. Tarrant, and G. M. Hansen, eds.), pp. 197–207. U.S. Dep. Agric., For. Serv., Pacific Northwest For. Range Exp. Stn., Portland, Oregon.

Younger, P. D., and Kaputska, L. A. (1983). $N_2(C_2H_2)$ase activity by *Alnus incana* ssp. *rugosa* (Betulaceae) in the northern hardwood forest. *Am. J. Bot.* **70,** 30–39.

16

Current and Potential Uses and Management of Casuarinaceae in the Tropics and Subtropics

H. G. Diem and Y. R. Dommergues
Biotechnologie des Systèmes Symbiotiques Forestiers Tropicaux
94736 Nogent sur Marne Cedex, France

I. Introduction

The family Casuarinaceae is a large group of perennials ranging from woody shrubs (e.g., *Allocasuarina nana*) to trees over 35 m tall (e.g., *Casuarina cunninghamiana*) and comprising some 90 species native to Australia, Malaysia, and Polynesia. All members of the family are characterized by highly reduced leaves and photosynthetic deciduous branchlets also called assimilatory branchlets (terminology recommended by Flores, 1978, and Johnson, 1982; the term cladode, which is sometimes used, is inappropriate). Casuarinaceae exhibit coniferlike traits, such as branchlets that resemble conifer needles, and conelike infructescences; a more careful examination reveals their basic angiospermous characteristics (Torrey and Berg, 1988).

Johnson (1980, 1982, 1988) recently divided the family Casuarinaceae into four genera: *Allocasuarina, Casuarina, Gymnostoma,* and *Ceuthostoma.* This classification is based on differences in chromosome numbers, stem morphology including the distribution of stomata on the branchlets, structure of male and female inflorescences and fruits, geographical occurrence, and fossil record. The validity of this classification was confirmed recently by a study of epidermal structures in five species of Casuarinaceae belonging to the three principal genera (Torrey and Berg, 1988). At a number of sites outside their native habitat, different species of *Casuarina* appear to have hybridized (Chen *et al.*, 1983).

Table 1 summarizes some of the major ecological requirements of the

Table I Characteristics of the Best-Known and Some Economically Promising Species of Casuarinaceae[a]

Genus and species	Climate	Altitude (m)	Rainfall[b] (mm)	Soil	Nodule[c]	Limitations
Casuarina cunninghamiana	Warm subhumid Warm semiarid, 50 days frost	0–800	500–1500	Acidic	+++	Not salt tolerant; not tolerant to highly calcareous soils
C. equisetifolia var. *equisetifolia*	Hot humid to hot subhumid	Up to 600	1000–2150	Tolerates salt, calcareous soil	+++	Not tolerant to weed competition; not resistant to insects; exhausts soil moisture
C. equisetifolia var. *incana*	Warm subhumid	0–100	1000–1500			
C. glauca	Warm humid Warm subhumid	0–900	900–1150	Tolerates salt, calcium clay	+++	Possibly weedy, profuse suckering
C. junghuhniana	Hot humid	0–3000		Various	++	Hybrids do not produce seeds
C. oligodon	Hot humid	Up to 2500		Various, including ultrabasic	++	Sensitive to salt, relatively exacting
C. obesa	Warm subhumid Warm semiarid Warm arid	0–300	200–500	Various	+++	Unreported

C. cristata	Warm arid Warm semiarid	0–350	175–275	Tolerates alkalinity, clay	+++	Susceptible to browsing
Allocasuarina littoralis	Humid/warm humid Warm subhumid Cool subhumid	0–1200	650–1250	Various, well drained, acid	0 to +++	Not tolerant to very wet soils
A. decaisneana	Arid	250–700	200–250	Deep, sandy	0 to +	Slow growing if not irrigated
A. verticillata	Wet subhumid Cool subhumid	0–800	600–900	Tolerates calcareous saline soils	++	Little natural regeneration
A. torulosa	Warm subhumid Warm humid	0–1100	950–2500	Various	+	Slightly more exacting than *A. littoralis*
A. luehmanniana	Warm subhumid	0–800	425–800	Various, tolerates salt, clay	0	Poor germination of seeds
Gymnostoma papuanum	Hot humid	0 to highlands		Various	+	Not reported
G. rumphiana	Hot humid	Up to 1000		Limestone, gullies	+	Slow growth
G. sumatrana	Hot humid	0–1000		Behind mangrove swamps	++	Slow growth

[a] After Doran and Hall (1983).
[b] Fifty percentile values (probability of one year in two).
[c] 0, no nodulation reported; +, sparse; ++, fair; +++, profuse nodulation.

species that are currently exploited or that are assumed to be among the most economically promising ones (Midgley *et al.*, 1983; National Research Council, 1984; Rosbrook, 1987).

II. Current and Potential Uses of Casuarinaceae

A. Casuarinaceae as a Source of Construction Wood and Fuelwood

The wood of many Casuarinaceae species, including *Casuarina equisetifolia, C. junghuhniana, Gymnostoma sumatrana,* and *G. rumphiana,* has the hardness and durability needed in roundwood used for piles, posts, poles, and rafters but splits too severely during drying to be popular as lumber. When no other wood was available, *Casuarina* has been used for making simple pieces of furniture, for example, chairs at the Desert Development Center in Egypt (M. H. El-Lakany, personal communication). In Egypt, *Casuarina* is commonly used for making particleboard. Certain species provide wood with specific qualities, for example, the red-brown heartwood of *Allocasuarina decaisneana* was used in constructing the railroad in Central Australia (Midgley *et al.*, 1983) because it is resistant to termites. In Malaysia, some species that are reputed to be stronger than teak are used for wheels and masts (Silvester, 1977).

Casuarinaceae stems and stumps are burnt as is or as charcoal. The caloric value (about 5000 kcal per kg) of *Casuarina equisetifolia* wood is significantly higher than that of many other well-known fast-growing N_2-fixing trees, such as *Acacia auriculiformis* and *Leucaena leucocephala.* Cones and photosynthetic branchlets shed from the trees provide cheap fuel that is widely used in India and China, a practice whose consequences are discussed later. *Casuarina* wood is excellent for making charcoal. Whereas most woods lose about three-quarters of their weight in this process, casuarinas only lose about two-thirds (National Research Council, 1984). Fuelwood production is vital since the average inhabitant in developing countries burns close to a ton of firewood a year, and in some 54 Third World countries the firewood supply has become critically low (Midgley *et al.*, 1984). Recent studies based on FAO and UN figures suggest that to meet the expected fuelwood and charcoal deficit in the tropics, 100 million ha will have to be planted, since the deficit is expected to double, or worse, between 1980 and 2000 (Robinson, 1987). *Casuarina equisetifolia* and related species are probably among the best candidates for this purpose. For instance, in Senegal, if an area of 30,000 ha of low fertility sandy and saline soils along the Atlantic coast were planted in *Casuarina equisetifolia* and *Casuarina glauca,* it would yield energy equivalent to one-tenth of the petroleum imports. Lastly, the wood of

Casuarina equisetifolia can be used to make paper pulp of satisfactory quality via neutral sulfite processes (National Research Council, 1984; J. Doat, personal communication).

B. Casuarinaceae as Components of Agroforestry Systems

A number of studies have shown that trees play a major role in sustaining soil fertility and crop production and that, among the woody perennials, Casuarinaceae could be introduced in agroforestry systems as windbreaks or protective plantations, or grown in mixed forests.

1. Windbreaks

Casuarina equisetifolia is extensively planted as windbreaks in many countries, for example, Guangdong Province of China, Senegal, and India. It is widely used in Tunisia, where its permeability to wind, a characteristic that is positively related to the windbreak effectiveness, is high (50–70%), thus reducing turbulence and potential evaporation by 29% at a distance of eight times the height of the windbreak (Paulsson, 1987).

Casuarina glauca is often used as windbreaks in Egypt, Tunisia, Israel, and Yemen, where its root suckering does not appear to be have the untoward effect sometimes reported (Geary, 1983). Information is lacking on the exact effect of *Casuarina* windbreaks on crop production, but as windbreaks they fit well into the rural scene not only by protecting the crops and apparently increasing their yield (El-Lakany, 1985) but also by providing the farmers with all the construction and fuelwood they require (in Egypt, the annual cash value of *Casuarina* wood is similar to that of the crops suited for cultivation on the same area).

Casuarina oligodon is planted around villages in the Highlands of Papua New Guinea as protection against strong winds (Midgley *et al.*, 1983). Other *Casuarina* species are also used as windbreaks.

2. Plantations for Dune Stabilization and Water Erosion Control

Because of its resistance to salt-laden winds, its low demand for nutrients, and its ability to fix N_2 actively, *Casuarina equisetifolia* has been successfully used to stabilize coastal sand dunes in many tropical and subtropical countries (Midgley *et al.*, 1983). In southeastern China (Guandong, Fujian, and Guangxi provinces), 1 million ha of *Casuarina equisetifolia* now prevent sand from drifting and causing large areas of arable lands to go to waste (Turnbull, 1983).

Some casuarinas, like *C. cunninghamiana*, cannot be used for stabilizing coastal dunes because of their intolerance to coastal conditions (Geary, 1983). Further research is needed to identify and breed species of Casuarinaceae that can protect agricultural land.

Casuarinas curtail not only wind but also water erosion, mainly because of their thick litter of interlocked branchlets that permanently blankets the soil. In addition, species that produce profuse suckers, such as *Casuarina glauca,* are efficient in controlling erosion by holding down the soil as the suckers spread (National Research Council, 1984).

3. Intercropping of Casuarinaceae with Annual or Perennial Crops

One classical example is that of Papua New Guinea, where *C. oligodon* and *Gymnostoma papuanum* (in the highlands) and *Casuarina equisetifolia* (in the lowlands) are intercropped with food crops or cash crops (coffee) or are used as shade trees or in rotation with crops (Thiagalingam, 1983; Bourke, 1985). In India, *Casuarina equisetifolia* plantations are associated with crops such as peanut, sesame, and various grain legumes (Kondas, 1983).

4. Mixed Forests

Spontaneous associations of non-N_2-fixing trees and Casuarinaceae have been reported, for example, in Australia, where *Eucalyptus* spp. grow with *Casuarina littoralis* (Turner *et al.*, 1978). In a limited number of cases artificially mixed forest including Casuarinaceae have been cultivated, for example, in North Sumatra, where *Casuarina junghuhniana* is associated with *Altinga excelsa* and *Podocarpus amarus* (Jafarsidik, 1986). *Casuarina equisetifolia* has been successfully associated with cashew and coconut in the coastal belt of India (Kumar, 1981).

C. Rehabilitation of Degraded Lands and Watersheds

1. Reclamation of Salt-Affected Soils

The development of forests on salt-affected soils seems justified and necessary for several reasons: resource mismanagement is accelerating land salinization around the world, the pressure on good soils for food production is increasing, and firewood and timber resources are dwindling rapidly. These soils can only be used for salt-tolerant woody species.

The choice is difficult because of variation in saline environments. There are four main categories: areas affected by seawater, coastal sand dunes, areas affected by irrigation programs, and areas affected by saline groundwater (Midgley *et al.*, 1984).

In arid areas of Pakistan where the soil had become exceedingly saline (conductivity: 2.7–2.8 mmho/cm) as a result of irrigation, of the ten species tested, *Casuarina equisetifolia* appeared to be the best; good results were also recorded with *Tamarix aphylla* (syn. *articulata*), a nonactinorhizal species renowned for its salt tolerance, *Prosopis juliflora*, and *Eucalyptus camaldulensis* (Malik and Sheikh, 1983).

Not all species of *Casuarina* are salt tolerant, and great variation exists among those that are. In a study of nine species of Casuarinaceae grown in liquid nutrient culture in a greenhouse, *Casuarina glauca*, *C. obesa*, and *Casuarina equisetifolia* var. *incana* were the most salt tolerant and could withstand up to 500 m*M* NaCl. *Casuarina cunninghamiana* and *C. cristata* tolerated up to 450 m*M* NaCl, *Allocasuarina verticillata* (syn. *C. stricta*) up to 350 m*M* NaCl, and *A littoralis* and *A. torulosa* up to 250 m*M* NaCl. *Allocasuarina decaisneana* was the most sensitive species with a survival threshold somewhere between 50 and 150 m*M* NaCl. A comparison of two provenances of *C. glauca* and *C. cunninghamiana* suggested that sufficient intraspecific variation exists to permit development of highly salt-tolerant clones (El-Lakany and Luard, 1982).

Neither *Frankia* growth nor the nodulation process per se seems to be greatly affected by salinity. Ng (1987) showed that growth of *Frankia* in the rhizosphere and nodulation and nodule development in *Casuarina equisetifolia* were not inhibited by increasing salt concentrations up to the 500 m*M* NaCl level.

2. Reclamation of Lands Wasted by Mining and Industrial Spoils

In a number of situations Casuarinaceae have been successfully introduced to recolonize lands spoiled by mine dumps and waste or by the removal of layers of arable soil. Casuarinas were also planted, together with *Acacia* and *Eucalyptus*, to revegetate lands suffering from fluoride emissions and salt-laden winds around a smelter located at Bell Bay, Tasmania (Mitchell, 1982). In arid and semiarid climates, casuarinas are also used in wastewater management, when wastewater is used to irrigate trees (Stewart and Flinn, 1984).

3. Rehabilitation of Watersheds

In many upland watersheds, deforestation has resulted in massive soil erosion, sedimentation of dams and reservoirs, and increased flooding of downstream arable lands and communities. This situation could be alleviated by developing plantations of fast-growing N_2-fixing trees. Casuarinaceae have not yet been used for this purpose but, if properly inoculated and tended, may well prove effective in rehabilitation of endangered areas.

D. Other Uses

Casuarinaceae are highly valued for landscaping. *Casuarina equisetifolia* is commonly grown along tropical and subtropical coastlines, where it is appreciated "for its shade and the pleasant sensation produced by the wind whispering through the foliage" (Geary, 1983). Together with other species, such as *Casuarina junghuhniana* in Indonesia, *Gymnostoma* spp. in Malaysia, and hybrids of *Casuarina* spp. in Thailand, it contributes to the beautification of parks and cities. In addition, species such as *Casuarina equisetifolia* tolerate air pollution by cars better than other ornamental trees do.

In contrast with a number of N_2-fixing trees, like *Leucaena leucocephala* or *Acacia mangium,* Casuarinaceae are generally not valued as forage because of the high tannin content in the plant tissue. Despite this, the nutritional value of *Casuarina equisetifolia* assimilatory branchlets is significant (El-Lakany, 1983), and *Allocasuarina verticillata* is lopped for fodder in times of drought (Midgley *et al.*, 1983).

III. Root Symbioses in Casuarinaceae

A. Actinorhizal Symbiosis

In 1883 all the vegetation on Krakatau Island was destroyed by exceptional volcanic eruptions. In 1896, 13 years later, coastal vegetation of this island included groves of large *Casuarina equisetifolia* along with other shrub species. Expeditions in 1919, 36 years later, found extensive forests dominated by *C. equisetifolia* up to 35 m tall and 1.65 m in circumference. This event, reported by Silvester (1977), shows that *C. equisetifolia* is an exceptional tree species capable of establishing in marginal sites under adverse environmental conditions. This ability is unquestionably due to the N_2-fixing activity induced by mutualistic symbiosis with the actinomycete *Frankia*.

1. Unique Characteristics of *Casuarina* Nodules

When *Casuarina* seedlings are excavated from soil containing *Frankia,* numerous small, multilobed, coralloid-type, amber or whitish nodules are found on their root systems. Since *Casuarina* nodules are perennial, layers of new nodule lobes successively develop from older lobes during the wet season, thus resulting in a cluster of hard and woody nodules that are generally observed on roots of mature *Casuarina* trees. Because of the numerous superposed layers of nodule lobes, some *Casuarina* nodules may reach a size of between 2 and 10 cm in diameter. In an *Allocasuarina verticillata* plantation established on a hydromorphic soil at Zerniza (Tunisia) we observed many nodules lying on the soil surface; some of them were 15–20 cm in diameter.

Casuarinas exhibit two distinct morphological types of nodules: species of the genus *Casuarina sensu-stricto* form the "*Myrica*-type" nodules characterized by the presence of a nodule root at the apex of each young nodule lobe; in contrast, species of the genus *Allocasuarina* form nodules of the "*Alnus*-type" that lack nodule roots. This expression of nodule morphological type is controlled by the host and not by the symbiotic microorganism (Zhang and Torrey, 1985a).

Nodules of *Casuarina* seem to be unique among actinorhizal nodules thus far described in that nodule cells contain only filamentous forms of *Frankia*. In their paper on the extranodular growth of *Frankia* of *Casuarina*, Diem *et al.* (1982a) reported that *Frankia* produced sporangia and vesicles as soon as the microorganism escaped from the mother nodule. This observation suggests that the formation of *Frankia* structures inside the nodule is under host control. For an in-depth discussion of nodule development and functioning of *Casuarina* symbioses refer to Torrey (1982).

2. Host Compatibility and Effectiveness of *Casuarina*-Compatible *Frankia*

a. Host Compatibility. Studies of *Frankia* and Casuarinaceae paradoxically started with the isolation, from *Casuarina* nodules, of *Frankia* strains incompatible with the host plant of origin (Diem *et al.*, 1982b; Zhang *et al.*, 1984; Chaudhary and Mirza, 1987). Up to now, all these *Casuarina*-incompatible strains have proved successful in nodulating members of Elaeagnaceae (Gauthier *et al.*, 1981; Diem *et al.*, 1982b; Baker, 1987), probably because plants of this family are promiscuous (Baker, 1987).

Since host-incompatible strains have been isolated from *Casuarina* nodules, it can be inferred that the endophytic population of *Frankia* of *Casuarina* nodules might be composed of a number of *Frankia* strains with different types of host compatibility.

After many disappointing attempts (Torrey, 1982), success in isolating

Casuarina-compatible strains was finally achieved by Diem *et al.* (1982c, 1983), Zhang *et al.* (1984), and Zhang and Torrey (1985b) using nodule pieces sterilized with OsO_4 and Q_{mod} medium (Lalonde *et al.*, 1981). By now, many pure-cultured isolates of *Frankia* have been obtained from nodules of *Casuarina* (Shipton and Burggraaf, 1983; Huang *et al.*, 1985; Jiang and Zhu, 1985; Chaudhary and Mirza, 1987), *Allocasuarina* (Puppo *et al.*, 1985; Zhang and Torrey, 1985a), and recently *Gymnostoma* (Racette and Torrey, 1989). A number of isolates have also been obtained in Australia (P. Reddell), China (Y.-H. Cao), Egypt (M. Girgis), and Senegal (L. Maggia). So far *Casuarina*-compatible strains have never been reported to nodulate plants of the family Elaeagnaceae. The problem of compatibility between *Frankia* and members of the genera *Casuarina* and *Allocasuarina* is also unclear especially with some species of *Allocasuarina* that nodulate irregularly. Cross-inoculation grouping within *Frankia* is discussed in detail by Torrey in Chapter 5.

b. Effectiveness. To date, only a few investigations have been made to estimate and compare the effectiveness of the various *Frankia* isolates in symbiosis with their host plant. Some studies have demonstrated marked variation in the effectiveness of N_2 fixation by different *Frankia*–Casuarina combinations (Reddell and Bowen, 1985; Reddell *et al.*, 1988; Sanginga *et al.*, 1988). Reddell and Bowen (1985) observed that the *Frankia* that originates from a given species is not necessarily the most effective *Frankia* on that species. Since effectiveness depends on both the *Frankia* source and the host, the effects on *Casuarina* species range from ineffectiveness to high effectiveness (Reddell and Bowen, 1985). Different levels of effectiveness according to species combinations and *Frankia* inocula were also shown by Sellstedt (1988).

It would be of great interest to compare the effectiveness of different *Casuarina–Frankia* combinations under marginal soil conditions, such as relatively high salinity or acidity. *Frankia* isolates from *Casuarina* seem to have a better tolerance to high concentrations of NaCl than do *Frankia* isolates from other actinorhizal plants *in vitro* at least with respect to growth and N_2 fixation (Dawson and Gibson, 1987).

3. *Frankia* Inoculant Technology

a. Crushed Nodules. Before it was possible to produce cultures of *Frankia, Casuarina* seedlings in nurseries were inoculated with crushed-nodule homogenates. This empirical method has too many limitations to be practical for widespread use (Torrey, 1982; Périnet *et al.*, 1985). The use of crushed-nodule homogenates as inoculants should be discouraged because it is a major cause of variation in *Casuarina* nodulation in nurseries (Reddell and Bowen, 1985) and requires large amounts of nodule material. In addition, it can seriously increase the dissemination of soil-borne pathogens like *Pseudomonas solanacearum*, a bacterium that causes *Casuarina* wilt in China (Liang and Chen, 1984).

b. Pure Cultures. By using pure cultures of *Frankia* the microbial identity of the inoculant can be strictly controlled to ensure that seedlings receive only effective *Frankia* strains. Further, the pure cultures of *Frankia* used as inoculants may contain both actively infecting hyphae and the resistant spores needed for extended storage of the inoculant.

In our laboratory, *Frankia* cultures prepared for *Casuarina* inoculation are grown until high concentrations of spores are obtained. The use of certain specific carbon sources (Jiang and Zhu, 1985) or amino acids (Diem *et al.*, 1988) can enhance the production of sporangia by *Frankia* isolates from *Casuarina*. Another possibility is incubating the *Frankia* cultures in a nitrogen-free liquid medium for 2 weeks, a simple technique that allows *Frankia* hyphae to produce three times more sporangia than hyphae grown on medium supplemented with combined nitrogen (Diem *et al.*, 1988).

We have developed a simple method for preparing *Frankia* inoculum that is technologically feasible in a modestly equipped laboratory. The technique is as follows:

Frankia stock cultures are embedded in a nutrient medium solidified with 1.2% (w/v) agar (Fig. 1A) and homogenized with sterile distilled water in a Waring blender for about 20 sec (Fig. 1B). Twenty milliliters of the homogenate is then used to inoculate a 1-liter Roux bottle containing 200 ml of the liquid nutrient medium and a stirring bar (Fig. 1C). After 10 days of incubation, the starter culture obtained in C (Fig. 1) is homogenized by placing the Roux bottle vertically onto a magnetic stirrer and stirring at high speed for a few minutes. The homogenized culture is then transferred into a 20-liter growth vessel (Fig. 1D,j) containing 12–15 liters of liquid nutrient medium and a stirring bar. The growth vessel is connected to a separatory bottle (Fig. 1D,o) and a collection bottle (Fig. 1D,q). After inoculation, the growth vessel and its satellite bottles (Fig. 1D,o and q) are placed in in the incubation room at 28°C. The growth vessel is placed on a magnetic stirrer. The stirring bar is kept rotating at the lowest speed needed to maintain *Frankia* cells in suspension in the whole volume of the liquid medium. After 10 days of incubation, the *Frankia* culture is siphoned into the separatory bottle (Fig. 1D,o) and allowed to settle, resulting in a concentrated culture. The culture medium (liquid phase) is then evacuated into bottle q (Fig. 1D) through a 1.0-mm-mesh screen (Fig. 1D,u), leaving the concentrated culture (s) in the

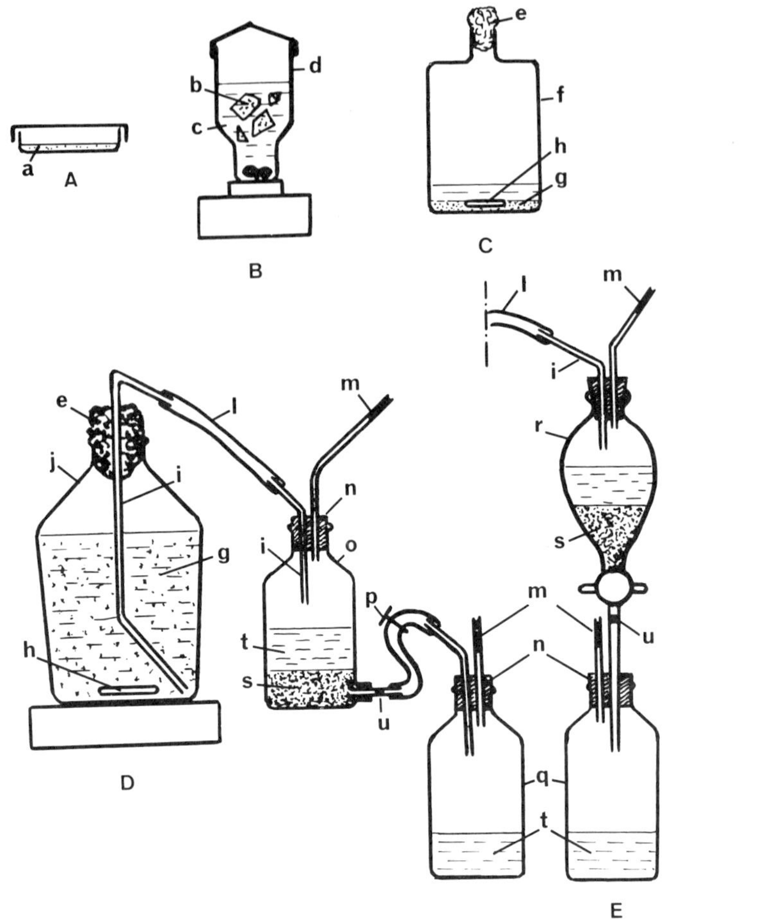

Figure 1 Proposed scheme for the mass production of *Frankia* inoculants for large-scale inoculation trials of *Casuarina*. a , *Frankia* cultured on agar nutrient medium; b , pieces of agar medium containing *Frankia* colonies; c , sterile water; d , Waring blender; e , cotton wool plug; f , Roux bottle; g , *Frankia* culture in liquid nutrient medium; h , stirring bar; j , growth vessel; i , glass tubing; k , magnetic stirrer; l , Tygon tubing; m , cotton wool filtered outlet tube; n , rubber stopper; o , separatory bottle; p , screw clamp; q , collection bottle; r , separatory funnel; s , concentrated suspension of *Frankia* cells; t , culture medium (liquid phase); u , 1.0-mm-mesh screen. The capital letters A,B,C,D,E refer to the different steps of the procedure (see text).

separatory bottle. The separatory bottle may be replaced by a separatory funnel (r) as shown in Fig. 1E.

c. Advantages of the Proposed Culture Technique.

—The *Frankia* stock culture maintained on a solid medium in a Petri dish is readily available and thus very convenient since it can initiate the starter culture at any time. The viability of *Frankia* colonies and the purity of the culture can easily be checked by observing the Petri dish under a dissecting microscope.
—This simple culture procedure is designed for mass culturing (up to 15 liters) and for harvesting *Frankia* with minimal labor, time, and risk of contamination.
—*Frankia* cells are continuously, gently, and regularly dispersed throughout the liquid medium, and do not suffer harmful fragmentation because of the low rotation speed of the stirring bar in the growth vessel during incubation.
—A dense suspension of *Frankia* cells can be harvested directly from the growth vessel by using the separatory and the collection bottles as described in Fig. 1D.
—*Frankia* issued from the culture technique described here is not a mass of fluffy hyphae as under stationary conditions. It is composed of distinct colonies that are relatively homogeneous in morphology and size (ca. 0.5 mm diameter).

d. Cultures Entrapped in Alginate Beads. As early as 1979, Dommergues and his team (see Diem *et al.*, 1988) proposed to entrap rhizobium in a polymeric matrix and use this formulation as a rhizobium inoculant for legume inoculation. Their concept has also been applied to *Frankia*. The *Frankia* cells are entrapped or grown (see Burleigh *et al.*, 1988; Diem *et al.*, 1988) in alginate beads. The principle of this method is as follows: microbial cells are aseptically mixed with 1–2% (w/v) alginate in an appropriate nutrient solution. The mixture is added dropwise into a gently stirred 0.1 *M* $CaCl_2$ solution at room temperature (28°C). Alginate beads are formed, then washed twice with sterile water, and incubated in a fresh nutrient liquid medium for 15–20 days. *Frankia* entrapped inside these beads are able to grow into colonies, with numerous sporangia produced in each center. As with rhizobium, *Frankia* cells directly entrapped in alginate beads may survive better if clays such as montmorillonite or kaolinite are added (10%) to the *Frankia*–alginate solution. Kaolinite enables alginate beads, which otherwise flatten when air-dried, to remain spherical. After being air-dried, the beads containing highly sporulating *Frankia* colonies can be stored for years without seeming to lose infectiveness. The inoculant quality can be checked by using the INT reduction method described by Prin *et al.* (1989). Before being used to inoculate *Casuarina* seedlings in the nursery, the beads are rehydrated in a phosphate buffer and crushed in water to obtain a pseudosolution. Another possibility would be to place the dried alginate beads directly in the soil.

The original technique for processing the *Frankia* inoculant in a polymeric matrix, for example, alginate beads (see Diem *et al.*, 1988) deserves careful attention because the resulting microbial inoculant meets the major requirements of a good practical inoculant: it is easy to store, transport, and apply. To date the bead inoculant production method for *Frankia* as well as for other beneficial microorganisms has only been used on an experimental basis. Industrial-scale production is currently being developed in France.

4. Inoculation with *Frankia* and N_2 Fixation in the Field

When a species of *Casuarina* is grown for the first time on a site that has never been planted before with a species of the same *Frankia*-compatible group, it is generally not nodulated. This has long been known in New Zealand (Torrey, 1982) and also observed on the sand dunes of Senegal. However, there are some exceptions. At Zerniza forest station in Tunisia, we noted that, in spite of the facts that the local flora did not include any actinorhizal plants and that foresters had not made any attempt to introduce nodule material, an *Allocasuarina verticillata* plantation established in 1970 had numerous large nodules.

In the first field inoculation trials, soil or leaf litter from around nodulated plants or crushed nodules collected directly from trees were applied to seedlings. This inoculation procedure was successful in different countries, for example, Egypt, Senegal, and Australia (Reddell *et al.*, 1988). However, in the future, this procedure should be replaced by inoculation with pure selected strains of *Frankia*.

The response to inoculation is spectacular when no specific *Frankia* are present and when no limiting factor inhibits the trees' N_2-fixing potential. The first inoculation experiment conducted at Dakar, Senegal, in field-simulating conditions showed that inoculation of *Casuarina equisetifolia* by a pure culture of *Frankia* (strain ORS 021001) increased plant biomass by 78%. Nitrogen fixation (measured by the isotopic dilution method) during the 6 months following transplantation was 3.07 ± 1.86 g per tree (Gauthier *et al.*, 1985). A recent field experiment carried out in Zimbabwe showed that inoculation with pure cultures of *Frankia* increased tree heights to between 50 and over 200% above the controls depending on the strains used (Reddell *et al.*, 1988).

In a field-simulating experiment carried out at the same time at the ORSTOM station in Dakar, Senegal, after inoculation of a clone (clone β) of *Casuarina equisetifolia* grown in a regularly irrigated and properly fertilized sandy N-deficient soil, the biomass doubled, as measured by dry weight or total N (Table II). The amount of N_2 fixed during the 2 years of growth estimated by the A value (AV) and total N difference (TND) methods were 85.4 (AV) or 84.8 (TND) g N_2 per tree. From these estimations, it can be calculated that at a planting density of 2000 trees per ha, clone β would fix about 85 kg N_2 per ha per year (Sougoufara *et al.*, 1989b).

When limiting factors are present, the increase of tree biomass resulting from inoculation may be less marked, but it is still highly significant. In the example given in Table II on the results of a relatively large scale field experiment at Notto, Senegal, to evaluate the response of *Casuarina equisetifolia* to the application of a dried inoculum made of alginate-entrapped cells of *Frankia* strain ORS 021001, the inoculation technique used was rated excellent, since all inoculated trees were nodulated, and the growth increase of the 2-year-old trees was 41% (dry weight) and 69% (total N). But the amount of N_2 fixed in 2 years of growth was only 11.8 g per tree. Extrapolating to a plantation of 2000 trees per ha, this gives about 12 kg N_2 fixed per ha per year. This N_2 fixation figure is seven times lower than that resulting from the Dakar experiment re-

Table II Response of 2-Year-Old *Casuarina equisetifolia* to Inoculation with *Frankia* Strain ORS 021001[a]

		Plant biomass (g/tree)			Estimates of N_2 fixation	
Inoculation	Height (cm)	Dry weight	Total N	Nodule biomass (g/tree)	AV	TND
		Field-simulating experiment (Dakar)				
Control	nd	12070	76.6	0		
Inoculated	nd	23851	161.1	220	85.4	84.8
		**	**			
		Field experiment (Notto)				
Control	291	2025	17.2	0		
Inoculated	313	2871	29.0	12	nd	11.8
	NS	**	**			

[a] In the field-simulating experiment set up at Dakar, Senegal, *Casuarina equisetifolia* was regularly irrigated and fertilized (inoculation was performed with a liquid inoculum). In the field experiment established at Notto, Senegal, trees were not irrigated except at the beginning of the experiment and fertilization was minimal (Sougoufara *et al.*, 1989a, and unpublished observations). Significance level: **: P = 0.01; NS: nonsignificant; nd: not determined.

ported and from an earlier field experiment (Dommergues, 1963) in a plantation along the coast at Malika, Cape Verde Peninsula. A calculation of the total N accretion of the latter ecosystem showed that approximately 50 kg of N_2 was fixed per year.

The discrepancies between the different estimations of N_2 fixation just reported are simply due to the fact that in the Dakar experiment there was no limiting factor. In the Malika experiment, drought had no effect since the plantation was established close to the sea and thus benefited from high atmospheric humidity, whereas at Notto the trees suffered from drought (Notto is 5 km inland) and very low rainfall during the trial (215 mm during the first year and 375 mm during the second). To summarize, it can be reasonably predicted that a young *Casuarina equisetifolia* plant can fix up to 45 g N_2 per year when growing in favorable conditions, and that this figure can fall to 6 g or less in constrained conditions.

B. Mycorrhizal Symbioses

Although the nodulation of *Casuarina* by effective *Frankia* strains is central to its successful establishment in adverse environments, Rose (1980) and Gardner (1986) suggest that the role of *Casuarina* as a pioneering plant is greatly assisted by the presence of mycorrhizae on the root system. Apart from mycorrhizae, typical proteoid roots have also been observed on the root systems of certain species of Casuarinaceae (Diem *et al.*, 1981; Reddell *et al.*, 1986). They are unique, tightly packed rows of rootlets probably structured to increase the ability of the host plant to absorb nutrients and thus better adapt to deficient soils.

The presence of ectomycorrhizae was reported on a number of Casuarinaceae growing in the field, especially *Allocasuarina* spp. (Reddell *et al.*, 1986), but the mycorrhizal fungi were not identified and their effect on growth of the Casuarinaceae host plant has not yet been evaluated. Bamber *et al.* (1980) and Thoen and B. Sougoufara at ORSTOM Centre, Dakar (personal communication), have found that under laboratory conditions some species of *Allocasuarina* may form mycorrhizae with the fungus *Pisolithus tinctorius*.

Because endomycorrhizal fungi are ubiquitous and not host specific, it is not surprising that they can infect roots from field-grown *Casuarina* (Rose, 1980; Gardner, 1986). Under greenhouse conditions, Diem and Gauthier (1982) found that double inoculation of *C. equisetifolia* with *Glomus mosseae* and *Frankia* (as crushed nodules) significantly improved plant growth and nodulation. After 6 months of growth, mycorrhizal nodulated plants contained twice as much nitrogen as seedlings that had only been inoculated with *Frankia*. Gardner (1986) found that nodulation of 3-month-old *Casuarina equisetifolia* was not stimulated by mycorrhizae, but the nodules on mycorrhizal plants were twice as efficient as those formed on nonmycorrhizal plants. When associated with endomycorrhizal fungi, roots of *Gymnostoma deplancheana* in New Caledonia bear along the root axis chains of spherical structures that Huguenin called "mycorrhizal nodules" (see Gauthier *et al.*, 1984). The so-called nodules might be derived from clusters of nondeveloping proteoid roots subsequently invaded by endomycorrhizal fungi.

Endomycorrhizae are generally known to improve the nutrient status of the host plant. In addition, we would like to emphasize that networks of endomycorrhizal hyphae can intermesh sand grains to form aggregates that subsequently help stabilize loose sand (Koske and Halvorson, 1981). This aspect is often overlooked but we believe that endomycorrhizae can benefit from their aggregating ability to play a significant role in stabilizing coastal sand dunes, thus synergistically increasing the effectiveness of *Casuarina* planted as shelterbelts to protect agricultural lands.

IV. Plant Selection and Breeding for Increased N_2 Fixation

The N_2 fixation potential of any symbiotic system depends on both the associated symbiont and the host plant. In the case of casuarinas, selection should seek to obtain plants with the following characteristics: ability to fix N_2 actively in association with *Frankia;* high tolerance of environmental constraints, especially drought, salinity, and wind; and the usual characteristics emphasized in tree breeding programs, including yield, shape, and pest resistance.

A. Species and Provenance Selection of Casuarinas

As in any genetic tree improvement program, the first stage should be devoted to testing the *Casuarina* population through species and provenance trials. The principal objective of provenance trials is to determine the extent of genetic variation between and within provenances through replicated field trials at different sites. Special attention should be given to the response to inoculation of the different provenances. This aspect of provenance selection was taken into account by Reddell *et al.* (1886, 1988), who showed that the response of *Casuarina cunninghamiana* to inoculation (performed with crushed nodules) significantly varied according to provenance.

B. Conventional Plant Breeding

The more elaborate conventional plant breeding approach is now being applied in Egypt. Elite trees have been selected and clonal seed orchards established. Progeny testing is being carried out at different sites (El-Lakany, 1983), but improvement programs have not yet taken into account the N_2-fixing capability of the plant.

C. Selection and Vegetative Propagation of Superior Phenotypes

A study carried out on *Casuarina equisetifolia* in Senegal (B. Sougoufara *et al.*, unpublished observations) has just shown that clonal selection followed by vegetative propagation is practicable, using the N_2-fixing potential as the criterion to rank the individual trees. The following criteria were used to make a comparative characterization of S, α, and β, three clones with different N_2-fixing potentials: biomass of nodules expressed as dry weight and total N, amount of N_2 fixed based on the total N difference (TND) or the isotope dilution (ID) method, and acetylene reduction activity. Nodule biomass, expressed on a dry weight or total N basis, was found to be a simple, reasonable criterion, at least for preliminary screening. TND is easier to calculate and cheaper than ID but can be misleading because clones (at least in the case of *Casuarina equisetifolia*) vary in their capacity to assimilate N from soil. Thus in the experiment reported in Table III, clone ranks were affected by the method of N_2 fixation assessment: $\beta = S > \alpha$ when the ID method was used, whereas $\beta > \alpha = S$ when the TND method was used.

Based on comparisons of *Casuarina equisetifolia* clones screened at 7 and 12 months, it was concluded that the host plants can be validly screened at an early stage, that is, at 7 months. In the case of young trees (7 months old), the acetylene reduction assay probably provides reliable results.

Progress has been made recently in methods of vegetative propagation of Casuarinaceae, for example, the method proposed by Duhoux *et al.* (1986) and based on the use of immature female inflorescences as explants, even in old trees.

V. Management Practices

There is a general tendency, not only among microbiologists but also among agronomists and foresters, to neglect the impact of management practices on the N_2-fixing ability and consequently on the growth of

Table III Comparison of Biomass and Nitrogen Fixation Characteristics of Three 12-Month-Old Uninoculated and Inoculated Clones (S, α, β) of *Casuarina equisetifolia* Grown at Dakar, Senegal[a]

Clone	Plant biomass: Dry weight	Plant biomass: Total N	NDFS	Nodule dry weight	Estimates of N_2 fixation[b]: ID	Estimates of N_2 fixation[b]: TND
			Uninoculated plants			
S	204c	1.51c	1.48c			
α	295b	2.18b	2.14b			
β	417a	3.31a	3.25a			
			Inoculated plants			
S	417c	4.54c	1.58c	11.5c	2.51ab	3.03b
α	534b	5.62b	3.87b	17.3b	1.84c	3.45b
β	943a	8.51a	4.36a	27.5a	3.27a	5.20a

[a]Comparisons were made according to plant biomass expressed as dry weight (g/plant) or total N (g/plant), N derived from soil (NDFS, g/plant), nodule dry weight (g/plant), and N_2 fixation measured by the direct isotope mehod (ID, g/plant) and the total N difference method (TND, g/plant) (B. Sougoufara *et al.*, unpublished observations). Soil used: Camberene soil (97% sand, 0.010% N); volume of containers where plants were grown: 15 liters. Same rate of ^{15}N urea was added to uninoculated and inoculated plants (0.232 g N/plant). Within each treament, data in each column followed by the same letter are not significantly different according to Newman–Keuls' test ($p = 0.05$).

[b]Estimates of N_2 fixation using isotope dilution (ID) and total N difference (TND) methods.

trees established in marginal conditions. The main objective of this section is to draw attention to important treatments and practices, such as spacing, fertilization, irrigation, and pest and disease control. In addition, data on yields of selected *Casuarina* plantations under current management systems are also given.

A. Spacing

Spacing is a vital factor in tree productivity and N_2 fixation. It differs according to final use (fuelwood or timber) and site characteristics. In most plantations of *Casuarina equisetifolia* and *Casuarina cunninghamiana* designed to produce pole-size timber, spacing is 2 × 2–3 m. Trees planted at the wider spacing (2 × 3 m) do not have to be thinned before reaching the targeted commercial size, which reduces labor costs, but probably also reduces the amount of N_2 fixed per unit area. In irrigated plantations the number of trees per hectare can reach 8000–10,000 (1 1.5 m spacing), but in rain-fed plantations, it is not possible to plant more than 1000–2500 trees per ha because of the competition for water, which can be especially acute in semiarid conditions (Midgley *et al.*, 1983). The optimum density for maximum wind protection is 2–4 rows every 1.5 m for main windbreaks and 1–2 rows at similar spacing for secondary windbreaks (El-Lakany, 1983).

B. Fertilization

Most investigations on the nutrient requirements of Casuarinaceae deal with the host plant partner and nearly all publications on this family are limited to *Casuarina equisetifolia.* Soil phosphorus status is crucial in the distribution of Casuarinaceae (Bowen and Reddell, 1987). Recent investigations clearly indicate that P stimulates nodulation more than plant growth (Sanginga *et al.*, 1988). But in spite of its essential role in the metabolic pathways of the host plant and the associated *Frankia,* less P is required for plant growth than potassium or magnesium (Verma *et al.*, 1987). In addition, *Casuarina equisetifolia* requires fairly large amounts of calcium, which explains why it grows poorly on soils low in Ca and quite well on alkaline soils (Yadav, 1983).

In the sandy soils of South-East China (H. G. Diem, unpublished observations) and Tamil Nadu, India (Kondas, 1983), which have very low nutrient reserves, all the litter under *Casuarina equisetifolia* in plantations is continuously removed. According to Turnbull (1983), up to 4 tons per ha of litter and branches can be obtained from older plantations in China each year for fuel. This practice, together with the uprooting of stumps, leads to nutrient deficiencies that hinder tree growth and its N_2-fixing ability. The decrease in *Casuarina equisetifolia* yield after one or two rotations, as observed in China and India, is the consequence of this chemical depletion of the soils. To restore fertility, P, Ca, K, and Mg will have to be added in proportions calculated according to the residual fertility status of the site. Further, the removal of litter should be discouraged or, at least, compensated by returning the ashes to the plantation sites. Fertilizer is also important to plant survival in eroded or wasted soils, for example, mine lands.

Acidity is a major constraint to nodulation, not because it decreases the viability of *Frankia* propagules but rather because it probably affects factors that control the susceptibility of the host plant to infection. In acid soils such as those in the coastal areas of South-East China, soil amendment around each tree would probably restore normal nodulation. Further research is required to determine the impact of high acidity on the nodulation process and the possible use of acid-tolerant *Frankia* strains to improve *Casuarina* nodulation.

C. Water Requirements and Irrigation

1. *Casuarina equisetifolia*

Casuarina equisetifolia does not have very specific water requirements. When annual rainfall is between 700 and 2000 mm, which is the normal range in its natural habitat, *Casuarina equisetifolia* grows well and seems to fix N_2 normally provided that the groundwater table does not sink to below a depth of 4–5 m during the dry season. In arid and semiarid climates, that is, where annual rainfall is less than 300 mm, *Casuarina equisetifolia* behaves like a typical littoral tree; it prospers along the coast, where the sea breeze bearing a spray of water supplements moisture shortages from the groundwater table. In this situation the trees often require some irrigation during the first years after planting (Yadav, 1983). In the hinterland or areas with less than 300 mm annual precipitation, *Casuarina equisetifolia* dries up if not regularly irrigated. The development of nodules can suffer from prolonged waterlogging, thus leading to the failure of the *Casuarina equisetifolia* plantations.

2. *Casuarina glauca*

In its natural habitat of Australia, *C. glauca* is common along the edges of swampy flats. However, this species can tolerate low atmospheric humidity, as was demonstrated by its successful development in desert or semidesert areas, for example, in the desert fringes of the Nile Delta, where it receives biweekly irrigation during summer (May–September) and supplementary irrigation for the rest of the year, depending on the rainfall (El-Lakany, 1983).

3. Drought-Tolerant Casuarinaceae

The family Casuarinaceae includes at least three species that are renowned for their ability to tolerate drought: *Casuarina obesa, C. cristata* subsp. *pauper,* and *Allocasuarina decaisneana.* The latter species is probably the best adapted to arid conditions but has not yet been fully tested. The list of drought-resistant Casuarinaceae will probably be lengthened in the future when the water requirements of the different species, provenances, and hybrids are better understood. It is interesting to note that during a severe drought in Thailand, *Casuarina equisetifolia* appeared to be less resistant than *C. junghuhniana* x *Casuarina equisetifolia,* which is surprising since *C. junghuhniana* is native to areas with 700 to 1500 mm rainfall and is reputed to be only moderately drought tolerant (National Research Council, 1984).

D. Pests, Diseases, and Their Control

Since N_2 fixation is optimal only when the host plant is healthy, it is important to know the Casuarinaceae pests and diseases, and treatments that can reduce or eliminate their damage to the trees. A detailed lists of pests and diseases affecting casuarinas was published in Browne's (1968) treatise and in a book edited by Midgley *et al.* (1983).

E. Yield

Casuarina equisetifolia lives for only 40–50 years. Its growth is rapid during the first 7 years, then gradually declines. Thus the species could be worked on a rotation of 7 to 15 years. Plantation yields vary considerably depending on soil fertility, moisture regimes, other site characteristics, and management practices such as irrigation, fertilization, thinning, and especially inoculation. The mean annual increment of *Casuarina equisetifolia* production is in the range of 4–12 m^3 per ha per year. The yield of wood is 40–120 tons per ha and about 40 tons per ha of litter and cones in a 10-year rotation (Midgley *et al.*, 1984). Compared to *Acacia mangium* or *Leucaena leucocephala*, *Casuarina equisetifolia* appears to be a moderately fast growing species. At some sites, however, *Casuarina equisetifolia* yields are much higher than those of other species. The preceding figures relate to plantations that could give much higher yields if management practices were good and if properly selected host trees were inoculated with selected *Frankia* strains. The natural yield of *Casuarina cunninghamiana* is probably higher, since Mendonza (1983) reported it to be in the range of 8 to 34 m^3 per ha per year in Argentina.

This book was in press when profuse aerial nodulation on *Casuarina cunninghamiana* trunks was reported by one of us (Y.D.) in a site characterized by high annual rainfall (3000–4000 mm). Aerial nodules would probably contribute more fixed N_2 than root nodules.

References

Baker, D. D. (1987). Relationships among pure cultured strains of *Frankia*based on host specificity. *Physiol Plant.* **70**, 245–248.

Bamber, R. K., Mullette, K., and Mackowski, C. (1980). Mycorrhizal studies. *In* "Research Report 1977–1978." Aust. For. Comm. N.S.W., Sydney, Australia.

Bourke, R. M. (1985). Food, coffee and casuarina: An agroforestry system from the Papua New Guinea highlands. *Agrofor. Syst.* **2**, 273–279.

Bowen, G. D., and Reddell, P. (1987). Nitrogen fixation in Casuarinaceae. *In* "Amelioration of Soil by Trees" (R. T. Prinsley and M. J. Swift, eds.), pp. 80–90. Commonw. Sci. Counc., London.
Browne, F. G. (1968). "Pests and Diseases of Forest Plantation Trees." Oxford Univ. Press (Clarendon), London and New York.
Burleigh, S., Baker, D., and Torrey, J. G. (1988). Mass inoculum production protocol for *Casuarina*. *Abstr. Int. Meet. Frankia Actinorhizal Plants, 7th.*
Chaudhary, A. H., and Mirza, M. S. (1987). Isolation and characterization of *Frankia* from nodules of actinorhizal plants of Pakistan. *Physiol. Plant.* **70,** 225–258.
Chen, Z. Z., Wang, T. T., and Yang, J. C. (1983). Identification and natural hybridity of two *Casuarina* species grown in Taiwan. *Nitrogen-Fixing Tree Res. Rep.* **1,** 22.
Dawson, J. O., and Gibson, A. H. (1987). Sensitivity of selected *Frankia* isolates from *Casuarina, Allocasuarina* and North American host plants to sodium chloride. *Physiol. Plant.* **70,** 272–278.
Diem, H. G., and Gauthier, D. (1982). Effet de l'infection endomycorhizienne (*Glomus mosseae*) sur la nodulation et la croissance de *Casuarina equisetifolia*. *C. R. Hebd. Seances Acad. Sci., Ser. C* **294,** 215–218.
Diem, H. G., Gueye, I., Gianinazzi-Pearson, V., Fortin, J. A., and Dommergues, Y. R. (1981). Ecology of VA mycorrhizae in the tropics: The semi-arid zone of Senegal. *Oecol. Plant.* **2,** 53–62.
Diem, H. G., Gauthier, D., and Dommergues, Y. R. (1982a). Extranodular growth of *Frankia* of *Casuarina equisetifolia*. *FEMS Microbiol. Lett.* **15,** 181–184.
Diem, H. G., Gauthier, D., and Dommergues, Y. R. (1982b). Isolation of *Frankia* from nodules of *Casuarina equisetifolia*. *Can. J. Microbiol.* **28,** 526–530.
Diem, H. G., Gauthier, D., and Dommergues, Y. R. (1982c). Isolement et culture *in vitro* d'une souche infective de *Frankia* isolée de nodules de *Casuarina*. *C. R. Hedb. Seances Acad. Sci., Ser. C* **295,** 759–763.
Diem, H. G., Gauthier, D., and Dommergues, Y. R. (1983). An effective strain of *Frankia* from *Casuarina* sp. *Can. J. Bot.* **61,** 2815–2821.
Diem, H. G., Duhoux, E., Simonet, P., and Dommergues, Y. R. (1988). Actinorhizal symbiosis biotechnology: The present and future. *Proc. Int. Symp. Biotechnol., 8th* **2,** 984–995.
Dommergues, Y. (1963). Evaluation du taux de fixation de l'azote dans un sol dunaire reboisé en filao (*Casuarina equisetifolia*). *Agrochimica* **186,** 335–340.
Doran, J. C., and Hall, N. (1983). Notes on fifteen Australian *Casuarina* species. *In* "Casuarina Ecology, Management and Utilization" (S. J. Midgley, J. W Turnbull, and R. D. Johnston, eds.), pp. 19–52. Commw. Sci. Ind. res. Org., Melbourne, Australia.
Duhoux, E., Sougoufara, B., and Dommergues, Y. (1986). Propagation of *Casuarina equisetifolia* through axillary buds of immature female inflorescences cultured *in vitro*. *Plant Cell Rep.* **3,** 161–164.
El-Lakany, M. H. (1983). A review of breeding drought resistant *Casuarina* for shelterbelt establishment in arid regions with special reference to Egypt. *For. Ecol. Manage.* **8,** 129–137.
El-Lakany, M. H. (1985). Biological effects of shelterbelts and windbreaks in arid regions. *Semin. Int. Brise-Vent* **IDRC-MR117e,f,** 104–110.
El-Lakany, M. H., and Luard, E. J. (1982). Comparative salt tolerance of selected *Casuarina* species. *Aust. For. Res.* **13,** 11–20.
Flores, E. M. (1978). The shoot apex of *Casuarina* (Casuarinaceae). *Rev. Biol. Trop.* **26,** 247–260.
Gardner, I. C. (1986). Mycorrhizae of actinorhizal plants. *MIRCEN J.* **2,** 147–160.
Gauthier, D., Diem, H. G., and Dommergues, Y. R. (1981). Infectivité et effectivité de

souches de *Frankia* isolées de nodules de *Casuarina equisetifolia* et d'*Hippophae rhamnoides*, *C. R. Hebd. Seances Acad. Sci., Ser. C* **293,** 489–491.

Gauthier, D., Diem, H. G., and Dommergues, Y. R. (1984). Tropical and subtropical actinorhizal plants. *Pesqui. Agropecu. Bras.* **19s/n,** 119–136.

Gauthier, D., Diem, H. G., and Dommergues, Y. R. (1985). Assessment of N_2 fixation by *Casuarina equisetifolia* inoculated with *Frankia* ORS 021001 using ^{15}N methods. *Soil Biol. Biochem.* **17,** 375–379.

Geary, T. F. (1983). Casuarinas in Florida, USA and some Caribbean islands. *In* "Casuarina Ecology, Management and Utilization" (S. J. Midgley, J. W. Turnbull, and R. D. Johnston, eds.), pp. 107–109. Commonw. Sci. Ind. Res. Org., Melbourne, Australia.

Huang, J. B., Zhao, Z. Y., Chen, G. X., and Liu, H. C. (1985). Host range of *Frankia* endophytes. *Plant Soil* **87,** 61–65.

Jafarsidik, Y. (1986). Hardwood forest tree plantations in Sumatra. *Indones. Agric. Res. Dev. J.* **8,** 7–11.

Jiang, J., and Zhu, B. (1985). Isolation, cultivation and infectivity of actinorhizal endophyte of *Casuarina equisetifolia*. *Kexue Tongbao (Engl. Transl.)* **30,** 833–838.

Johnson, L. A. S. (1980). Notes on Casuarinaceae. I. *Telopea* **2,** 83–84.

Johnson, L. A. S. (1982). Notes on Casuarinaceae. II. *J. Adelaide Bot. Gard.* **6,** 73–87.

Johnson, L. A. S. (1988). Notes on Casuarinaceae. III: The new genus *Ceuthostoma*. *Telopea* **3,** 133–137.

Kondas, S. (1983). *Casuarina equisetifolia*—A multipurpose tree crop in India. *In* "Casuarina Ecology, Management and Utilization" (S. J. Midgley, J. W. Turnbull, and R. D. Johnston, eds.), pp. 66–76. Commonw. Sci. Ind. Res. Org., Melbourne, Australia.

Koske, K. E., and Halvorson, W. L. (1981). Ecological studies of vesicular-arbuscular mycorrhizae in a barrier dune. *Can. J. Bot.* **59,** 1413–1422.

Kumar, P. H. (1981). Problems and prospects of establishing a plantation forestry with *Casuarina*, cashew and coconut in the coastal belt of India. *Riv. Agric. Subtrop. Trop.* **75,** 317–323.

Lalonde, M., Calvert, H. E., and Pine, S. (1981). Isolation and use of *Frankia* strains in actinorhizae formation. *In.* "Current Perspectives in Nitrogen Fixation" (A. H. Gibson and W. E. Newton, eds.), pp. 296–299. Aust. Acad. Sci., Canberra, Australia.

Liang, Z. C., and Chen, X. H. (1984). Selection of clones of casuarina for resistance to bacterial wilt. *J. South China Agric. Coll.* **5,** 53–59.

Malik, M., and Sheikh, M. I. (1983). Planting of trees in saline and waterlogged areas. Part I. Plant testing at Azakhel. *Pak. J. For.* **33,** 1–17.

Mendonza, L. A. (1983). Growth and uses of *Casuarina cunninghamiana*. *In* "Casuarina Ecology, Management and Utilization" (S. J. Midgley, J. W. Turnbull, and R. D. Johnston, eds.), pp. 53–54. Commonw. Sci. Ind. Res. Org., Melbourne, Australia.

Midgley, S. J., Turnbull, J. W., and Johnston, R. D. (eds.) (1983). "Casuarina Ecology, Management and Utilization." Commonw. Sci. Ind. Res. Org., Melbourne, Australia.

Midgley, S. J., Turnbull, J. W., and Hartney, V. J. (1984). Fuelwood species for salt-affected sites. *ADAB Res. Dev. Semin., Forage Fuel Prod. Salt-Affected Wasteland* 36 pp.

Mitchell, J. (1982). Revegetation of lands surrounding the Bell Bay smelter. *In* "Fluoride Emissions. Their Monitoring and Effects on Vegetation and Ecosystems" (F. Murray, ed.), pp. 207–215. Academic Press, New York.

National Research Council (1984). "Casuarinas: Nitrogen-Fixing Trees for Adverse Sites." Natl. Acad. Press, Washington, D. C.

Ng, B. H. (1987). The effects of salinity on growth, nodulation and nitrogen fixation of *Casuarina equisetifolia*. *Plant Soil* **103,** 123–125.

Paulsson, K. (1987). "Effects of Windbreaks on Evapotranspiration and Windspeed in Sidi

Bouzid, Central Tunisia. Part 2. Study of Four Windbreaks," Work. Pap. 43, p. iii and 29 pp. Int. Rural Dev. Cent., Swed. Univ. Agric. Sci., Sweden.

Périnet, P., Brouillette, J. G., Fortin, J. A., and Lalonde, M. (1985). Large scale inoculation of actinorhizal plants with *Frankia*. *Plant Soil* **87,** 175–183.

Prin, Y., Neyra, M., and Diem, H. G. (1989). Evaluation of *Frankia* growth and viability using total protein and INT reduction activity estimations. *Abstr. Int. Meet.* Frankia *Actinorhizal Plants, 7th.*

Puppo, A., Dimitrijevic, L., Diem, H. G., and Dommergues, Y. R. (1985). Homogeneity or superoxide dismutase patterns in *Frankia* strains from Casuarinaceae. *FEMS Microbiol. Lett.* **30,** 43–46.

Racette, S., and Torrey, J. G. (1989). The isolation, culture and infectivity of a *Frankia* strain from *Gymnostoma papuanum* (Casuarinaceae). *Plant Soil* **118,** 165–170.

Reddell, P., and Bowen, G. D. (1985). *Frankia* source affects growth, nodulation and nitrogen fixation in *Casuarina* species. *New Phytol.* **100,** 115–122.

Reddell, P., Bowen, G. D., and Robinson, A. D. (1986). Nodulation of Casuarinaceae in relation to host species and soil properties. *Aust. J. Bot.* **34,** 435–444.

Reddell, P., Rosbrook, P. A., Bowen, G. D., and Gwaze, D. (1988). Growth responses in *Casuarina cunninghamiana* plantings to incoulation with *Frankia*. *Plant Soil* **108,** 78–86.

Robinson, P. J. (1987). The dependence of crop production on trees and forest land. *In* "Amelioration of Soils by Trees" (R. T. Prinsley and M. J. Swift, eds.), pp. 104–120. Commonw. Sci. Counc., London.

Rosbrook, P. A. (1987). A *Frankia* source selection programme to improve nitrogen fixation by Casuarinaceae. *In* "Amelioration of Soils by Trees" (R. T. Prinsley and M. J. Swift, eds.), pp. 91–94. Commonw. Sci. Counc., London.

Rose, S. L. (1980). Mycorrhizal associations of some actinomycete nodulated nitrogen-fixing plants. *Can. J. Bot.* **58,** 1449–1454.

Sanginga, N., Danso, S. K. A., and Bowen, G. D. (1988). Nodulation and growth response of *Allocasuarina* and *Casuarina* species to phosphorus fertilization. *Abstr. Int. Meet.* Frankia *Actinorhizal Plants, 7th.*

Sellstedt, A. (1988). Nitrogenase activity, hydrogen evolution and biomass production in different *Casuarina* symbioses. *Plant Soil* **105,** 33–40.

Shipton, W. A., and Burggraaf, A. J. P. (1983). Aspects of the cultural behaviour of *Frankia* isolates in relation to infectivity and nitrogen fixation (acetylene reduction). *Plant Soil* **61,** 2783–2792.

Silvester, W. B. (1977). Dinitrogen fixation by plant associations excluding legumes. *In* "A Treatise on Dinitrogen Fixation" (R. W. F. Hardy and A. H. Gibson, eds.), Sect. IV, pp. 141–189. Wiley, New York.

Sougoufara, B., Diem, H. G., and Dommergues, Y. R. (1989a). Response of field-grown *Casuarina equisetifolia* to inoculation with *Frankia* strain ORS 021001 entrapped in alginate beads. *Plant Soil* **118,** 133–137.

Sougoufara, B., Duhoux, E., and Dommergues, Y. R. (1989b). Choix des critères de sélection chez un arbre fixateur d'azote de N_2: *Casuarina equisetifolia*. *Oecol. Plant.* **10,** 215–214.

Stewart, H. T., and Flinn, D. W. (1984). Establishment and early growth of trees irrigated with wastewater at four sites in Victoria, Australia. *For. Ecol. Manage.* **8,** 243–256.

Thiagalingam, K. (1983). Role of casuarina in agroforestry. *In* "Casuarina Ecology, Management and Utilization" (S. J. Midgley, J. W. Turnbull, and R. D. Johnston, eds.), pp. 175–179. Commonw. Sci. Ind. Res. Org., Melbourne, Australia.

Torrey, J. G. (1982). *Casuarina:* Actinorhizal nitrogen-fixing tree of the tropics. *In* "Biological Nitrogen Fixation Technology for Tropical Agriculture" (P. H. Graham and S. C. Harris, eds.), pp. 427–439. CIAT, Cali, Colombia.

Torrey, J. G., and Berg, R. H. (1988). Some morphological features for generic characterization among the Casuarinaceae. *Am. J. Bot.* **75,** 844–874.

Turnbull, J. W. (1983). The use of *Casuarina equisetifolia* for protection of forests in China. *In* "Casuarina Ecology, Management and Utilization" (S. J. Midgley, J. W. Turnbull, and R. D. Johnston, eds.), pp. 55–57. Commonw. Sci. Ind. Res. Org., Melbourne, Australia.

Turner, J., Kelly, J., and Newman, L. A. (1978). Soil nutrient–vegetation relationships in the Eden area, N.S.W. II. Vegetation–soil associations. *Aust. For.* **41,** 223–231.

Verma, V.P.S., Tandon, V. N., and Rawat, H. S. (1987). Biomass production and plant nutrient distribution in different aged plantations of *Casuarina equisetifolia* in Puri, Orissa. *Indian For.* **113,** 273–280.

Yadav, J. S. P. (1983). Soil limitations for successful establishment and growth of casuarina plantations. *In* "Casuarina Ecology, Management and Utilization" (S. J. Midgley, J. W. Turnbull, and R. D. Johnston, eds.), pp. 138–166. Commonw. Sci. Ind. Res. Org., Melbourne, Australia.

Zhang, Z., and Torrey, J. G. (1985a). Studies of an effective strain of *Frankia* from *Allocasuarina lehmanniana* of the Casuarinaceae. *Plant Soil* **87,** 1–16.

Zhang, Z., and Torrey, J. G. (1985b). Biological and cultural characteristics of the effective *Frankia* strain HFPCcI3 (Actinomycetales) from *Casuarina cunninghamiana* (Casuarinaceae). *Ann. Bot. (London)* **56,** 367–378.

Zhang, Z., Lopez, M. F., and Torrey, J. G. (1984). A comparison of cultural characteristics and infectivity of *Frankia* isolates from root nodules of *Casuarina* species. *Plant Soil* **78,** 79–80.

17

Actinorhizal Plants in Pacific Northwest Forests

David E. Hibbs and Kermit Cromack, Jr.
Department of Forest Science
Oregon State University
Corvallis, Oregon, U.S.A.

I. Introduction

A. The Region

The forests of the Pacific Northwest are extensive and diverse. They contain representatives of most of the world's conifer genera. Geographically, they are constrained in a relatively narrow north–south trending belt between the Pacific Ocean and the high-elevation mountain ranges that occur 20 to 200 km inland. These forests extend from 42° to 60°N latitude. Climate ranges from cool maritime to cold alpine. Summers are dry, and annual precipitation decreases north to south. The geology includes sedimentary and volcanic rock, and the northern portions of the region have been heavily glaciated. Historically, extensive fire has been the major disturbance in the south while wind has been of primary importance in the north. For more detail about the region, see Franklin and Dyrness (1973) for an in-depth review of the climate, geology, vegetation types, and vegetation dynamics of Oregon and Washington. Much of their discussion of the *Picea sitchensis, Tsuga heterophylla, Abies amabilis,* and *Tsuga mertensiana* zones applies to the Canadian and southern Alaskan coast as well.

B. Actinorhizal Species

Three genera of actinorhizal plants occur in the region, *Purshia, Alnus,* and *Ceanothus. Purshia* species are found east of the Cascade Mountains

The Biology of *Frankia* and Actinorhizal Plants

and occasionally on dry sites in the Pacific Northwest. East of the Cascades, *P. tridentata* can fix a modest amount of nitrogen (Dalton and Zobel, 1977).

Red alder (*A. rubra*) is an early successional species found on mesic sites in the *Picea sitchensis* and *Tsuga heterophylla* zones (Franklin and Dyrness, 1973). North of about 50° latitude or above 800 m elevation, red alder is first mixed with and, with increasingly cooler climate, replaced by the shrubby Sitka alder (*A. viridis* ssp. *sinuata*). White alder (*A. rhombifolia*) occurs in riparian zones in the drier southern portions of the region (mixed-evergreen and interior valley zones). One other species of alder, the nonnative *A. glutinosa*, has been planted in a very limited set of research plots.

Species of the genus *Ceanothus* are restricted almost entirely to early successional sites in Oregon (Conard *et al.*, 1985), although they occur on drier sites as far north as Vancouver Island (Binkley, 1986). As a group, they are more stress tolerant than *Alnus*. They occur primarily as early successional shrubs on cold or drought-prone sites. Snowbrush (*C. velutinus*) regenerates from buried seed and inhabits middle-elevation sites in the Cascade Range and higher elevations in the Coast Range (*Tsuga heterophylla* and mixed-conifer zones). Redstem (*C. sanguineus*) overlaps snowbrush in range and habitat characteristics. However, it is more drought and cold tolerant and extends up the Cascade Range north of the Canadian border. Deerbrush (*C. integerrimus*) is the most drought tolerant of the three more common *Ceanothus* species. In the Pacific Northwest, deerbrush reaches its best development in the Siskiyou Mountains (mixed-evergreen and mixed-conifer zones). Deerbrush is an important wildlife browse species. Of minor importance in the region are *C. cuneatus*, *C. thyrsiflorus*, *C. cordulatus*, and *C. prostratus*.

Two genera of leguminous plants are also found in the forests of the region: *Lupinus* and *Cytisus*. The lupines are restricted to high elevation or dry sites. Scotch broom (*C. scoparius*) is a nonnative legume that is spreading northward as a roadside and forest plantation weed on mesic sites. It cannot be considered an important nitrogen fixer in the region (Helgerson *et al.*, 1979). Lupines are present but not common west of the Cascade crest, so they also are probably not important nitrogen fixers. East of the Cascade Mountains, lupines can fix 10 to 15 kg/ha/year (Cromack *et al.*, 1979).

C. Nitrogen Fixation Estimates

Estimates of nitrogen fixation rates vary greatly by species and region (Table I). Red alder shows the most consistently high levels of fixation,

Table I Examples of Estimates of Annual Nitrogen Fixation by Pacific Northwest Actinorhizal Plants[a]

Annual N fixation (kg/ha/year)	Location	Stand age (years)	Reference
Red alder			
23	Or. Coast Range	2	Kim *et al.* (1989)
320	Or. Coast Range	2–15	Newton *et al.* (1968)
70	NW. Oregon	5	Bormann and Gordon (1984)
80	W. Washington	1–4	DeBell and Radwan (1979)
62	W. Washington	2–4	Tripp *et al.* (1979)
85	W. Washington	1–38	Cole *et al.* (1978)
100	W. Washington	10–40	Bormann and DeBell (1981)
42	NW. Washington	1–23	Binkley (1982)
65	British Columbia	1–23	Binkley (1982)
130	British Columbia	20	Binkley (1981)
Sitka alder			
2	British Columbia	20	Binkley (1981)
Ceanothus velutinus			
8	SW. Oregon	5–6	Kim (1987)
69	SW. Oregon	11	Kim (1987)
32	NW. Oregon	11	McNabb *et al.* (1979)
70–108	Or. Cascades	1–10	Youngberg and Wollum (1976)
42–100	Or. Cascades	1–12	Binkley *et al.* (1982)
0–20	Or. Cascades	1–15	Zavitkovski and Newton (1968)
80	Or. Cascades	17	Cromack *et al.* (1979)
101	Or. Cascades	17	McNabb and Cromack (1983)
Ceanothus sanguineus			
24–50	British Columbia	1–12	Binkley and Husted (1983)
Ceanothus integerrimus			
<1	SW. Oregon	5–6	Kim (1987)

[a]This list is not exhaustive.

although snowbrush (*C. velutinus*) and Sitka alder can equal it on sites without temperature or moisture stress. The values in Table I need to be interpreted with caution because they comprise a diversity of estimation methods: ^{15}N, accretion, acetylene reduction, and chronosequences (sequences of similar sites of different successional ages).

D. Importance of Actinorhizal Plants

The historic importance of actinorhizal plants to nitrogen budgets and forest productivity in the Pacific Northwest is uncertain. In recent history, logging activities with ground-based equipment and preparation of harvested land for replanting have increased the area occupied by *Alnus* and *Ceanothus*. This spread following disturbance has led to speculation that similar increases in abundance of these species would have occurred following the extensive fires or blowdowns that occurred throughout the presettlement history of the region. These nitrogen-fixing pioneer species would then have been succeeded by the mature forest conifers. If these nitrogen fixers were common following disturbances, they surely would have played an important role in maintaining or rebuilding nitrogen pools.

Two observations cause some uncertainty about this hypothesized importance in both the red alder- and snowbrush-dominated portions of the region (the *Picea sitchensis* and *Tsuga heterophylla* zones; Franklin and Dyrness, 1973). First, chronosequence studies of red alder succession in the Coast Range (Henderson, 1970; Carlton, 1988) show that conifer regeneration under alder is very rare. Instead, most alder stands appear to succeed to salmonberry (*Rubus spectabilis*)-dominated communities (Newton *et al.*, 1968; Hibbs, 1987b). How long these shrub communities would be stable is uncertain. This successional sequence, however, clearly delays conifer colonization by more than the 100-year life span of red alder and several decades or, perhaps, centuries of shrub dominance. Fire may be required to remove the shrubs and restart succession on another pathway.

The second observation comes from tree-age distribution studies in old growth Douglas-fir (*Pseudotsuga menziesii*) stands in the Oregon Cascade Mountains (Franklin and Hemstrom, 1981). There are at least two extensive fires that can be documented, about 250 years ago and 500 plus years ago. Stands resulting from the fire 500 plus years ago are uneven-aged. Conifer regeneration occurred over a period of 200 years. This could have resulted from the slow invasion of an already established plant community, perhaps including actinorhizal plants. It could also have resulted from a lack of seed source for Douglas-fir. The lack of signs of early growth suppression in dominant trees argues for the latter. Stands resulting from the fire 250 years ago are even-aged. Even-aged stands result from immediate colonization following a disturbance. Immediate conifer invasion precludes the possibility of a period of red alder domination. Because Douglas-fir can establish in and succeed snowbrush, this successional pathway is a possibility.

E. Defining the Bounds to Management

Two long-term studies have stimulated interest in actinorhizal plants and fueled the controversy that surrounds their use in forestry. Miller and Murray (1978) described a 4-year-old Douglas-fir plantation on a nitrogen-deficient soil in southwestern Washington (Wind River Experimental Forest) that was interplanted with red alder from a coastal seed source. By age 48, the mixed stand contained almost 7% more conifer wood volume than the pure conifer stand. Conifer diameter increased 23%; height increased by 24%. In addition, the mixed stand contained almost as much alder wood volume as conifer volume. The increase in forest productivity with mixed culture is clear. As the next example shows, however, the where and how of achieving such increases is not clear.

In 1935–1936, researchers established plots in an 8-year-old stand of mixed conifer/red alder at Cascade Head Experimental Forest (Berntsen, 1961). Soil nitrogen levels are high in this area. The four treatments were (1) control (mixed species, unthinned), (2) pure conifer, thinned, (3) pure alder, thinned, and (4) pure alder, unthinned (this plot originally

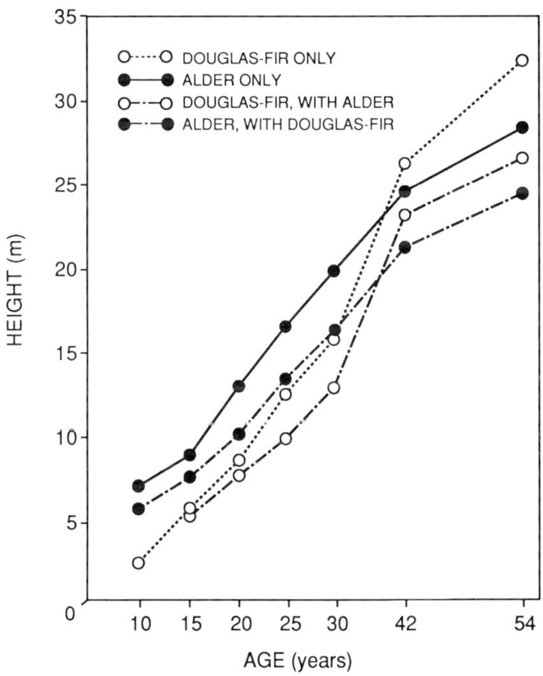

Figure 1 Height growth of red alder and Douglas-fir grown in mixture and in pure-species plots at Cascade Head Experimental Forest (S. Greene, unpublished observations). Note that there were some differences in height growth potential among plots at the beginning of the study.

had a low conifer complement). These treatments were unreplicated, and initial height measurements indicate that there were some site quality differences among the plots (Fig. 1). Table II documents the change in volume with time in these treatments. Berntsen (1961) and S. Greene (unpublished observations) do not separate these volumes by species. The early rapid growth of alder is clear. Through age 30, Douglas-fir was shorter than alder (Fig. 1) and the mixed stand underproduced relative to either species alone (Table II). By age 42, the Douglas-fir in the mixed plot grew through the alder canopy and then stand growth increased. In contrast, Miller and Murray (1978) found an almost immediate increase in Douglas-fir growth following alder planting at Wind River even though the Douglas-fir height did not surpass that of alder until age 30 (Douglas-fir). The ability of Douglas-fir to survive under and eventually grow through the dense alder canopy in both examples comes as a surprise to foresters.

The differences in stand development and productivity between Wind River and Cascade Head Experimental forests have made it clear that factors related to soil fertility and the spacing and timing of establishment may all be important to the production and outcome of pure-species stands and mixing species (see also Binkley, 1983). In the next sections, we will review the research in these areas. Then, we will discuss how these findings are or could be implemented in forest management.

II. Soil Building/Fertility

A. Increased Soil Nitrogen

In the Pacific Northwest, red alder clearly has the capacity to add significant amounts of N to soil (Table I) and also to increase soil organic matter

Table II Wood Volume (m^3/ha) in Cascade Head Mixed Alder/Conifer Study[a]

Treatment	Age at measurement (years)					
	15	20	25	30	42	54
Alder/conifer mix	48	121	206	278	434	518
Conifer, thinned	15	71	185	331	595	770
Alder, thinned	85	170	234	303	357	420
Alder, unthinned	174	240	294	361	420	406

[a] Minimum tree diameter measured was 4 cm. From Berntsen (1961) and S. Greene (unpublished observations).

(Tarrant and Miller, 1963; Bormann and DeBell, 1981). Other alder species, including Sitka alder, can also increase soil N and organic matter (Van Cleve *et al.*, 1971; Binkley *et al.*, 1984). In Europe, *Alnus glutinosa* has been shown to have similar capacity (Wild, 1988), as has *A. incana* ssp. *rugosa* in the eastern United States (Voight and Steucek, 1969).

Red alder has the capability of fixing N over a wide range of stand densities and ages, ranging from 2-year-old stands having 10,000 stems/ha to a 58-year-old stand having 250 stems/ha (Kim *et al.*, 1990). Estimates of N fixation by red alder range from 22 kg/ha/year in 2-year-old alder (Kim *et al.*, 1990) to over 300 kg/ha/year in vigorous, mature stands (Zavitkovski and Newton, 1968). Bormann and Gordon (1984), for 5-year-old red alder, report fixation rates of 62 kg/ha/year with 2000 stems/ha ranging up to a maximum rate of 85 kg/ha/year with 10,000 stems/ha. Heilman and Ekuan (1982) recently demonstrated the ability of red and Sitka alders to fix large amounts of nitrogen on coal mine spoils in western Washington.

Alnus is quite versatile as a successional genus. Alders can occur as pioneer species in primary succession following deglaciation (Lawrence, 1958; Davis, 1981), landslides (Ugolini, 1968), and volcanic eruptions (J. Means, personal communication), and they can occur in secondary succession following a variety of forest disturbances (Franklin and Dyrness, 1973). Alder can occur as a dominant species early in succession. In other cases, it is an understory species in mature conifer stands (Van Cleve and Viereck, 1981; Simard, 1989). Substantial N fixation can occur when red alder is an understory species growing after heavy thinning of conifer stands, if light is adequate (Berg and Doerksen, 1975).

Addition of N to soil by alder species has been shown to increase significantly both total soil N and available soil N (Binkley *et al.*, 1984; Sollins *et al.*, 1984; Binkley, 1986). Increased available N is a key to increased growth of economically important conifers such as Douglas-fir (Shumway and Atkinson, 1978).

B. Soil Organic Matter

Increases in soil organic matter as high as 20% have been found under alder (Tarrant and Miller, 1963; Franklin *et al.*, 1968; Bormann and DeBell, 1981; Binkley *et al.*, 1982; Binkley, 1983). A chronosequence study documented that snowbrush can also increase soil organic matter (Binkley *et al.*, 1982). Soil organic matter is the primary storage medium for soil N, and the relatively decay-resistant nature of humified organic matter means that chemically bound organic N will be released fairly slowly (Stevenson, 1982; Wild, 1988). Soil organic matter has a high cation-exchange capacity and retains cations such as Ca^{2+}, K^{+}, Mg^{2+}, NH_4^{+}, and micronutrients. Retention of NH_4^{+} is especially important since NH_4^{+} is the dominant form of mineralized N in many forest soils (Binkley, 1986). Alder also has the capacity to acidify soil (Lawrence, 1958). Lowering soil pH can minimize volatilization losses of N as NH_4^{+} from soils that normally have a high pH (Ledgard *et al,* 1984).

Addition of soil organic matter has been shown to increase aggregate formation and improve soil tilth (Oades, 1988; Wild, 1988). Soil aggregate structure is important to maintaining soil porosity and good aeration of the soil (Wild, 1988). Stable soil aggregates also protect soil organic matter from decomposition due to physical occlusion (Oades, 1988). In addition to contributing to improved soil structure, addition of soil organic matter to soil can decrease soil bulk density (Wild, 1988). Several studies have shown that both red and Sitka alder have decreased soil bulk density as well as increased soil N and organic matter (Tarrant and Miller, 1963; Binkley *et al.*, 1984; Binkley, 1986).

III. Use in Forest Production

This review of the interactions between actinorhizal plants and soil properties reflects the broad ecological interest that exists regarding these species. The interests in forest management are more narrow. Where they occur naturally in forest plantations, the various species of *Alnus* and *Ceanothus* are treated as undesirable weeds. They are removed from plantations to improve conifer survival and growth.

Only in the last 10 years has interest begun to develop in the management of red alder, both in pure stands and in mixed culture with conifers. There is also a small interest in Sitka alder as an interplanted species in conifer plantations. The very limited interest that exists for *Ceanothus* species, as discussed earlier, has focused on its short-term nitrogen input as an early-successional shrub in plantations. Evidence for its effects on conifer regeneration is mixed. *Ceanothus* can be competitive with Douglas-fir (Peterson *et al.*, 1988); it also may provide protection from animal damage during seedling establishment (Youngberg *et al.*, 1979). Considering red alder's overwhelming economic importance, as a weed and as a crop, the rest of this discussion on forest management will focus on red alder. We will return at the end, however, to review interplanting actinorhizal shrubs in conifer plantations.

A. Reasons for Management

Red alder is planted and managed for three reasons. It is a disease-resistant crop in areas infected with conifer root diseases. It is a valuable pulp and timber species. As a nitrogen fixer, it can improve the growth of associated species.

Laminated root rot (*Phellinus weirii*) is a pathogenic fungus of Douglas-fir and many other conifers in the Pacific Northwest. An estimated 32 million cubic feet of Douglas-fir is lost each year in Oregon and Washington to the disease (Childs and Shea, 1967). Red alder, like all hardwoods, is immune. Thus, it can be an alternative crop in infected areas.

Nelson *et al.* (1978) review the research regarding the interactions between red alder and the fungus. First, it is well known that alder changes soil properties by adding nitrate, fatty acids, and phenolic compounds, by reducing pH, and by increasing the activity of microbial antagonists to *P. weirii*. Second, in the laboratory, some of these changes can inhibit fungal growth. In the field, however, it has been difficult to separate direct effects of alder from the indirect one of removing the host. Following harvest of an infected stand, fungal distribution becomes increasingly limited as first Douglas-fir roots and then stumps decay. In 50 years, the incidence of infected stumps can be reduced to 25% of the original number (Hansen, 1976). This reduction in fungal incidence has not been directly associated with the presence of alder.

Most debates about the alternatives of growing red alder or a conifer as a timber crop are abstract (Atkinson and Hamilton, 1978; Tarrant *et al.*, 1983). They involve comparisons of alder yield table estimates (Worthington *et al.*, 1961; Chambers, 1974) with Douglas-fir yield tables or growth models. For red alder, they also have involved estimates of growth response to thinning. The Cascade Head study (Table I) provides one side-by-side comparison. Here, it appears that alder can outproduce conifers only in short rotations. Atterbury (1978) came to a similar conclusion for high site quality land in the Coast Range.

B. Short Rotation Management

Two types of short rotation management systems have been proposed and explored in a limited way. DeBell (1975; DeBell *et al.*, 1978) has examined very short rotation (2 years) coppice systems and found yields on the order of 5.6 oven-dry tons per ha per year. DeBell and co-workers are continuing to explore the effects of spacing, fertilizer, and rotation length. Recent work has shown that resprouting after cutting is vigorous at this age (Harrington, 1984; Hoyer and Belz, 1984; DeBell and Turpin, 1989). The number of times alder can be recut is unknown. In one study, Harrington and DeBell (1984) found 54.8% of the plants dead after four 2-year cutting cycles. Technical questions remain regarding efficient planting and harvesting systems.

DeBell *et al.* (1978) also discuss a short pulpwood log management system. Mean annual increment in red alder stands peaks at age 10 to 15, depending on site quality and stocking (Worthington *et al.*, 1961; DeBell, 1972; Zavitkovski and Stevens, 1972). With short rotations, spacing control, and weed competition reduction, DeBell *et al.* (1978) estimate that yields almost double those given in normal yield tables can be obtained, thus, producing a 15-cm-dbh (diameter at 1.3 m) tree in 10 years on a site index 37 m at 50 years.

C. Sawlog Management

Red alder has a long history of studies in density management for sawlog production (Warrack, 1949, 1964; Berntsen, 1961, 1962; Hibbs *et al.*, 1989). Although stocking guides do exist (Hibbs, 1987a; Hibbs and Carlton, 1989), these studies have not systematically addressed density and timing of spacing. Because of the high value of knot-free wood, questions about the effects of density and timing of thinning on wood quality are of great economic importance. Research being conducted by DeBell and Hibbs will provide most of the answers to these questions.

DeBell *et al.* (1978) estimated alder thinning response from the response of other species and predicted that 30-cm-dbh trees could be grown in 27–28 years in plantations (Atkinson *et al.*, 1979; Tarrant *et al.*, 1983). Hibbs *et al.* (1989) thinned a 14-year-old stand and found that diameter growth rate doubled to 1 cm per year for at least the first 5 years after thinning. This growth rate could result in a 30-cm-dbh tree at 29 years. In spite of this early rapid growth, the most recent economic comparison of red alder and Douglas-fir management shows alder trailing Douglas-fir (Tarrant *et al.*, 1983). Recent increases in alder value may have changed this comparison (R. F. Tarrant, personal communication).

D. Mixed-Species Plantations

To utilize the nitrogen-fixing capacity of red alder to increase the production of other tree species, management systems that either mix or alternate alder with another species have been considered.

Alternating alder and conifer crops has received the least experimental attention. A recent study of red alder and Douglas-fir at one location in Washington has shown that each species grows better following the other than following itself in sequential rotations (Cole, 1988).

In a theoretical analysis of alternating crops, Tarrant *et al.* (1983) estimated the fertilizer effect of several alder rotation lengths on subsequent Douglas-fir growth. They applied management costs and product values and compared alternate rotations of several lengths with conifer culture, with and without chemical fertilizer. Among the tested alternatives, they found the greatest present net worth with 45-year, fertilized Douglas-fir rotations. A close second, however, was alternating 28-year alder and 45-year Douglas-fir rotations.

In mixed-culture systems, the Wind River (Miller and Murray, 1978) and Cascade Head (Berntsen, 1961) studies described earlier appear to delimit the extremes of results. Atkinson and Hamilton (1978) calculate the number of alder trees needed to supply all the nitrogen needs of a conifer crop. From this, they calculate the cost of fertilization by alder and compare this figure with the cost of application of chemical fertilizer. They conclude that mixing 500 alder per ha in the conifer plantation for 25 years will provide fertilizer at less than half the cost of chemical fertilization. Unfortunately, they do not include in this cost comparison any of the conifer growth loss that will occur as a result of shading.

Newton *et al.* (1968) have analyzed the height growth patterns of red alder and Douglas-fir over a range of site conditions and proposed a set of adjustment periods necessary to successful mixing of these two species. They divided site conditions into wet and dry, and into Coast, Valley, and Cascade. In comparing height growth curves, they show that delays in alder establishment of 4 to 8 years are necessary to prevent overtopping of the Douglas-fir. The longest delays are required on Coastal wet sites.

Mixing red alder and black cottonwood (*Populus trichocarpa*) in a 2-year coppice rotation has been tested (DeBell and Radwan, 1979). Yields of mixed plantings yielded 19,510 kg/ha (dry weight, minus leaves), 71% higher than pure cottonwood and 55% higher than alder. The yield increase was confined to the cottonwood component of the mix, indicating a nitrogen response. Alder growth was not reduced in mixture. In longer rotations, early-dominant alder is eventually overtopped by the cottonwood (Pezeshki and Oliver, 1985).

Mixing actinorhizal shrubs in conifer plantations has the short-term benefit of nitrogen input without the long-term hazard of conifer overtopping and suppression that can occur when red alder is used. Sitka alder, redstem (*C. sanguineus*), and snowbrush (*C. velutinus*) have been considered for mixed plantations. In all cases, high shrub density is expected to result in some crop tree growth reduction (Harrington and Deal, 1982; Conard *et al.*, 1985; Peterson *et al.*, 1988). Harrington and Deal (1982) examined height growth of Sitka alder over a range of site conditions and compared this growth with that of Douglas-fir. Their method of analysis (harmonized growth curves for each species by site quality) allows a land manager to compare the different competitive regimes created by using different conifer stock types or delays in Sitka alder establishment. They conclude that mixtures can increase conifer growth and that, on sites of poorer quality, Douglas-fir should be given a head start to reduce competitive growth loss. Binkley *et al.* (1984) examined one natural example of Sitka alder in a Douglas-fir plantation on Vancouver Island, B.C., and found that the presence of Sitka alder increased conifer biomass growth by 40%.

Shrubs can increase conifer seedling survival and growth on some sites through environmental amelioration. On an Oregon Cascades site, Youngberg *et al.* (1979) found that increased Douglas-fir survival and growth resulted from the reduced soil and air temperature, increased soil moisture in the top 15 cm, and reduced animal browsing, all in association with snowbrush. They did not discount the nitrogen fertilization effect. Rather, it could not be separated from the other environmental changes. At another location in the Oregon Cascades, Peterson *et al.* (1988) came to a different conclusion. They removed snowbrush cover in eight plantations and found that Douglas-fir seedling growth improved. One could suggest that the snowbrush had increased the soil N pool by this time, and its removal also removed its competitive effect on the Douglas-fir.

IV. Utilization of Red Alder

A. The Alder Resource

Information on the red alder resource (Bassett and Oswald, 1981a,b, 1982; Gedney *et al.*, 1986a,b, 1987; Chester, 1988) is at least 10 years old, has large sampling errors, and omits a critical part of the resource dynamics. Hence, this overview of the resource and its dynamics should be viewed with caution.

Forest stands classified as being primarily alder cover more than 1.2 million ha in the Pacific Northwest. Most of this land is held in private ownership. The alder volume is 238 million cubic meters, with almost half of this volume being found in stands that are less than 50% alder by volume. Growth of alder (NPP minus mortality) in Oregon and Washington averages 3.2%. Harvest removals (Fig. 2) for Oregon and Washington are estimated at 1.9 million cubic meters, about 25% of growth. However, this removal estimate does not include whole tree chipping, a serious omission (Beachy and McMahon, 1985). Harvest levels in British Columbia are low but growing.

There are two weaknesses in this picture. First, harvest rates have increased greatly since these figures were compiled. Second, the age-class distribution is not even. Most stands are in the 30- to 60-year-old

Figure 2 Red alder logs being delivered to a mill in the Oregon Coast Range.

age range. When these stands are harvested, they are replanted to conifers, not alder. As a result, the area of alder forest is decreasing. Both of these factors cause the harvest to represent an ever-increasing portion of growth.

B. Alder Products

The highest value use for red alder is found in the furniture and cabinet industry. Resch (1980) states that the manufacturing process adds $1,186 per thousand board feet to the value of the raw material. Alder lumber is prized for its even texture, moderate density, strength, and excellent working characteristics (turning, gluing, finishing, staining, bending).

Alder has found a ready place in the paper industry (Hrutfiord, 1978). Both the domestic and export markets currently compete strongly with the sawlog industry for the resource.

Probably the oldest use of red alder is for firewood. The demand is strong in urban areas, but competition with pulp chip and sawlog markets is reducing its availability. Alder could play a role in regional biomass energy programs (Smith, 1978).

Alder has occasionally been turned for veneer, both core stock in structural plywood and face veneer for interior paneling. Alder could be used in structural flake boards (Maloney, 1978).

V. Long-Term Productivity

Maintenance of long-term site productivity is a topic of considerable interest in both agriculture and forestry (Wild, 1988; Perry and Maghembe, 1989). Interest in long-term site productivity stems, in part, from the perspective of 150 years of experience in agriculture, where data exist for productivity of experimental plots under a variety of treatments (Wild, 1988). Changes in yield have occurred under fertilizer or manure treatments in contrast to unamended controls (Wild, 1988). Under the very long time perspective of millenia, there is evidence of productivity declines in some forest ecosystems. For example, on the deeply weathered soil occurring in Australian dune ecosystems, productivity may be limited by availability of essential nutrients such as phosphorus (Walker *et al.*, 1981). In this case, total ecosystem productivity declined by a factor of approximately 10 between the youngest and oldest dune ecosystems, paralleling a decline in soil phosphorus.

Although actinorhizal species are important as a source of nitrogen, their presence would not prevent long-term limitations of other nutrients. These secondary limits can result in productivity declines and retrogressive succession in ecosystems like the Australian dune example (Walker *et al.*, 1981). Although phosphorus is generally not limiting in the Pacific Northwest, there is evidence that nitrogen-fixing understory plants can recycle greater amounts of phosphorus in litter and thus influence phosphorus cycling on phosphorus in litter and thus influence phosphorus cycling on phosphorus-limited sites. This has been seen in eucalypt forests of Australia (O'Connell, 1986). Thus, actinorhizal plants may be able to mitigate some of the effects of limitations on nutrients other than nitrogen.

In the Pacific Northwest, greater retention of anions such as PO_4^- and SO_4^- can occur in soils colonized by red alder (Johnson *et al.*, 1986). This greater retention is apparently due to a greater soil content of amorphous iron and aluminum and less of the crystalline forms of these elements. This results in greater retention of phosphorus and sulfur from organic ligand formation (Johnson *et al.*, 1986). Therefore, these elements would be more available when alder is or has recently been present on a site.

In long-term studies, there is an opportunity to confuse anthropogenic and autogenic changes. The current levels of industrial emissions of nitrogen and sulfur in many parts of the world can result in nutrient imbalance or other soil effects such as increased soil acidity. This ultimately would decrease forest productivity (Waring and Schlesinger, 1985). However, such results from cases of environmental pollution will have to be distinguished from cases of natural soil acidification such as found under nitrogen-fixing species like alder (Crocker and Major, 1955; Franklin *et al.*, 1968), including effects of acidification on nutrient cycling and production.

VI. Nontimber Values

A. Riparian Areas

Red alder is a frequent dominant in riparian areas because of its tolerance of moister sites (Minore, 1970; Minore and Smith, 1971) and the seedbeds created by stream erosional and depositional processes. Most streambanks today are composed of intermixed patches or trees of alder and conifers. This forest cover provides shade and large woody debris to the stream.

Both natural succession and riparian area buffer strip practices tend to increase understory shrub density and decrease tree regeneration (Hibbs, 1987b), particularly in alder-dominated reaches. This has led to a growing social concern that the very regulations that were intended to preserve riparian area values for fish, wildlife, and water quality through limiting human impact may be accelerating its degeneration.

B. Wildlife

The seeds, twigs, and foliage of *Ceanothus* species are an important food source for domestic stock and wildlife (Hickey and Leege, 1970; Leege, 1979; Conard *et al.*, 1985). Redstem (*C. sanguineus*) and deerbrush (*C. integerrimus*) are the most important food sources. Up to 80% of the annual growth of deerbrush may be consumed in a year by elk and deer (McCulloch, 1955; Thilenius and Hungerford, 1967). Snowbrush is a poor browse species.

The foliage, buds, and twigs of *Alnus* are eaten by deer and elk (Dayton, 1931; Cowan, 1945), although other common species rank as high or higher in preference. Alder in riparian zones is collected by beaver for food and structural purposes (Maser *et al.*, 1981). Mountain beaver (*Aplodontia rufa*) show little preference for alder (Crouch, 1968; Voth and Black, 1973).

VII. Summary

Actinorhizal plants play an important role in the nitrogen budgets of the Pacific Northwest. All the actinorhizal species are pioneer species that colonize newly disturbed sites. They are commonly found in riparian areas, avalanche tracks, and dry, fire-prone slopes and ridges. Recent forest management activity has increased their occurrence greatly. Where they occur, they can improve soil fertility by increasing soil nitrogen, organic matter, and cation-exchange capacity. They can also be an important wildlife food source.

In forestry, red alder is harvested as a timber crop. Some deliberate regeneration of red alder is now being practiced. Red alder and other of the region's actinorhizal plants have been considered for mixed culture with conifers. Such mixtures are theoretically possible but are currently not practiced. Mixed culture with shrubs is clearly easier than mixed culture with red alder. However, red alder offers the double benefit of nitrogen fixation and a timber crop.

Acknowledgments

Contribution No. 2555, Forest Research Laboratory, Oregon State University, Corvallis, OR. This work was supported in part by USDA Grant 84-CRCR-1-1434.

References

Atkinson, W. A., and Hamilton, W. I. (1978). The value of red alder as a source of nitrogen in Douglas-fir/alder mixed stands. *U.S. Dep. Agric., For. Serv. Gen. Rep.* **PNW-70,** 337–351.

Atkinson, W. A., Bormann, B. T., and DeBell, D. S. (1979). Crop rotation of Douglas-fir and red alder: A preliminary biological and economic assessment. *Bot. Gaz. (Chicago), Suppl.* **140,** S102–S107.

Atterbury, T. (1978). Alder characteristics as they affect utilization. *U.S. Dep. Agric., For. Serv. Gen. Tech. Rep.* **PNW-70,** 71–81.

Bassett, P. M., and Oswald, D. D. (1981a). Timber resource statistics for southwest Washington. *U.S. Dep. Agric., For. Serv. Resour. Bull.* **PNW-91,** 24 pp.

Bassett, P. M., and Oswald, D. D. (1981b). Timber resource statistics for the Olympic Peninsula, Washington. *U.S. Dep. Agric., For. Serv., Resour. Bull.* **PNW-93,** 31 pp.

Bassett, P. M., and Oswald, D. D. (1982). Timber resource statistics for the Puget Sound Area, Washington. *U.S. Dep. Agric., For. Serv., Resour. Bull.* **PNW-96,** 31pp.

Beachy, D. L., and McMahon, R. O. (1985). "Economic Value of the Pacific Northwest Hardwood Industry in 1985," a report sponsored by the Western Hardwood Association, 53 pp. For. Res. Lab., Oregon State Univ., Corvallis, Oregon.

Berg, A., and Doerksen, A. (1975). "Natural Fertilization of a Heavily Thinned Douglas-Fir Stand by Understory Red Alder," Res. Note 56, 3 pp. For. Res. Lab., Oregon State Univ., Corvallis, Oregon.

Berntsen, C. M. (1961). Growth and development of red alder compared with conifers in 30-year-old stands. *U.S. Dep. Agric., For. Serv., Res. Pap.* **PNW-38,** 20 pp.

Berntsen, C. M. (1962). A 20-year growth record for three stands of red alder. *U.S. Dep. Agric., For. Serv. Res. Pap.* **PNW-36,** 9 pp.

Binkley, D. (1981). Nodule biomass and acetylene reduction rates of red alder and Sitka alder on Vancouver Island, B.C. *Can J. For. Res.* **11,** 281–286.

Binkley, D. (1982). Nitrogen fixation and net primary production in a young Sitka alder stand. *Can. J. For. Res.* **12,** 281–284.

Binkley, D. (1983). Ecosystem production in Douglas-fir plantations: Interaction of red alder and site fertility. *For. Ecol. Manage.* **5,** 215–227.

Binkley, D. (1986). "Forest Nutrition," 290 pp. Wiley, New York.

Binkley, D., and Husted, L. (1983). Nitrogen accretion, soil fertility, and Douglas-fir nutrition in association with redstem ceanothus. *Can. J. For. Res.* **13,** 122–125.

Binkley, D., Cromack, K., Jr., and Fredrikson, R. L. (1982). Nitrogen accretion and availability in some snowbrush ecosystems. *For. Sci.* **28,** 720–724.

Binkley, D., Lousier, J. D., and Cromack, K., Jr. (1984). Ecosystem effects of Sitka alder in a Douglas-fir plantation. *For. Sci.* **30,** 26–35.

Bormann, B. T., and DeBell, D. S. (1981). Nitrogen content and other soil properties related to age of red alder stands. *Soil Sci. Soc. Am. J.* **45,** 428–432.

Bormann, B. T., and Gordon, J. C. (1984). Stand density effects in a young red alder plantation: Production, photosynthate partitioning and nitrogen fixation. *Ecology* **65,** 394–402.

Carlton, G. C. (1988). "The Structure and Dynamics of Red Alder Communities in the Central Coast Range of Western Oregon," M.S. thesis, 173 pp. Oregon State Univ., Corvallis, Oregon.

Chambers, C. J. (1974). Empirical yield tables for predominantly alder stands in western Washington. *Wash., Dep. Nat. Resour., DNR Rep.* **31,** 70 pp.

Chester, R. A. (1988). "Management and Use of Red Alder in the Vancouver Forest Region," M.S. thesis, 268 pp. Simon Fraser Univ., Vancouver, British Columbia, Canada.

Childs, T. W., and Shea, K. R. (1967). Annual losses from diseases in Pacific Northwest forests. *U.S. Dep. Agric., For. Serv. Resour. Bull.* **PNW-20,** 19 pp.

Cole, D. W. (1988). "Forest Ecosystem Investigations at the Thompson Research Center: Soil Processes, Nutrient Cycling, Atmospheric Deposition," 20 pp. Coll. For. Resour., Univ. Washington, Seattle, Washington.

Cole, D. W., Gessel, S. P., and Turner, J. (1978). Comparative mineral cycling in red alder and Douglas-fir. *U.S. Dep. Agric., For. Serv. Gen. Tech. Rep.* **PNW-70,** 327–336.

Conard, S. G., Jaramillo, A. E., Cromack, K., Jr., and Rose, S. (1985). The role of the genus *Ceanothus* in western forest ecosystems. *U.S. Dep. Agric., For. Serv., Gen. Tech. Rep.* **PNW-182,** 72 pp.

Cowan, I. M. (1945). The ecological relationships of the food of the Columbian black-tailed deer, *Odocoileus hemionus columbianus* (Richardson), in the coast forest region of southern Vancouver Island, British Columbia. *Ecol. Monogr.* **15,** 109–139.

Crocker, R. L., and Major, J. (1955). Soil development in relation to vegetation and surface age at Glacier Bay, Alaska. *J. Ecol.* **43,** 427–448.

Cromack, K., Jr., Delwiche, C. C, and McNabb, D. H. (1979). Prospects and problems of nitrogen management using symbiotic nitrogen fixers. *In* "Symbiotic Nitrogen Fixation in the Management of Temperate Forests" (J. C. Gordon, C. T. Wheeler, and D. A. Perry, eds.), pp. 210–223. For. Res. Lab., Oregon State Univ., Corvallis, Oregon.

Crouch, G. L. (1968). Clipping of woody plants by mountain beaver. *J. Mammal.* **49,** 151–152.

Dalton, D. A., and Zobel, D. B. (1977). Ecological aspects of nitrogen fixation by *Purshia tridentata*. *Plant Soil* **48,** 57–80.

Davis, M. B. (1981). Quaternary history and stability of forest communities. *In* "Forest Succession: Concepts and Application" (D. C. West, H. H. Shugart, and D. B. Botkin, eds.), pp. 132–153. Springer-Verlag, Berlin and New York.

Dayton, W. A. (1931). Important winter browse plants. *U.S. Dep. Agric., Misc. Publ.* **101,** 214 pp.

DeBell, D. S. (1972). "Potential Productivity of Dense, Young Thickets of Red Alder," For. Res. Note 2, 7 pp. Crown Zellerbach Corp., Camas, Washington.

DeBell, D. S. (1975). Short-rotation culture of hardwoods in the Pacific Northwest. *Iowa State J. Res.* **49,** 345–352.

DeBell, D. S., and Radwan, M. A. (1979). Growth and nitrogen relations of coppiced black cottonwood and red alder in pure and mixed plantings. *Bot. Gaz. (Chicago), Suppl.* **140,** S97–S101.

DeBell, D. S., and Turpin, T. C. (1989). Control of red alder by cutting. *U.S. Dep. Agric., For. Serv., Res. Pap.* **PNW-RP-414.** 10 pp., in press.

DeBell, D. S., Strand, R. F., and Reukema, D. L. (1978). Short-rotation production of red alder: Some options for future forest management. *U.S. Dep. Agric., For. Serv., Gen. Tech. Rep.* **PNW-70,** 231–244.

Franklin, J. F., and Dyrness, C. T. (1973). Natural vegetation of Oregon and Washington. *U.S. Dep. Agric., For. Serv. Gen. Tech. Rep.* **PNW-8,** 417 pp.

Franklin, J. F., and Hemstrom, M. A. (1981). Aspects of succession in the coniferous forest of the Pacific Northwest. *In* "Forest Succession: Concepts and Application" (D. C. West, H. H . Shugart, and D. B. Botkin, eds.), pp. 212–229. Springer-Verlag, Berlin and New York.

Franklin, J. F., Dyrness, C. T., Moore, D. G., and Tarrant, R. F. (1968). Chemical soil properties under coastal Oregon stands of alder and conifers. *In* "Biology of Alder" (J. M. Trappe, J. F. Franklin, R. F. Tarrant, and G. M. Hansen, eds.), pp. 157–172. U.S. Dep. Agric., For. Serv., Pacific Northwest For. Range Exp. Stn., Portland, Oregon.

Gedney, D. R., Bassett, P. M., and Mei, M. A. (1986a). Timber resource statistics for non-federal forest land in northwest Oregon. *U.S. Dep. Agric., For. Serv., Resour. Bull.* **PNW-140,** 26 pp.

Gedney, D. R., Bassett, P. M., and Mei, M. A. (1986b). Timber resource statistics for non-federal forest land in southwest Oregon. *U.S. Dep. Agric., For. Serv. Resour. Bull.* **PNW-138,** 26 pp.

Gedney, D. R., Bassett, P. M., and Mei, M. A. (1987). Timber resource statistics for non-federal forest land in west-central Oregon. *U.S. Dep. Agric., For. Serv. Resour. Bull. PNW-143,* 26 pp.

Hansen, E. M. (1976). Twenty year survival of *Phellinus weirii* in Douglas-fir stumps. *Can. J. For. Res.* **6,** 123–128.

Harrington, C. A. (1984). Factors influencing initial sprouting of red alder. *Can. J. For. Res.* **14,** 357–361.

Harrington, C. A., and Deal, R. L. (1982). Sitka alder, a candidate for mixed stands. *Can. J. For. Res.* **12,** 108–111.

Harrington, C. A., and DeBell, D. S. (1984). Effects of irrigation, pulp mill sludge and repeated cutting on growth and yield of black cottonwood and red alder. *Can J. For. Res.* **14,** 844–849.

Heilman, P., and Ekuan, G. (1982). Nodulation and nitrogen fixation by red alder and Sitka alder on coal mine spoils. *Can J. For. Res.* **12,** 992–997.

Helgerson, O., Wheeler, C. T., Perry, D. A., and Gordon, J. C. (1979). Annual nitrogen fixation in Scotch broom (*Cytisus scoparius*). *In* "Symbiotic Nitrogen Fixation in the Management of Temperate Forests" (J. C. Gordon, C. T. Wheeler, and D. A. Perry, eds.), p. 477. For. Res. Lab., Oregon State Univ., Corvallis, Oregon.

Henderson, J. A. (1970). "Biomass and the Composition of the Understory Vegetation in Some *Alnus rubra* Stands in Western Oregon," M.S. thesis, 64 pp. Oregon State Univ., Corvallis, Oregon.

Hibbs, D. E. (1987a). The self-thinning rule and red alder management. *For. Ecol. Manage.* **18,** 273–281.

Hibbs, D. E. (1987b). Management of riparian zone vegetation in western Oregon. *In* "Managing Oregon's Riparian Zone for Timber, Fish and Wildlife," Tech. Bull. 514, pp. 57–63. Nat. Counc. Pap. Ind. Air Stream Improv., New York.

Hibbs, D. E., and Carlton, G. (1989). Comparing volume- and diameter-based stocking guides using red alder. *West J. Appl. For.*, **4**(4), 113–115.

Hibbs, D. E., Emmingham, W., and Bondi, M. (1989). Thinning red alder: Effects of method and spacing. *For. Sci.*, **35**(1), 16–29.

Hickey, W. O., and Leege, T. A. (1970). Ecology and management of redstem ceanothus—A review. *Idaho Dep. Fish Game, Wildl. Bull.* **4,** 18 pp.

Hoyer, G. E., and Belz, D. (1984). Stump sprouting related to time of cutting red alder. *Wash., Dep. Nat. Resour., DNR Rep.* **45,** 17 pp.

Hruitfiord, B. F. (1978). Red alder as a pulpwood species. *U.S. Dep. Agric., For. Serv., Gen. Tech. Rep.* **PNW-70,** 135–138.

Johnson, D. W., van Miegroet, H., and Cole, D. W. (1986). Factors affecting anion retention in four forest soils. *Soil Sci. Soc. Am. J.* **50,** 776–783.

Kim, D. Y. (1987)."Seasonal Estimates of Nitrogen Fixation by *Alnus rubra* and *Ceanothus* Species in Western Oregon Forest Ecosystems," M.S. thesis, 71 pp. Oregon State Univ., Corvallis, Oregon.

Kim, D. Y., Cromack, K., and Hibbs, D. E. (1990). Seasonal estimates of nitrogen fixation by red alder in western Oregon forest ecosystems. Submitted for publication.

Lawerence, D. B. (1958). Glaciers and vegetation in southeastern Alaska. *Am. Sci.* **46,** 81–122.

Ledgard, S. F., Feney, J. R., and Simpson, J.R. (1984). Variations in natural enrichment of ^{15}N in the profiles of some Australian pasture soils. *Aust. J. Soil Res.* **22,** 155–164.

Leege, T. A. (1979). Growth and dieback of redstem (*Ceanothus sanguinius*) in Idaho. *Northwest Sci.* **49,** 58–64.

Maloney, T. M. (1978). Alder: One of tomorrow's important structural materials? *U.S. Dep. Agric., For. Serv., Gen. Tech. Rep.* **PNW-70,** 125–132.

Maser, C., Male, B. R., Franklin, J. F., and Dyrness, C. T. (1981). Natural history of Oregon coast mammals. *U.S. Dep. Agric., For. Serv., Gen. Tech. Rep.* **PNW-133.** 496 pp.

McCulloch, C. Y., Jr. (1955). Utilization of winter browse on wilderness big game range. *J. Wildl. Manage.* **19,** 206–215.

McNabb, D. H., and Cromack, K. (1983). Dinitrogen fixation by a mature *Ceanothus velutinus* (Dougl.) stand in the western Oregon Cascades. *Can. J. Microbiol.* **29,** 1014–1072.

McNabb, D. H., Geist, J. M., and Youngberg, C. T. (1979). Nitrogen fixation by *Ceanothus velutinus* in northeastern Oregon. *In* "Symbiotic Nitrogen Fixation in the Management of Temperate Forests" (J. C. Gordon, C. T. Wheeler, and D. A. Perry, eds.), pp. 481–482. For. Res. Lab., Oregon State Univ., Corvallis, Oregon.

Miller, R. E., and Murray, M. D. (1978). The effects of red alder on Douglas-fir growth. *U.S. Dep. Agric., For. Serv., Gen. Tech. Rep.* **PNW-70,** 283–306.

Minore, D. (1970). Seedling growth of eight Northwestern tree species over three water tables. *U.S. Dep. Agric., For. Serv., Res. Note* **PNW-115,** 8 pp.

Minore, D., and Smith, C. E. (1971). Occurrence and growth of four Northwestern tree species over shallow water tables. *U.S. Dep. Agric., For. Serv., Res. Note* **PNW-160,** 9 pp.

Nelson, E. E., Hansen, E. M., Li, C. Y., and Trappe, J. M. (1978). The role of red alder in

reducing losses from laminated root rot. *U.S. Dep. Agric., For. Serv., Gen. Tech. Rep.* **PNW-70,** 273–282.
Newton, M., el Hassen, B. A., and Zavitkovski, J. (1968). Role of red alder in western forest succession. *In* "Biology of Alder" (J. M. Trappe, J. F. Franklin, R. F. Tarrant, and G. M. Hansen, eds.), pp. 73–83. U.S. Dep. Agric., For. Serv., Pacific Northwest For. Range Exp. Stn., Portland, Oregon.
Oades, J. M. (1988). The retention of organic matter in soils. *Biogeochemistry* **5,** 35–70.
O'Connell, A. M. (1986). Effects of legume understory on decomposition and nutrient content of eucalypt forest litter. *Plant Soil* **92,** 235–248.
Perry, D. A., and Maghembe, J. (1989). Ecosystem concepts and current trends in forest management: Time for reappraisal. *For. Ecol. Manage.* **26,** 123–140.
Peterson, T. D., Newton, M., and Zedaker, S. M. (1988). Influence of *Ceanothus velutinus* and associated forbs on the water stress and stemwood production of Douglas-fir. *For. Sci.* **34,** 333–343.
Pezeshki, S. R., and Oliver, C. D. (1985). Early growth patterns of red alder and black cottonwood in mixed species plantation. *For. Sci.* **31,** 190–200.
Resch, H. (1980). Utilization of red alder in the Pacific Northwest. *For. Prod. J.* **30,** 21–26.
Shumway, J., and Atkinson, W. A. (1978). Predicting nitrogen fertilizer response in unthinned stands of Douglas-fir. *Commun. Soil Sci. Plant Anal.* **9,** 529–539.
Simard, S. W. (1989). "Competition among Lodgepole Pine Seedlings and Plant Species in a Sitka Alder-Dominated Shrub Community in the Southern Interior of British Columbia," M.S. thesis, 143 pp. Oregon State Univ., Corvallis, Oregon.
Smith, N. J. (1978). Red alder as a potential source of energy. *U.S. Dep. Agric., For. Serv., Gen. Tech. Rep.* **PNW-70,** 139–156.
Sollins, P., Spycher, G., and Glassman, C. A. (1984). Net nitrogen mineralization from light and heavy fraction forest floor organic matter. *Soil Biol. Biochem.* **16,** 31–37.
Stevenson, F. J. (1982). "Humus Chemistry," 443 pp. Wiley, New York.
Tarrant, R. F., and Miller, R. E. (1963). Accumulation of organic matter and soil nitrogen beneath a plantation of red alder and Douglas-fir. *Soil Sci. Soc. Am. Proc.* **27,** 231–234.
Tarrant, R. F., Bormann, B. T., DeBell, D. S., and Atkinson, W. A. (1983). Managing red alder in the Douglas-fir region: Some possibilities. *J. For.* **81,** 787–792.
Thilenius, J. F., and Hungerford, K. E. (1967). Browse use by cattle and deer in northern Idaho. *J. Wildl. Manage.* **31,** 141–145.
Tripp, L. N., Franklin, J. F., and Heilman, O. E. (1979). Seasonal and diurnal patterns and rates of nitrogen fixation by young red alder. *For. Sci.* **25,** 372–380.
Ugolini, F. C. (1968). Soil development and alder invasion in a recently deglaciated area of Glacer Bay, Alaska. *In* "Biology of Alder" (J. M. Trappe, J. F. Franklin, R. F. Tarrant, and G. M. Hansen, eds.), pp. 115–140. U.S. Dep. Agric., For. Serv., Pacific Northwest For. Range Exp. Stn., Portland, Oregon.
Van Cleve, K., and Viereck, L. A. (1981). Forest succession in relation to nutrient cycling in the boreal forest of Alaska. *In* "Forest Succession: Concepts and Application" (D. C. West, H. H. Shugart, and D. B. Botkin, eds.), pp. 185–210. Springer-Verlag, Berlin and New York.
Van Cleve, K., Viereck, L. A., and Schlenter, R. L. (1971) Accumulation of nitrogen in alder (*Alnus*) ecosystems near Fairbanks, Alaska. *Arct. Alp. Res.* **3,** 101–114.
Voight, G. K., and Steucek, G. L. (1969). Nitrogen distribution and accretion in an alder ecosystem. *Soil Sci. Soc. Am. J.* **33,** 946–949.
Voth, E. H., and Black, H. C. (1973). A historic technique for determining feeding habits of small herbivores. *J. Wildl. Manage.* **37,** 223–231.
Walker, J., Thompson, C. H., Fergus, I. F., and Tunstall, B. R. (1981). Plant succession and soil development in coastal sand dunes of subtropical eastern Australia. *In* "Forest

Succession: Concepts and Application" (D. C. West, H. H. Shugart, and D. B. Botkin, eds.), pp. 107–131. Springer-Verlag, Berlin and New York.

Waring, R. H., and Schlesinger, W. H. (1985). "Forest Ecosystems: Concepts and Management," 340 pp. Academic Press, Orlando, Florida.

Warrack, G. C. (1949). Treatment of red alder in the coastal region of British Columbia. *B.C., For. Serv., Res. Note* **14,** 7 pp.

Warrack, G. C. (1964). "Thinning Effects in Red Alders," paper presented to the annual meeting of the Pacific Northwest Hardwood Association, Port Angeles, Washington, 7 pp.

Wild, A. (1988). "Russells's Soil Conditions and Plant Growth," 991 pp. Wiley, New York.

Worthington, N. P., Johnson, F. A., Staebler, G. R., and Lloyd, W. J. (1961). Normal yield tables for red alder. *U.S. Dep. Agric., For. Serv., Res. Pap.* **PNW-36,** 32 pp.

Youngberg, C. T., and Wollum, A. G. (1976). Nitrogen accretion in developing *Ceanothus velutinus* stands. *Soil Sci. Soc. Am. J.* **40,** 109–112.

Youngberg, C. T., Wollum, A. G., and Scott, W. (1979). *Ceanothus* in Douglas-fir clear-cuts: Nitrogen accretion and impact on regeneration. *In* "Symbiotic Nitrogen Fixation in the Management of Temperate Forests" (J. C. Gordon, C. T. Wheeler, and D. A. Perry, eds.), pp. 224–233. For. Res. Lab., Oregon State Univ., Corvallis, Oregon.

Zavitkovski, J., and Newton, M. (1968). Ecological importance of snowbrush, *Ceanothus velutinus,* in the Oregon Cascades. *Ecology* **49,** 1134–1145.

Zavitkovski, J., and Stevens, R. D. (1972). Primary productivity of red alder ecosystems. *Ecology* **53,** 235–242.

18

Current and Potential Uses of Actinorhizal Plants in Europe

C. T. Wheeler and I. M. Miller
Department of Botany
University of Glasgow
Glasgow G12 8QQ, Scotland, United Kingdom

I. Introduction

Records of the regard with which actinorhizal plants, in particular alders, have been held in Europe can be traced back to the Roman Empire and poetry from writers such as Virgil have often been quoted to extol the virtues of these plants (Tarrant and Trappe, 1971). Those interested in early accounts of native and introduced actinorhizal plants, specifically of Britain but including wide reference to other countries in Europe and elsewhere, would benefit from consulting Loudon's *Arboretum et Fruticetum Britannicum* of 1838, where intriguing accounts of the culture, management, and uses of these and of other trees and shrubs of temperate climates are to be found. It is clear from these volumes that the genus *Alnus,* in past as in present times, offered the largest number of useful species for temperate/warm temperate climates. Its utility as a source of dyes, to provide wood for supports for construction work over water, and for water pipes, turnery, cabinet manufacture, barrel production, baskets, poles, eating vessels, clogs, and shoe soles is reported. It was recommended that alders be planted to form hedges in moist meadows and to prevent erosion of riverbanks; the improvement of impoverished soils following the growth of these species is also implied. One early practice was the spreading of leafy alder branches over the fields in summer so that the soil is improved for crop production the subsequent spring following their decay. Concoctions were prepared from the leaves to treat throat infections and, in common with *Myrica gale* (bog myrtle or sweet gale), the young shoots were used to deter flea infestations of houses. Although alder is not a fuelwood of quality for domestic heating, it was deemed satisfactory for industrial furnaces and gave charcoal suitable for the manufacture of high-quality gunpowder and for the operation of smelters and furnaces. Large alder coppices were maintained in Europe from the seventeenth to the twentieth century for these purposes.

Attractive horticultural varieties, such as the cut-leaved alder (*A. glutinosa* var. *laciniata*), are mentioned and specimen nursery price lists for seed and plants are given—a wide range of indigenous and exotic species of alder and of other genera of actinorhizal plants could be obtained cheaply from European nurserymen. A quote from a Mr. Boutcher suggests that not all growers were happy with their purchases, however: "Vast quantities [of alders] were brought from Holland to Scotland at a considerable price and planted in large tracts of moist land from which no returns, suitable to the labour and expense, has been received" (Loudon, 1838). This gentleman went on later to characterize the alder as "an ugly, melancholy tree." Irrespective of such unfortunate experience, Loudon (1838) provides perspective on the utilization of alder in these earlier centuries by his comment that "the oak falls for the greater purpose of man; and the alder is ready to supply a variety of his smaller wants." At that time, of course, no one was aware of the wider scientific implications of nitrogen fixation in the root nodules, although an early report of the presence of nodules on the root system of alder had appeared (Meyen, 1829).

Loudon's (1838) account is particularly interesting as an indication of the extensive range of exotic actinorhizal plants already readily available in Europe 150 years ago. This early importation of seed and perhaps rooted plants and the worldwide development of shipping and trade, which often involved the deposition of soil ballasts from sailing ships at ports, must have facilitated extensive exchange of microorganisms between countries. Modern observations of nodulation of exotic plants in countries where the natural occurrence of an appropriate strain of *Frankia* might not be expected should not be considered unusual therefore, although if it is desired to maximize symbiotic N fixation, screening of the nodulating strains may be profitable since it is conceivable that these may not always be the best adapted to the local conditions (Hooker and Wheeler, 1987). In the next section, the actinorhizal plants that are indigenous to Europe are detailed, together with observations on some of the more important introduced species.

II. European Actinorhizal Species

As noted in the foregoing, the range of actinorhizal species found in Europe is large because of the long history of explorers and ecclesiasts who set out in previous centuries to colonize lands or to expound Christianity and who returned with new, interesting, or economically useful plants to their countries. These plants were established on the estates of wealthy patrons or in the botanical or herbal gardens being formed in increasing numbers in monasteries throughout Europe from the sixteenth century onward. Therefore, by 1838 the only actinorhizal genera not mentioned in Loudon as plants indigenous or introduced to Britain are the South American rhamnaceous genera *Kentrothamnus* and *Talguena* and also the genus *Datisca*. *Datisca cannabina* is listed in "Flora Europaea" as an indigen of Crete and was reported to be nodulated in Italy by Trotter (1902). The root system of species other than those that might furnish root crops was rarely commented on prior to the beginning of the twentieth century, however, so that in general it is impossible to provide precedence for observation of the nodulation status of most introductions. Indeed, as already pointed out, the number of introduced actinorhizal species that may now be found in Europe is so great that little purpose would be served by listing them here.

Actinorhizal species that are indigenous to Europe are fewer in number and are listed in Table I. The Rhamnaceae and Casuarinaceae are not represented, although species of the latter family are planted widely for amenity purposes in southern Europe. The only indigenous genus of the Rosaceae is *Dryas*, but there are no records for nodulation of European species (Elkington, 1971; Bond, 1983). Although they are important pioneer plants in their natural habitats, species of this genus are utilized for horticultural purposes only in providing plants for rock gardens. The other four actinorhizal families contain between them five genera that are indigenous to Europe. However, relatively few of the large numbers of actinorhizal plants that have been introduced into Europe have been used for other than ornamental, horticultural purposes. Consequently, examination of the plant or seed lists of many modern European nurseries (e.g., Hillier & Sons, Winchester, England; Vilmorin-Andrieux, Paris; Den Ouden & Son, Boskoop, The Netherlands; Siemer & Mettler, Wittingen, West Germany) shows a ready availability of stock from most actinorhizal genera, including a range of specially selected horticultural varieties.

Species of the genus *Ceanothus* are particularly favored by the gardening public, although their nodulation status in Europe is not clear. For

Table I Distribution in Europe of Indigenous Actinorhizal Species

Species	Geographical areas[a]	Natural habitat
Coriariaceae		
Coriaria myrtifolia L.	S.W. Europe (F, P, E, Bal, I, G)	Dry woods, hedges, rocky places
Rosaceae		
Dryas octopetala L.	Most European countries except B, NL, DK, G	Neutral and basic soils, mountains, but low altitudes in north
Datiscaceae		
Datisca cannabina L.	Crete	Streambanks
Myricaceae		
Myrica gale L.	N.W. Europe, extending to N.W. E, central D, and N.W. Ru	Bogs, acid peatland, wet heaths
Myrica faya Ait.	E, P, Can, Mad, Az	Sandy soils, woodland, to 1500 m in Canaries
Betulaceae		
Alnus glutinosa L. Gaertn.	Most of Europe (see Fig. 1)	Wet or swampy ground, riversides; tree or shrub at higher altitude in mountain valleys
A. incana L. Moench.	N.E. to central Europe and southern mountains (see Fig. 1)	Riverside, floodplains, mountain valleys; tolerates drier soil than *A. glutinosa*
ssp. *incana*	Throughout range	
ssp. *kolaensis*	Arctic Europe	
[*A. glutinosa* × *A. incana*	A fairly common hybrid]	
A. cordata (Lois.)	I, Cors	Thickets, woodland, neutral to basic soils
A. viridis (Chaix.) D.C.		
ssp. *viridis*	Mountains of central Europe and Balkans	Thickets
ssp. *fruticosa*	N.E. Ru	
ssp. *suavolens*	Cors	
Elaeagnaceae		
Hippophae rhamnoides L.	Coastal and local distribution (see Fig. 1)	Sand dunes, cliffs, river gravel, dry river beds, and valleys in mountains
Elaeagnus angustifolia L.	Native of temperate Asia; widely naturalized from S. Europe to CZ and central Ru	

[a]Key to countries: Bal, Balearic Isles; B, Belgium; Cors, Corsica; F, France; P, Portugal; E, Spain; DK, Denmark; D, Germany; CZ, Czechoslovakia; NL, Netherlands; I, Ialy; Ru, Russia; G, Greece; Can, Canaries; Mad, Madeira; Az, Azores. From Tutin *et al.* (1968).

example, specimens of *C. foliosus* cv. 'Gloire de Versailles' failed to nodulate in the author's garden in Scotland but the Japanese introduction, *Elaeagnus pungens* var. *maculata*—selected as "shrub of the year" in Britain in 1983!—nodulated well. In terms of numbers of plants produced and sold annually, the horticultural industry constitutes an important economic outlet for many actinorhizal species. However, the ability of woody plants to form nitrogen-fixing root nodules is not a feature that stimulates sales to gardeners to a significant extent at present and consequently further analysis of this market will not be undertaken here. Many horticultural varieties are listed in Hillier and Son's *Manual of Trees and Shrubs*, by Bean (1970), and by Mitchell and Jobling (1984). Ashburner (1986) provides an interesting account of the origins, location, and growth of many species and varieties of *Alnus* found in Britain. The occurrence of *Alnus cordata* and *Alnus incana* in Europe is considered in detail by Gamisans (1983) and Schwabe (1985).

Species that are currently most widely used for a variety of purposes other than horticulture come from the genera *Alnus, Hippophae,* and *Elaeagnus*—in particular *A. glutinosa, A. incana, A. cordata, A. viridis, Hippophae rhamnoides,* and *Elaeagnus angustifolia*. The normal geographical range of the first five species is shown in Fig. 1, while that of other species is given in Table I. Most species are known to be nodulated widely throughout their range and readers should refer to Bond (1983) for further information on this aspect. Introduced species for which a wider role has been developed outside of horticulture are fewer in number. Important species include a few alders, for example, the red alder, *Alnus rubra* (tree) and *A. incana* ssp. *rugosa* (shrub), and shrubs such as *Elaeagnus umbellata*. Indigenous European species have been used on a wider scale, sometimes following introduction into European countries where they do not form part of the native flora. The realized and potential application of these species is primarily in forestry, which has been considered recently in a European context (Mikola *et al.*, 1983), in land management, and in some countries for fruit production. These and other minor uses are considered in the following section.

III. Utilization for Land Management and Forestry

A. Land Reclamation and Stabilization

While a number of countries in Eastern Europe have maintained an interest in the utilization of some actinorhizal species for fuel, for the manufacture of wood products, and for fruit, these uses in Western European countries have declined during the twentieth century and the main outlets now are for the reclamation of industrial wastelands and

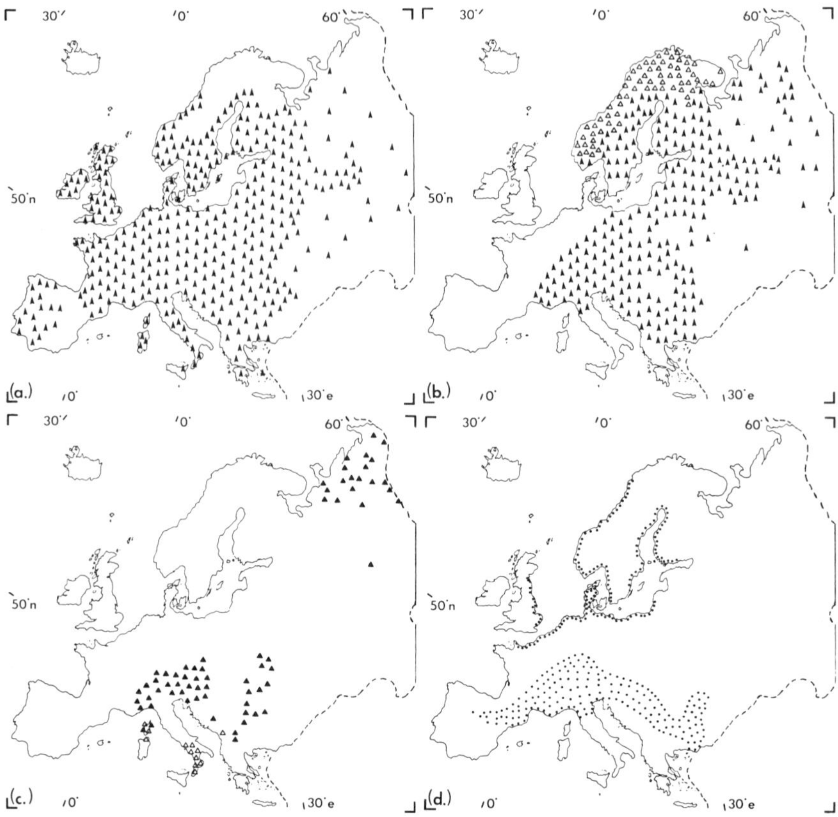

Figure 1 The distribution in Europe of economically important, indigenous actinorhizal species. (a) *Alnus glutinosa*. (b) *Alnus incana* ssp. *incana* (▲) is found throughout the range of the species; where it and *A. incana* ssp. *kolaensis* coexist is indicated (△). (c) *Alnus cordata* (△) and *Alnus viridis* (▲). (d) *Hippophae rhamnoides*. Redrawn from Jalas and Suominen (1972) and from Pearson and Rogers (1958).

for land stabilization. Those features of actinorhizal plants that have made them successful pioneers in their natural habitats, coupled with the ability of different species to grow on soils varying widely in properties from boggy to dry and from acid to alkaline (Dixon and Wheeler, 1983), suit them for these purposes in particular.

In addition to the possession of actinorhizal root nodules, many species are known to be ecto- and/or endomycorrhizal (Daft *et al.*, 1985), so that they are able to colonize a range of nutrient-impoverished lands. Generally, they possess well-developed root systems that are often capable of extensive and deep penetration of the substratum. In common with other pioneer species, however, they have medium to high requirements for light so that they can be shaded out fairly readily by other tall fast-growing species. Many researchers have compared the ability of appropriate actinorhizal species to grow on different wasteland sites but, compared with studies conducted in North America, few have determined the nodulation status of the actinorhizal outplants or have examined in detail the changes with time in the nutrient status of the site following planting. Usually, observations have been confined to the general appearance and quality of new vegetation, often together with measurements of top growth (e.g., height and stem diameter) of the planted species but with little attempt to quantify the contribution of microbial associations to the revegetation process.

The industrial wastelands found in Britain, which are listed by Gemmell (1977), are typical of those present in Europe. They comprise mainly coal mine spoils and ash from coal-burning power stations, metalliferous mine spoils and smelter wastes, china clay and quarry wastes, and chemical wastes. Land utilization problems also are created by gaseous pollutants generated by industrial activity. Problems for the revegetation of wastelands arise from many sources, including nutrient deficiencies, in particular of nitrogen and phosphate, extremes of pH, toxicity of heavy metals and cations, and the physical structure of the substratum. Alleviation of ground compaction is essential to ensure good growth of alders on reclamation sites (Binns and Fourt, 1980).

Actinorhizal species differ widely in their tolerance of pH but some adjustment of pH toward neutrality generally improves plant growth, as evidenced by the improved nodulation and growth of *A. incana* ssp. *incana*, the gray alder, on acid mine spoils and peatlands following application of lime (Huss-Danell, 1986). In addition to correction of acidity, calcium ions antagonize the toxicity of many heavy metal cations and can improve tolerance of pollutants (Gemmell, 1977; Grimstad, 1986).

Alnus glutinosa, the European common alder or black alder, and to a lesser extent the gray alder and *Hippophae rhamnoides* (sea buckthorn) are tolerant or semitolerant of a range of soil pollutants, including boron, cadmium, lead, and zinc. *Elaeagnus angustifolia* (oleaster) is generally reported as highly tolerant of pollutants and more tolerant than *E. umbellata* of some ions such as boron in alkaline fuel ash. However, both oleaster and sea buckthorn failed on acid shale on which black alder showed satisfactory growth (Carpenter and Hensley, 1978; Hodgson and Buckley, 1975; Greszta, 1982). Both black and gray alder form good barriers for catching industrial dusts (Supuka, 1980; Ullrich, 1980).

Reports for the resistance of particular species to different gaseous pollutants are somewhat inconsistent. For example, black alder was reported as more resistant to fluorine toxicity than some other broad-leaved species and also relatively insensitive to SO_2 (Bernatzky, 1978; Jamrich, 1984). By contrast, other studies have found this species to be damaged readily by SO_2 and also to be especially susceptible to ammonia fumes (Ewert, 1979). Both black and gray alder are highly sensitive to nitrogen trioxide but black alder tolerated nitrous oxide, which stimulated stem elongation in the first year of seedling growth (Bernatzky, 1978; Freer-Smith, 1984). Black alder is now regarded as one of the more sensitive trees to air pollution, along with other species that show limited tolerance of shade, and may be used as a pollution indicator (Ehrenberg, 1979; Freer-Smith, 1984; Umbach and Davis, 1984). Oleaster generally is considered to be more tolerant of air pollutants (Smirnov, 1983, 1987). It is clear, however, that assessment of the susceptibility of a species, particularly to gaseous pollutants, is rather dependent on the experimental system of the investigator and probably also on genetic variability in the species studied.

In the European studies mentioned earlier, the susceptibility of the symbiotic relationships of actinorhizal plants to damage by pollutants is generally inferred rather than measured directly. For example, evaluation of top growth has resulted in black alder becoming the most widely planted species on coal mine spoils in Britain, whereas in the years between 1966 and 1976 it was used in the reclamation of almost 80% of spoil heaps. Gray alder may be more suitable for some dry sites (Jobling and Stevens, 1980). There has been some decline in planting of alders in drier areas of Britain because of check or dieback in young stands, attributed mainly to drought (Evans, 1984). Trials have also shown *Alnus* species to be among the best choices for the reclamation of other spoils, such as alkaline magnesian limestone wastes, where both black and gray alder performed well despite a high pH, or on acidic shales and clays, where black and red alder and *A. viridis* (the green alder) were all productive (Richardson and Evans, 1987; Vann *et al.*, 1988). The last species was praised particularly for its bush form and attractive foliage.

Some results, typical of those presented by many European researchers utilizing actinorhizal plants for land reclamation, are illustrated in Fig. 2. Compared with pure stands, after 5 years of growth on alkaline, pulverized fuel ash of low nitrogen content, a significant increase was shown in mean tree height of *Picea sitchensis* (Sitka spruce) in mixture with oleaster. Intermixing had less effect on the growth of the conifer on a site with higher nitrogen content, thus suggesting, but by no means proving, that growth improvement was attributable to nitrogen fixation by the companion species (Hodgson and Buckley, 1975). The failure of intermixes with black alder or *Robinia pseudoacacia* (false

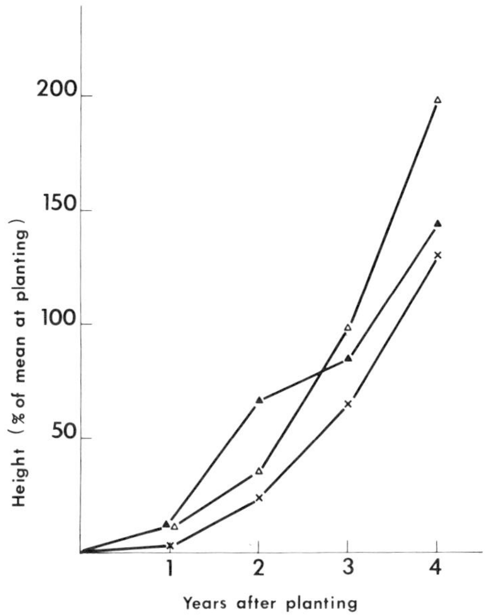

Figure 2 The effect on the growth of *Picea sitchensis* (Sitka spruce) on pulverized fly ash of intermixing with *Elaeagnus angustifolia* or *Alnus glutinosa*. Pure spruce, ×; spruce + alder, ▲; spruce + *Elaeagnus*, △. The mean height of spruce at planting was 1.54 m. Data adapted from Hodgson and Buckley (1975).

acacia or black locust) to increase significantly the height of the conifer was attributed to competition for light with these taller-growing species.

The importance of good nodulation of outplants of alder for reclamation work is now well recognized by most concerned organizations. However, there is often less appreciation among nurserymen that other species, such as sea buckthorn or oleaster, also form nitrogen-fixing symbioses. Clearly, more public education is needed in this area. Many growers have observed that *Frankia* already present in nursery soil can give good nodulation without further treatment and therefore need to be convinced that the growth and quality of their outplants are likely to be improved by inoculation. Some of the advantages of inoculation that have been demonstrated for the growth of seedlings of red alder in a forestry nursery will be discussed later.

Alnus, Elaeagnus, and *Hippophae* are also the genera that have been used most widely for land stabilization in Europe. Many unstable areas also have nutrient-impoverished soil, so that benefits are derived from both soil binding by the root systems as well as nitrogen fixation and mycorrhizal scavenging of nutrients by the actinorhizal plant root systems. Again, the authors are not aware of any substantial publications, in a European context, that detail the contribution of root nodules to the soil fertility of sites stabilized through the use of actinorhizal plants.

Although better adapted to wetter sites, *Alnus* species have been planted with some success on sand dunes as well as on eroded lands (Kohnke, 1941). However, the soil-binding properties of the rhizomatous root system of sea buckthorn have made this species of particular value for stabilization of light, sandy, and gravel soils. Consequently, it has been planted widely in Europe for shelterbelt formation and stabilization of sand dunes, for stabilization of erosion gullies and ravines, as hedges, and for planting along roadways (Flueckiger *et al.*, 1979; Green, 1981; Schiechtl, 1983; Grigor'ev, 1985; Jensen, 1986; Myakushko *et al.*, 1986). Species of *Elaeagnus* and black or gray alder have also been used for some of these purposes, for example, gray alder for landslide and erosion control in the Austrian Tyrol (Stucki, 1981; Florineth, 1982). Problems can arise from uncontrolled planting of vigorous colonizers such as sea buckthorn, however, for the formation of dense, thorny thickets will suppress other vegetation as well as be unacceptable environmentally to the public (Green, 1981). In addition to the discomfort caused by thorns, other undesirable effects are indirectly attributable to sea buckthorn. On sand dunes near Oostvorne, The Netherlands, infestation of buckthorn by the moth *Eupractis chryssorhoeo* defoliated the thickets, after which the moth spread to the foliage of other park and shade trees. Direct larval contact and wind dispersal also caused an epidemic of caterpillar dermatis among people using the beach areas! (Kniest and Hoffman, 1984).

Resistance to damage by deicing salt is an important criterion for the selection of species for planting on roadside shoulders and embankments in Europe. Oleaster and sea buckthorn are among the most resistant species to salt applied either to the roots or to the foliage and are recommended widely for planting along roads (Thompson and Rutter, 1982).

B. Forestry

Alnus is the only European actinorhizal genus with trees of a size suitable for large timber. Pure or mixed alder plantations may be managed as high forest or as coppice but currently comprise only a minor component of the forested areas of Europe. For example, in Britain, where forest and woodlands comprise 8–9% of the land area, alder covers some 10,000 ha or about 0.5% of the total forested land (Evans, 1984; Green, 1981).

1. Stand Growth and Nitrogen Fixation

Alders, as nitrogen-fixing pioneer species, usually show rapid juvenile growth so that maximum current annual increments of woody biomass in species such as red, black, and gray alders are achieved at about 10 to 12 years (Miller, 1983; Matthews, 1987). In the studies quoted by these authors, annual yields of dry biomass varied between 10 and 14 tons ha^{-1} $year^{-1}$ but there is large variability in biomass production between sites. For example, black alder yielded only 0.8 to 2.16 tons ha^{-1} $year^{-1}$ on wet acid forest land in France (Teissier du Cros, 1988). Estimates of annual rates of nitrogen fixation in *Alnus* stands vary widely and their accuracy is open to criticism on methodological grounds. Therefore, it is difficult to provide unequivocal predictions of relationships between age, biomass production, and rates of nitrogen fixation for stands of particular species. The relationships between these parameters shown in Fig. 3 have been derived from published data for current annual biomass accretion by red and gray alder (Binkley, 1981; Miller, 1983), and for rates of nitrogen fixation for different stands of several European and North American alders (Tarrant and Trappe, 1971; Silvester, 1974; Binkley, 1981; Miller, 1983). The figure shows that maximum fixation occurs early in the life of a stand, thus supporting rapid growth of the tree before sufficient mineralizable litter is produced to permit the development of a fully operational nitrogen cycle. The rotation age to maximize dry-matter production in such stands is around year 20 (Miller, 1983), by which time optimum input of fixed nitrogen should have been achieved.

Changes in woody biomass only are shown, but roots probably account for some 20 to 25% of the total weight of the tree and in dense stands (0.5-m spacing) of 4-year-old gray alder, nodules comprised 4.5 to 6% of the biomass of the root system (Miller, 1983; Saarsalmi *et al.*, 1985). This estimate (250–290 kg nodules ha^{-1}) is of the same order as found for other alder species of different ages and growing at different densities on a variety of sites, for example, red alder, 5 to 7 years old: 50 to 175 kg ha^{-1}; 15 to 20 years: 390 kg ha^{-1}; black alder, 6 years old: 80 to 130 kg ha^{-1}; 20 year old: 450 kg ha^{-1} (Zavitkovski and Newton, 1968; Zavitkovski and Stevens, 1972; Akkermans and van Dijk, 1976; Bormann and Gordon, 1984; Binkley, 1981; Wheeler *et al.*, 1986). Of course, only a fraction of this nodule biomass, the younger nodule lobes, is active in nitrogen fixation.

2. Afforestation of Degenerated Forest Soil and Peatlands

Alders are sometimes planted on clear-cut areas to improve the soil prior to replanting. Mikola *et al.* (1983) report experiments of Hausser on the

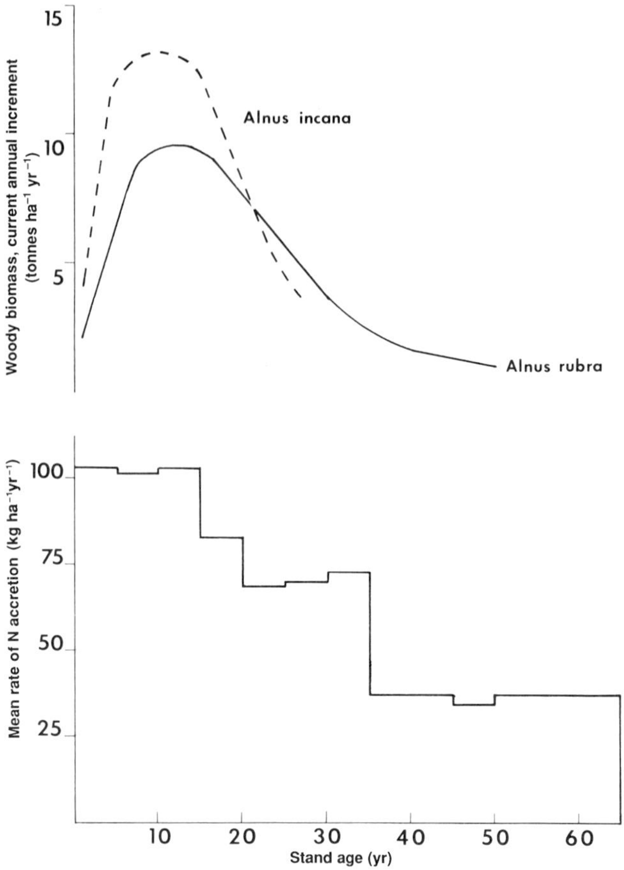

Figure 3 Nitrogen accretion and woody biomass increments in stands of alder. Mean rates of nitrogen accretion are derived from data for several stands of different-aged alder species. (Tarrant and Trappe, 1971; Silvester, 1974; Binkley, 1981; Miller, 1983). The current annual increments in woody biomass for *Alnus incana* (dashed line) and *Alnus rubra* (solid line) are adapted from Miller (1983).

utilization of gray alder to improve degraded soil in the Black Forest, West Germany, where litter removal had been practiced for centuries. Substantial application of calcium and phosphorus fertilizer was required to establish the alders. After 5 to 8 years, the height of both alder and spruce increased three to four times compared with control plots, while the nitrogen content of the spruce needles increased by 20%. In northern Sweden, there are also large tracts of previously productive forest land that are now impoverished of organic matter and nutrients as a result of unfavorable forest practices in the past and forest fires. Huss-

Danell and Lundmark (1988) applied leaf litter to small plots to evaluate the probable effects on soil humus content, pH, and mineral content of replanting areas with gray alder. Addition of >1000 kg ha^{-1} of dry leaves (equivalent to the litterfall in a dense stand of gray alder) every year for 6 years more than doubled the humus layer, increased the soil pH, and increased by more than four times the total and mineralizable soil nitrogen. This study, together with preliminary assessment of plant growth on field sites, permitted recommendation of gray alder for planting on degenerate soils in warmer, mesic areas, but *Lupinus nootkatensis* survived better on cooler, most northerly sites. Although the outplants used in these experiments were nodulated, it is interesting that *Frankia* was common in these degenerate soils, which were devoid of *Alnus* at least in recent years (Huss-Danell and Frej, 1986).

Gray alder has been used very successfully in Finland for improvement of the fertility of peat-bog bottoms, left after the removal of peat for fuel (Mikola, 1975). Swedish experience has also shown gray alder to be a promising species for growing on ombrotrophic *Sphagnum* peats following treatments such as draining and liming (Granhall *et al.*, 1983; Rytter *et al.*, 1989). However, experience with a number of species of *Alnus* for afforestation of acid peat in Britain and Ireland generally has been disappointing, with the production of short-lived plants of poor stature (Zehetmayer, 1954; MacDonald *et al.*, 1957; McVean, 1962; Matthews, 1987). Of the species tested, gray alder grew best on upland *Molinia* peats and *Juncus* flushes (MacDonald *et al.*, 1957).

Glasshouse pot experiments to study the effects of inoculation with *Frankia* on the growth of black and red alders on acid peat and acid brown earths have demonstrated the necessity for additional phosphorus, following which superior growth was shown by black alder (Sheppard *et al.*, 1989). Poor mycorrhizal development is a common feature of the growth of alders on peat and phosphorus fertilization can improve growth (McVean, 1962; Rytter *et al.*, 1989).

Sheppard *et al.* also obtained evidence of adaptation of *Frankia* to particular soils. While the growth of all strains tested was stimulated by aqueous extracts of mineral soils, inhibitory effects of peat extracts were less apparent with isolates from nodules from peat than from mineral soils. Liming, phosphorus fertilization, and selection of the most appropriate combination of species and *Frankia* strains should all be considered if it is proposed to plant alders on peat sites. Greatly enhanced nodulation of *Alnus* has also been observed following addition of clay to gravel substrates, in part due to adsorption of inhibitory root secretions. Development of this finding could have practical application for improved nodulation on some soils (Smolander *et al.*, 1989).

3. Admixture with Broad-Leaved and Coniferous Species

Utilization of alders as a "nurse," particularly for broad-leaved species, is a longstanding practice in Europe. For example, Carr (1866) refers to the use of gray alder as a "nurse" for oak in the Scottish Highlands. This species was successful in mixtures with beech and ash (MacDonald *et al.*, 1957), and in Finland, Virtanen (1957) showed a 17% increase in N content of the needles and a 30% increase in the mean height of pine grown on a nutrient-poor sandy soil in mixture with gray alder. Similar results have been obtained by Schalin (1966) for pine mixed with black or gray alder. In a 14-year-old stand, average growth of pine in the mixed stand was 1 m, and basal area was 70% greater than in pure pine stands. Beneficial effects of mixture with black alder have also been observed for poplars (Courrier and Garbaye, 1981).

However, the benefits for growth of the main species in such admixtures are not always so obvious and depend not only on the nutrient status, particularly nitrogen availability, of the site but also on competition between the component species of the mixed plantation. Thus, growth assessments 26 years after planting Norway spruce and larch on an upland site in England showed that benefits from mixture with black alder were not significantly different from those where *Pinus sylvestris* (Scots pine) was used as a "nurse" (Lines, 1982). Scot pine is heavily mycorrhizal, so that the growth benefits derived from the additional nutrients taken up by the mycorrhizal associations and from the shelter provided by the companion species here outweighed the effect of additional nitrogen from mineralization of alder litter.

Malcolm *et al.* (1985) also failed to find significant differences in growth of Sitka spruce 23 years after establishment of pure and mixed plantations with red alder on a surface water gley soil. Alder roots penetrated much more deeply into the subsoil than those of spruce, however, and it was suggested that an observed increase in the phosphorus content of the upper soil horizons might have resulted from surface deposition in litter of minerals exploited by the deep-rooting alder. The nitrogen content of the upper soil also increased by 585 kg ha^{-1}, which approximates the standard 150 kg ha^{-1} of fertilizer nitrogen applied at 5-year intervals in forestry practice to nitrogen-deficient stands. In this experiment, two factors mitigated against the appearance of significant beneficial "nurse" effect on spruce growth. First, the soil at planting was not especially nitrogen deficient so that the effect on spruce growth of additional nitrogen from the alders would be less obvious than on nutrient-impoverished soils. Second, at 10 years some of the spruce were being suppressed by the alders, so that growth was less at 23 years than it might otherwise have been. Shrubby nitrogen fixers, which would have provided an early input of nitrogen prior to being shaded out by the main species, might have been more beneficial for spruce growth on this site.

Attention to planting densities, relative growth rates of the main and "nurse" species, and the nutrient content of the site are all necessary to assess the value of mixed plantation practices. These factors have been considered for the growth of red alder in Chapter 17. Beneficial effects arising from causes other than improved nutrition of the main species may stem from interplanting, however. For example, Altmann (1940) noted a reduction in insect attacks in spruce stands in Nuremberg in mixtures of spruce and alder. Birch was also effective but was more susceptible to browsing by deer. Exudates from the roots and nodules of sea buckthorn and black alder are reputed to inhibit growth of *Heterobasidium annosum;* admixtures with pines susceptible to this fungus have been suggested for use in Russian forestry (Kobets and Andreeva, 1983). Such findings are similar to observations of inhibition of *Poria weirii* by root exudates of red alder (Li *et al.*, 1968).

4. Coppicing, Short Rotation Forestry, and Biomass Production

Interest in the use of *Alnus* species for biomass production has been maintained in Eastern Europe for many years and there has recently been a resurgence of research on this problem in Western Europe, largely due to the increases in oil prices and suggestions of impending oil shortages in the 1970s and early 1980s. These factors stimulated efforts to reduce dependence on oil imports, through both increased biomass production and the more effective use of nitrogen-fixing species. Although these pressures have temporarily abated, new concerns of overproduction of food within the European Common Market have provided further impetus to seek alternative uses for farm land. Recent estimates (Hummel *et al.*, 1988) suggest that 15 million ha will have to be taken out of food production in the EEC if agricultural surpluses are to be brought under control. Short rotation forestry may constitute the best use of up to 5 million ha. Currently, market prices for pulp and biomass render viable production of about 12 dry tons ha^{-1} year^{-1} within 100 km of major outlets such as pulp mills or within 10–15 km of small outlets. Only species that coppice readily give adequate yields on short rotations less than 15 years—a requirement met by many nonnitrogen-fixing European species, such as poplars, willows, and chestnut, as well as exotics such as eucalypts.

Critical comparison of biomass yields of short rotation plantations in Europe is difficult because of the empirical nature of much of the published data (Cannell, 1988). However, yield figures for alders suggest that they can meet the preceding EEC productivity criteria when planted on appropriate sites and with suitable management such as optimum planting densities. For example, Dennington *et al.* (1983) recommend densities no greater than 10,000 tree ha^{-1} for short rotation coppices. Thus, 15-year-old gray alder in Finland produced mean annual woody biomass of about 10.7 ton ha^{-1} and a 22-year-old stand in England produced about 16 tons (Ovington, 1956; Miller, 1983). Biomass production in a 7-year stand of gray alder on a raised sphagnum peat bog in central Sweden was 10 to 11 tons ha^{-1} $year^{-1}$ (Rytter *et al.*, 1989). The mean annual production per hectare of a 12-year stand of black alder on nitrogen-poor pulverized fuel ash in Wales was 11.2 tons compared with 7 tons from *Populus canadensis* on the same site (Dennington *et al.*, 1983). In terms of annual current increment in volume per hectare, Ehrenberg (1979) gives 6.0 to 9.8 m^3 for 40-year-old black alder in south Sweden, 10 to 50 m^3 for gray alder coppice, and 6 to 16 m^3 for 15- to 20-year-old coppice of *A. cordata* (Italian alder). The interest in the utilization of alder species for biomass production in a European context is shown by recent reports of investigations of management practices, yield, species, and provenance trials that are under way in virtually all of the ten EEC countries reporting to the Commission's current programs on biomass research and development (Hummel *et al.*, 1988).

Some encouraging economic projections for conversion of agricultural land into biomass production have been made. For example, the financial return in Austria from the production of wood chips from the intensive silviculture, on a 17-year rotation, of *Populus* or gray alder was estimated to be equal to or better than that from cattle pasture (Kniessel, 1985). However, it is likely that subsidies will be required, at least in the initial years of establishment, to make short rotation forestry viable, and it must be noted that some highly pessimistic reports of the economic viability of the practice have been published. For example, Blom (1986) calculated for The Netherlands that harvested yields of 40 tons ha^{-1} would be required to balance returns from sales with production and transport costs! Much depends on the price the Community is prepared to pay for the development of alternative land use strategies and on future global movements in energy costs.

Biomass productivity can obviously be increased by selection of the most appropriate species and provenance for particular sites and methods of cultivation. However, there is a need for long-term studies to quantify cycles of nutrients and of biomass in coppiced stands and to determine how patterns of nodulation and nitrogen fixation are affected by repeated coppicing. In addition, attempts should be made to optimize the symbiotic relationships in coppiced actinorhizal species by selection of, and inoculation with, superior strains of microsymbiont.

Variation in the effectivity of *Frankia* for nodulation and symbiotic nitrogen fixation has been noted for several actinorhizal species and controlled environment tests of the effectivity of strains nodulating *Alnus* have been published (Normand and Lalonde, 1982; Hooker and Wheeler, 1987). However, relatively few data from the evaluation of strains under field conditions are available. In collaborative tests carried out in a tree nursery by the Forestry Commission, the University of Glasgow, and The Institute for Terrestrial Ecology, a superior *Frankia* strain, isolated from the nodules of red alder growing in Scotland, supported good plant growth and a high rate of nitrogen accretion in the host plant species. Outplants inoculated with this strain have continued to show the benefits of inoculation through improved growth in height, particularly when transplanted to nitrogen-deficient sites. In addition, inoculation with *Frankia* ensured early nodulation of alder seedlings, which has permitted outplanting at year 1 instead of in year 2 as recommended previously by the Forestry Commission (Evans, 1984).

One example that illustrates the enormous possibilities for improvement of alders by selection and breeding programs, and that is of particular importance for applied work, will be given here. Verweij (1983) examined host plant ecotypes of black alder and identified two types. One was a tall-growing (to 25 m) forest ecotype that, although not showing rapid initial growth, increased in height for a long period and was a good competitor with other species. A second type was identified in alder groves and showed characteristics more typical of pioneer species as well as rapid growth for the first 8 years, which then slowed so that height at maturity was about 10 m. This ecotype was better nodulated, produced more adventitious and lateral roots, and flowered more abundantly than the forest type. the forest ecotype clearly is better suited for high forests and for the production of larger timber sizes, whereas the alder grove ecotype is more suited to biomass production in short rotations and for soil improvement. Such observations have also been made for alder stands in Germany.

IV. Commercial Products

A. Timber

The wood of alders, as in ancient times, continues to be used on a minor scale for the production of turnery and for boxes and crates. The wood is diffuse and porous, of medium to light density and strength, and is moderately useful for bending. It reddens on exposure to air and is usually durable under water (Evans, 1984; Matthews, 1987). Alder wood has been used commonly in Europe for fuel (Elgar, 1982), although it burns quickly and is often utilized as charcoal. Large-scale utilization of wood from actinorhizal plants for fuel, however, is probably unlikely until the price of other fuel sources increases dramatically.

The uniformity, strength, and length of fibers from wood of common and gray alder have led to its utilization for plywood production, for particle and fiber boards, and for pulping, usually in blends with coniferous wood (Alestalo and Hentola, 1967; Farmer, 1972; Morin, 1974; Surminski, 1980; Haarlaa and Kaarkkainen, 1982). Krzysik (1977) evaluated the industrial use of *A. incana* in Poland and noted that prior to 1970, a harvest of 10,000 m^3 year $^{-1}$ was used mainly for charcoal, burnt directly for fuel, or used in drainage and ditch lining. The development of industrial procedures for processing fiber and particle board increased the use of gray alder wood tenfold. Its utilization for these purposes was found to be profitable for industrialists, but the low price of the wood made its supply unattractive to foresters. Unfortunately, the timber does not possess outstanding qualities in comparison with other broad-leaved species and alders generally tend to suffer from the widely held view that they are weed species—all features that combine to suppress market prices. These difficulties should not be insurmountable as the efficiency of production increases with the development of better growing practices, the cultivation of elite lines selected from breeding programs, and increased efficiency in harvesting techniques and transportation.

B. Food Products

The main products for human consumption from actinorhizal plants are fruits, primarily from sea buckthorn, and also a number of alcoholic beverages. The latter represents a minor use, although the pleasure derived from the moderate consumption of the products of fermentation is an important ancillary attraction of these plants! Typical examples are the liqueurs "Tyrni," made from buckthorn berries in Finland, and "Els," a drink flavored in part with alder leaves and produced in the Low Countries. Sea buckthorn berries are also used for cider production in Germany (Albrecht and Koch, 1981). Alcoholic drinks have long been flavored with aromatic bog myrtle leaves and were reputed to be drunk by marauding Vikings in previous centuries. Bog myrtle beer can still be purchased locally in Denmark and there are recipes for its production (Mabey, 1972).

Cultivation of sea buckthorn and the harvesting of its berries constitute an important agricultural industry in Eastern Europe, particularly in Russia. The species is dioecious, so that it is necessary to plant rows of male and female bushes for fruit production. Although it is productive on relatively poor soil, the soil must be light, well drained, and sheltered but with plenty of light for good yield. High-yielding and thornless varieties have been developed, but to the authors' knowledge, data on the effects on yield of inoculation with *Frankia* are not available. Recent accounts of the techniques for cultivation of the species for fruit production have been published (Albrecht and Koch, 1981; Albrecht *et al.*, 1984; Von Beldean and Leahu, 1985).

The value of the fruit lies in its high content of vitamins (vitamins C, E, and F and also carotene) and of unsaturated aliphatic acids, particularly linoleic and linolenic acids. Vitamin C content can reach 0.3%, vitamin E 0.2%, and carotene up to 0.1% in juice oil (Zhmyrko *et al.*, 1978). *Hippophae* berry oil, "Siberian Balsam," has been used in the USSR for treatment of skin burns, inflammations, and ulcers and might be considered as an alternative to the oil of evening primrose (*Oenothera erythrosepala*), which is currently in vogue in Western Europe as a natural medicine. In China, a range of teas and jellies is sold by the China National Tea Import and Export Corporation and the "Sha Wei Kang" drink was designated as the official beverage for the Chinese delegations to the Asian games in Korea. This company also planned the production of a buckthorn health drink to be sold in ring-pull cans (Anonymous, 1987)! Sea buckthorn is currently under investigation as part of a novel crops initiative at SCRI Invergowrie, Scotland (M. Cormack, personal communication), whence have come suggestions for its use as a coloring and flavoring for ice cream. Such a product has been deemed "elegant and subtle of taste" (Grigson, 1988).

C. Chemical Extractants

Alders continue to have some use in Eastern Europe as a source of tannins (Khvorost *et al.*, 1986). They are also a source of a wide variety of chemicals, particularly flavonoids (Krawiarz and Chalupka, 1980). Utilization for the production of alcohols and other industrial chemicals must again await changes in petrochemical economics. The authors are aware of industrial programs in progress in Britain and elsewhere to screen *Frankia* cultures for products of medical or agricultural importance, but do not know of any public pronouncements in this field as yet.

V. Future Prospects

It is obvious from the foregoing discussion in this and other chapters that there are substantial opportunities for greater and more effective use of many actinorhizal species, often with the prospect of a satisfactory financial return. Utilization of *Alnus*, *Elaeagnus*, and *Hippophae* for land reclamation and stabilization is now a widely accepted practice. Actinorhizal plant researchers have a duty to publicize how the quality and rate of land reclamation programs may be improved through the exploitation of superior genotypes of host plant and microsymbiont and to try to ensure that the products of their research are available for purchase and utilization! Further research is required to optimize forestry practices such as intermixes and to provide economic analyses of the expected returns from these procedures. Research into biomass production, particularly from alders, must continue both for ecological reasons and as insurance against future scarcity of fossil fuels.

Within Europe, there is a great need for more extensive, properly documented data on biomass yields and nutrient cycling in short rotation and coppiced plantations so that informed comparisons can be drawn concerning the relative advantages of the utilization of nitrogen-fixing and non-nitrogen-fixing species. Cannell (1988) notes that partitioning of photosynthates to wood can be increased dramatically by decreasing rates of fine root turnover and that this can often be achieved by improving rates of nutrient or water supply. Data on fine root turnover in root-nodulated species are notable for their absence, yet lower rates of root turnover following improved nitrogen nutrition could be a feature of well-nodulated actinorhizal plants. Root development in plants relying on fixation for their nitrogen is usually less than in plants supplied with mineral nitrogen. Basic research on the molecular genetics of *Frankia* symbioses may eventually suggest ways to nodulate alternative biomass species such as poplars, or rosaceous fruit species such as apples or raspberries. Finally, the biosynthetic capacity of *Frankia* has been explored little but could open new horizons for research as screening programs proceed.

We have discussed the various products that can be obtained from the berries of sea buckthorn. We would like to conclude with some lines from the poem "Sea Buckthorn," written by Helen B. Cruickshank, which praise in Scots the visual splendor that this shrub brings to its environment when in fruit:

Saut[1] *an' cruel winds to shear it,*
Nichts[2] *o' haar*[3] *an' rain,*

Ye micht think the sallow buckthorn
Ne'er a hairst[4] could hain[5]
But amang the sea-bleach'd branches,
Ashen grey as pain,
Thorn-set orange berries cluster,
Flamin' beauty-fain.[6]

References

Akkermans, A. D. L. and van Dijk, C. (1976). The formation and nitrogen-fixing activity of root nodules of *Alnus glutinosa* under field conditions. *In* "Symbiotic Nitrogen Fixation in Plants" (P. S. Nutman, ed.), pp. 511–520. Cambridge Univ. Press, Cambridge, England.

Albrecht, H.-J., and Koch, H.-J. (1981). The cultivation of sea-buckthorn for fruit-processing industry. *Gartenbau* **28,** 175–177.

Albrecht, H.-J., Gerber, J., Koch, H.-J., and Wolf, D. (1984). Experience in growing sea buckthorn. *Gartenbau* **31,** 242–244.

Alestalo, A., and Hentola, Y. (1967). Alder in sulphate pulping. *Papperi Puu* **49,** 25–27.

Altmann, T. (1940). Die Erles als Vorwald. *Forstarchiv.* **16,** 92–96.

Anonymous (1987). A range of sea buckthorn products. *Food Market. Technol.* pp. 15–16.

Ashburner, K. (1986). *Alnus*—A survey. *Plantsman* **8,** 170–188.

Bean, W. J. (1970). "Trees and Shrubs Hardy in the British Isles." Murray, London.

Bernatzky, A. (1978). "Tree Ecology and Preservation," 357 pp. Elsevier, Amsterdam.

Binkley, D. (1981). Nodule biomass and acetylene reduction rates of red alder and Sitka alder on Vancouver Island, B.C. *Can. J. For. Res.* **11,** 281–286.

Binns, W. O., and Fourt, D. F. (1980). Surface workings and trees. *Occas. Pap.—U.K., For. Comm.* **10,** 60–75.

Blom, J. (1986). *In* "Energy from Biomass," 3rd E. C. conference, Vienna. (W. Halz, J. Coombs, and D. O. Hall, eds.). Elsevier, London.

Bond, G. (1983). Taxonomy and distribution of non-legume nitrogen-fixing systems. *In* "Biological Nitrogen Fixation in Forest Ecosystems: Foundations and Applications" (J. C. Gordon and C. T. Wheeler, eds.), pp. 55–87. Nijhoff/Dr. W. Junk, The Hague.

Borman, B. T., and Gordon, J. C. (1984). Stand density effects in young red alder plantations: productivity, photosynthate partitioning and N fixation. *Ecology* **65,** 394–402.

Cannell, M. G. R. (1988). The scientific background. *In* "Biomass Forestry in Europe: A Strategy for the Future" (F. C. Hummel, W. Palz, and G. Grassi, eds.), pp. 83–140. Elsevier, Amsterdam.

Carpenter, P. L., and Hensley, D. L. (1978). Accumulation of cadmium in *Alnus glutinosa* L. *HortScience* **13,** 363.

[1]Saut: salty.
[2]Nichts: nights.
[3]Haar: mist.
[4]Hairst: crop, harvest.
[5]Hain: give, produce.
[6]Reprinted by permission of Bayley and Ferguson, Ltd., publishers, Glasgow, Scotland.

Carr, R. (1866). On the silver alder, its value as a nurse plant for oak. *Trans. Highl. Agric. Soc. Scotl. 4th Ser.* **1,** 381–385.

Courrier, G., and Garbaye, J. (1981). The silviculture of mixed stands. An example of the beneficial effect of alder on growth of poplars. *Rev. For. Fr.* **33,** 289–292.

Cruikshank, H. B. (1932). Sea Buckthorn. The Collegiate Choir, no. 441. Bayley and Ferguson Ltd., Glasgow.

Daft, M. J., Clelland, D. M., and Gardner, I. C. (1985). Symbiosis with endomycorrhizas and nitrogen-fixing organisms. *Proc. R. Soc. Edinburgh, Sect. B* **85,** 283–298.

Dennington, V. N., Chadwick, M. J., and Chase, D. S. (1983). Energy cropping on derelict and waste land. *J. Environ. Manage.* **16,** 241–260.

Dixon, R. O. D., and Wheeler, C. T. (1983). Biochemical, physiological and environmental aspects of symbiotic nitrogen fixation. *In* "Biological Nitrogen Fixation in Forest Ecosystems: Foundations and Applications" (J. C. Gordon and C. T. Wheeler, eds.), pp. 107–171. Nijhoff/Dr. W. Junk, The Hague.

Ehrenberg, C. (1979). Mini-monograph on *Alnus*. *In* "Technical Consultation on Fast-Growing Plantation Broadleaved Trees for Mediterranean and Temperate Zones." *F.A.O. [Tech. Rep.] FO–FGP* **79-6/3,** 142–186.

Elgar, N. (1982). Growing wood for fuel. *Q. J. For.* **76,** 53–55.

Elkington, T. T. (1971). *Dryas octopetala* L. *J. Ecol.* **59,** 887–905.

Evans, J. (1984). Silviculture of broadleaved woodlands. *Bull.—U.K., For. Comm.* **62.**

Ewert, E. (1979). Phytotoxicity of ammonia. *Hercynia* **16,** 75–80.

Farmer, R. H. (1972). "Handbook of Hardwoods," pp. 21–22. Her Majesty's Stationery Office, London.

Florineth, F. (1982). Biological engineering measures used by the office for torrent control in southern Tyrol. *For. Abstr.* **43,** abstr. no. 1788.

Flueckiger, W., Oertli, J. J., Flueckiger-Keller, H., and Braun, S. (1979). Premature senescence in plants along a motorway. *Environ. Pollut.* **20,** 171–176.

Freer-Smith, P. H. (1984). The response of six broad leaved trees during long-term exposure to SO_2 and NO_2. *New Phytol.* **97,** 49–61.

Gamisans, J. (1983). The Corsican alder, *Alnus cordata*, in its natural environment. *Rev. For. Fr.* **35,** 187–197.

Gemmell, R. P. (1977). "Colonisation of Industrial Wasteland," Inst. Biol. Stud. 80. Arnold, London.

Granhall, U., Ericsson, T., and Clarholm, M. (1983). Dinitrogen fixation and nodulation by *Frankia* in *Alnus incana* as affected by inorganic nitrogen in pot experiments with peat. *Can. J. Bot.* **61,** 2956–2963.

Green, B. (1981). "Countryside Conservation." Allen & Unwin, London.

Greszta, J. (1982). Accumulation of heavy metals by certain tree species. *In* "Urban Ecology" (R. Bornkamm, J. A. Lee, and M. R. D. Seeward, eds.), pp. 161–166. Blackwell, Oxford.

Grigor'ev, A. G. (1985). Trees and shrubs for landscaping the west coast of the Crimea. *Byull. Gos. Nikitsk. Bot. Sada* **57,** 23–26.

Grigson, J. (1988). The life and loves of a long-forgotten fruit. *Observer* **10267,** 36.

Grimstad, S. O. (1986). Addition of peat, liming, fertiliser application and choice of species for revegetation in a highly polluted industrial area. *Meld. Nor. Landbrukshoegsk.* **65,** 20 pp. [quoted in *For. Abstr.* **49,** abstr. no. 3159 (1988)].

Haarlaa, R. H., and Kaarkkainen, M. (1982). Fibre length in black alder. *Silva Fenn.* **16,** 343–350.

Hillier and Sons (1988). "Manual of Trees and Shrubs." Hillier and Sons, Winchester, Hampshire, England.

Hodgson, D. R., and Buckley, G. P. (1975). Establishment of trees and shrubs on pulverised fuel ash. *In* "The Ecology of Resource Degradation and Renewal" (M. J. Chadwick and G. T. Goodman, eds.), pp. 305–330. Blackwell, Oxford.

Hooker, J. E., and Wheeler, C. T. (1987). The effectivity of *Frankia* for nodulation and nitrogen fixation in *Alnus rubra* and *Alnus glutinosa*. *Physiol. Plant.* **70,** 333–341.

Hummel, F. C., Palz, W., and Grassi, G. (eds.) (1988). "Biomass Forestry in Europe: A Strategy for the Future." Elsevier, Amsterdam.

Huss-Danell, K. (1986). Growth and production of leaf litter nitrogen by *Alnus incana* in response to liming and fertilisation on degenerated forest sites in northern Sweden. *Can. J. For. Res.* **16,** 847–853.

Huss-Danell, K., and Frej, A.-K. (1986). Distribution of *Frankia* in soils from forest and afforestation sites in northern Sweden. *Plant Soil* **90,** 407–418.

Huss-Danell, K., and Lundmark, J.-E. (1988). Growth of nitrogen-fixing *Alnus incana* and *Lupinus* spp. for restoration of degenerated forest soil in northern Sweden. *Stud. For. Suec.* **181,** 1–20.

Jalas, J., and Suominen, J. (eds.) (1972). "Atlas Flora Europaeae," Vol. 1. Committee for Mapping the Flora of Europe and Soc. Biol. Fenn. Vanamo, Helsinki.

Jamrich, V. (1984). Possibilities of classifying the resistance of woody plants to fluorine intoxication by means of visual damage symptoms. *Folia Dendrol.* **11,** 83–107.

Jensen, F. A. (1986). Living hedges. *Gartner Tidende* **102,** 1507–1509.

Jobling, J., and Stevens, F. W. R. (1980). Establishment of trees on regraded colliery spoil heaps. U.K. For. Comm., Occas. Paper **7.**

Khvorost, O. P., Vyetrov, P. P., and Serbin, A. H. (1987). Optimisation of the process of extraction of tannic substances from the compound fruit of the European alder *For. Abstr.* **83,** abstr. no. 78760.

Kniessel, F. (1985). Evaluation of the operational economics of field plantations of agricultural land use. *Allg. Forstztg.* **96,** 293–294.

Kniest, F. M., and Hoffman, J. R. (1984). Brown-tail moth, *Eupractis chryssorhoea*, an indigenous pest of parks and public in The Netherlands. *Great Lakes Entomol.* **17,** 111–112.

Kobets, E. V., and Andreeva, I. N. (1983). The influence of non-legume N fixers on *Heterobasidium annosum*. *Lesn. Khoz.* **4,** 55–56.

Kohnke, H. (1941). The black alder as a pioneer tree on sand dunes and eroded land. *J. For.* **39,** 333–334.

Krawiarz, K., and Chalupka, W. (1980). Physiology of growth and development. *In* "Olsze (*Alnus* Mill.)" (S. Bialobok, ed.), Monogr. Popularn. 8, pp. 165–200. Polska Akad. Nauk, Inst. Dendrol., Warsaw, Poland.

Krzysik, F. (1977). Wood of *Alnusincana* as a raw material. *Sylwan* **11,** 19–22.

Li, C. Y., Lu, K. C., Trappe, J. M., and Bollen, W. B. (1968). Enzyme systems of red alder in relation to infection by *Poria weirri*. *In* "Biology of Alder" (J. M. Trappe, J. F. Franklin, R. F. Tarrant, and G. M. Hansen, eds.), pp. 241–249. U.S. Dep. Agric., For. Serv., Pacific Northwest For. Range Exp. Stn., Portland, Oregon.

Lines, R. (1982). Mixture experiments. Rep. For. Res.—U. K., For Comm. pp. 13–14.

Loudon, J. C. (1838). "Arboretum et Fruticetum Britannicum," Vols. III and IV. Longman, Orme, Brown, Green and Longmans, London.

Mabey, R. (1972). "Food for Free." Collins, Glasgow.

MacDonald, J., Wood, R. F., Edwards, R. V., and Aldhouse, J. R. (1957). Exotic forest trees in Britain. *Bull.—U.K., For. Comm.* **30,** 144–146.

Malcolm, D. C., Hooker, J. E., and Wheeler, C. T. (1985). *Frankia* symbiosis as a source of nitrogen in forestry: A case study of symbiotic nitrogen fixation in a mixed *Alnus–Picea* plantation in Scotland. *Proc. R. Soc. Edinburgh, Sect. B* **85,** 263–282.

Matthews, J. D. (1987). The silviculture of alders in Britain. *Oxford For. Inst., Occas. Pap.* **34,** 29–38.

McVean, D. N. (1962). The establishment of alder on peatland and its possible role in afforestation. *Ir. For.* **19,** 81–84.

Meyen, J. (1829). Ueber das Herauswachsen parasitischer Gewachse aus den Wurzeln anderer Pflanzen. *Flora Allg. Bot. Z.* **12,** 49–63.

Mikola, P. (1975). Afforestation of bogs after industrial exploitation of peat. *Silva Fenn.* **9,** 101–115.

Mikola, P., Uomala, P., and Malkonen, E. (1983). Application of biological nitrogen fixation in European silviculture. *In* "Biological Nitrogen Fixation in Forest Ecosystems: Foundations and Applications. (J. C. Gordon and C. T. Wheeler, eds.), pp. 279–294. Nijhoff/Dr. W. Junk, The Hague.

Miller, H. G. (1983). "Nutrient Cycling in Alder," Rep. 2. Int. Energy Agency, Stockholm.

Mitchell, A., and Jobling, J. (1984). "Decorative Trees for Town and Garden." For. Comm. Her Majesty's Stationery Office, London.

Morin, M. J. (1974). NSSC pulping of young European black alder. *Tappi* **55,** 133–135.

Myakushko, V. K., Kersenko, V. M., and Bedritskii, A. S. (1986). *Hippophae rhamnoides* in stands of gulley and ravine systems. *Lesn. Khoz.* **10,** 30–34.

Normand, P., and Lalonde, M. (1982). Evaluation of *Frankia* strains isolated from provenances of two *Alnus* species. *Can. J. Microbiol.* **28,** 1133–1142.

Ovington, J. D. (1956). The form, weight and productivity of tree species grown in close stands. *New Phytol.* **55,** 289–304.

Pearson, M. C., and Rogers, J. A. (1958). *Hippophae rhamnoides* L. *J. Ecol.* **50,** 501–503.

Richardson, J. A., and Evans, M. E. (1987). Establishing trees and shrubs on magnesian limestone waste. *J. Environ. Manage.* **24,** 157–167.

Rytter, L., Slapokas, T., and Granhall, U. (1989). Woody biomass and litter production of fertilised grey alder plantations on a low-humified peat bog. *For. Ecol. Manage.,* **28,** 161–176.

Saarsalmi, A., Palmgren, K., and Levula, T. (1985). Biomass production and nutrient and water consumption in an *Alnus incana* plantation. *Folia. For.* **628,** 1–24.

Schwabe, A. (1985). Monographie Alnus incana—reicher waldgesellschaften in Europa. Variabilitat und ahnlichkeiten einer azonal verbreiten gesellschaftsgruppe. *Phytocoenology* **13,** 197–302.

Schalin, I. (1966). *Alnus incana* in forestry practice. *Metsa Pun.* **9,** 362–370.

Schiechtl, H. M. (1983). Woody plants along motorways which exhibit long term resistance to salt. *Gart. Landsch.* **93,** 876–882.

Sheppard, L. J., Hooker, J. E., Wheeler, C. T., and Smith, R. I. (1989). Glasshouse evaluation of the growth of *Alnus rubra* and *Alnus glutinosa* on peat and acid brown earths soils when inoculated with four sources of *Frankia*. *Plant Soil* **110,** 187–198.

Silvester, W. B. (1974). Ecological and economic significance of the non-legume symbioses. *Proc. Int. Symp. Nitrogen Fixation, 1st* **2,** 489–506. Washington State Univ. Press, Pullman, Washington.

Smirnov, I. A. (1983). Resistance of woody plants to gases in an arid climate. *Sov. J. Ecol.* **14,** 156–158.

Smirnov, I. A. (1987). Fume resistant trees and shrubs. *Lesn. Khoz.* **4,** 65–67. [abstracted in *For. Abstr.* **49,** abstr. no. 7282 (1988)].

Smolander, A., Nurmiaho-Lassila, E.-L., and Sundman, V. (1988). Effect of clay mineral (montmorillonite) on the nodulation of *Alnus* and on the nitrogenase activity of *Frankia* in pure culture. *Symbiosis,* **6,** 37–52.

Stucki, B. (1981). Vegetation cover and the threat of landslides on the NW slope of the Buchserhorn, Switzerland. *For. Abstr.* **43,** abstr. no. 2393.

Supuka, J. (1980). Barrier properties of woody plants against industrial dust. *Acta Dendrobiol.* **3,** 247–297.

Surminski, J. (1980). Technical properties of alderwood and possibilities of its utilisation. *In* "Olsze (*Alnus* Mill.)" (S. Bialobok, ed.), Monogr. Popularn. 8, pp. 325–341. Polska Akad. Nauk, Inst. Dendrol., Warsaw, Poland.

Tarrant, R. F., and Trappe, J. M. (1971). The role of *Alnus* in improving the forest environment. *Plant Soil, Spec. Vol.* pp. 335–348.

Teissier du Cros, E. (1988). In "Biomass Forestry in Europe: a strategy for the future" (F. C. Hummell, W. Palz, and G. Grassi, eds.). Elsevier, London.

Thompson, J. R., and Rutter, A. J. (1982). Planning motorway planting in relation to deicing salt. *In* "Urban Ecology" (R. Bornkamm, J. A. Lee, and M. R. D. Seward, eds.), pp. 332–333. Blackwell, Oxford.

Trotter, A. (1902). Intorno a tubercoli radicali di *Datisca cannabina*. *Bull. Soc. Bot. Ital.* pp. 50–52.

Tutin, T. G., Heywood, V. G., Burges, N. A., Moore, D. M., Valentine, D. H., Walters, S. M., and Webb, D. A. (1968). "Flora Europaea," Vols. 1 and 2. Cambridge Univ. Press, Cambridge, England.

Ullrich, T. (1980). Determination of the dust-catching capacity of different broad leaved species. *Arch. Naturshutz Landschaftsforsch.* **20,** 43–59.

Umbach, D. M., and Davis, D. D. (1984). Severity and frequency of SO_2-induced leaf necrosis on seedlings of 57 tree species. *For. Sci.* **30,** 587–596.

Vann, A. R., Brown, L., Chew, E., Denison Smith, G., and Miller, E. (1988). Early performance of *Alnus* on derelict land in the industrial Pennines. *Q. J. For.* **82,** 165–170.

Verweij, J. A. (1983). Selection and Breeding of Alder. Landschapsbouw "de Dorschkamp," Annu. Rep. 142, pp. 43–44. Wageningen, The Netherlands.

Virtanen, A. I. (1957). Investigations on nitrogen fixation by the alder without combined nitrogen. *Physiol. Plant.* **10,** 164–169.

Zavitkovski, J., and Newton, M. (1968). Effect of organic matter and combined nitrogen on nodulation and nitrogen fixation in red alder. *In* "Biology of Alder" (J. M. Trappe, J. F. Franklin, R. F. Tarrant, and G. H. Hansen, eds.), pp. 209–223. U.S. Dep. Agric., For. Serv., Pacific Northwest For. Range Exp. Stn., Portland, Oregon.

Zavitkovski, J., and Stevens, R. D. (1972). Primary productivity of red alder ecosystems. *Ecology* **53,** 235–242.

Zehetmayer, J. W. L. (1954). Afforestation of upland heaths. *Bull.—U.K., For. Comm.* **32,** 102 103.

Zhmyrko, T. G., Gigienova, E. I., and Umarv, A. U. (1979). Vitamins from the oils of *Hippophae rhamnoides* fruits. *Biol. Abstr.* **68,** abstr. no. 18015.

Index